THE BASICS OF STRUCTURAL ANALYSIS

Michael R. Botwin
California Polytechnic State University
San Luis Obispo

George J. Murnen
The University of Toledo

Engineering Press, Inc. **San Jose, California 95103-0001**

Printed in the United States of America

Library of Congress Cataloging in Publication Data

Botwin, Michael R., date.
The basics of structural analysis.

Bibliography: p. 271
Includes index.
1. Structures, Theory of. I. Murnen, George, J., date. II. Title.
TA645.B654 1986 624.1'7 85-20540
ISBN 0–910554–35–8

Engineering Press, Inc. P.O. Box 1 San Jose, California 95103-0001

CONTENTS

PREFACE

This text introduces the necessary elements toward enabling the student to perform successful structural analysis.

Designed to serve as an introductory textbook and not a comprehensive source of reference, the material is practically oriented. While theory is naturally included, we emphasize problem solving. The focus is on those techniques most basic to the analysis of particular structural configurations. With the illustration of many figures and fully developed examples, our approach encourages the student to understand and know how to apply a *few* techniques well, rather than being superficially exposed to many techniques.

In conceiving of structural analysis as one phase in the analysis–design continuum, we integrate statics and mechanics of materials into *structural analysis*; further, we indicate where structural analysis can be best used in the process of design. Many structural analysis texts deal only with deflections and internal stress resultants (moments, axial and shear forces); we also consider the stresses *caused* by the moments and forces. With this additional information, the student is introduced to how structural members are best designed to withstand the stresses to which they are subjected.

The first chapter reviews statics and strength of materials. These concepts are carried into Chapters 2, 3 and 4 where trusses, beams and frames are treated. Experience has shown that students have as much trouble with statics as they have with "new" material in structural analysis. This review (or the first exposure for some) attempts to bring all students to the same level of understanding before introducing new material.

Chapters 5 and 6 examine the moment-area method—for beams and frames—and virtual work—primarily for trusses. Both determinate and indeterminate structures are handled in a unified manner.

Chapters 7, 8 and 9 cover, in turn, slope-deflection, moment-distribution and the stiffness matrix method. These are actually variations of the same method. (For those who have never studied matrices, an appendix reviews the principles involved.)

The text concludes with a look at approximate methods and treatments of moving loads. From the point of view of preliminary design, approximate methods are indispensable. Finally, there is a selection of answers to problems posed at the end of each chapter.

The authors thank their wives, Caroline and Patricia, for their patience and understanding during the writing of this text. The authors are greatly indebted to two former students: Mark Ennis, for the drafting of the example problems, and Rick Beall, for the figure illustrations; working with both was a pleasure. Denise Grote typed much of the manuscript; in addition to her professional skills, we acknowledge her continuous good spirits.

Michael Botwin
George Murnen

1

INTRODUCTION

Structural analysis—what is it? How does it fit into the educational spectrum? Structural analysis is a natural extension of basic engineering mechanics—statics and mechanics of materials—to problems that structural engineers commonly meet. While structural analysis can be seen as an end in itself, we favor the notion that analysis is a prelude to design.

In statics we are basically concerned with finding reactions at supports or connections. That is, given configuration of a structure and the loads it is subjected to, the equations of equilibrium are used to find the magnitude and direction of forces and/or couples that develop at supports and/or connections in order to maintain equilibrium. Our study of structural analysis relies heavily on statics.

Knowledge of statics is imperative when exploring the inner workings of structural members. In mechanics of materials, interest is focused on quantities that occur inside the member. These quantities—stress and strain—depend not only upon structural configuration and loadings, as in statics, but upon cross-sectional characteristics of the structural member such as area and moment of inertia. Further, strain and deformation are directly related to properties of the material used, e.g., Young's modulus and Poisson's ratio.

In this text we shall encounter three of the loading patterns introduced in mechanics of materials: uniaxial tension, uniaxial compression including buckling, and bending. We shall, further, combine these loadings to study in detail three important structural devices: trusses, beams, and simple frames, investigating primarily the same quantities studied in mechanics of materials; that is, we shall calculate internal quantities like stresses and stress resultants (forces and couples) as well as strains and deformations. Our study of deformations will allow us to resolve statically indeterminate problems.

Applications of this analysis are many. They, however, can be divided into 1) purely analytical and 2) as the basis of design. Let's look at two examples.

As an instance of the "purely analytical," let's say the owner of an old office building is interested in converting the building into a warehouse. He asks your consulting office to see if the building will be "safe." This involves, among many considerations, checking if the steel floor beams can withstand the larger loads that this conversion will bring, by checking the stresses and deflections of existing beams being subjected to presupposed loads. The beams will be considered "safe" if these stresses and deflections are not greater than those allowed by the local building codes.

Design is a broad term meaning different things to different people. Say you're working as a junior structural engineer in a consulting office. A client wishes to build a new warehouse for his wholesale hardware concern. He has the site and ideas about required space for the warehouse and office portions of the building. He wants the warehouse "designed." At first glance the options available and necessary considerations are overwhelming.

Focusing on just the roof structure a multitude of configurations come to mind: a trussed roof, a beam-column system, open-web joist-column system, a rigid frame or even an air-supported pneumatic structure. On closer examination these choices multiply. For example, for the truss roof we can use a Pratt, a Howe or a Fink truss among many others. Each can be designed at any slope and at any number of feet between trusses. And trusses can be fabricated of steel, timber, light metal, or—occasionally—concrete. And we've just scratched the surface. The options for a junior engineer would be bewildering. However, the experienced senior man who has "designed" many similar structures eliminates most of these options by "inspection." He intuitively knows the important design considerations; he knows the loads the structure will be subjected to; he is familiar with the local labor market and workers' degrees of skill. He understands construction techniques and knows the feasibility of certain techniques for this particular job. He has at his fingertips good ball-park figures for the relative cost of different materials and how these material costs interact with labor costs.

From these and other restrictions imposed by his colleagues, the architect and the mechanical engineer, the senior consultant will arrive at, say, six good design schemes. All of these will have to be "checked out." In so doing the senior man asks you, the junior engineer to "design a 30° tubular steel Pratt truss of 40′ span with these loads . . ." What he's asking primarily is for you to arrive at sizes of the structural members. Generally, these sizes are found by equating the actual stress in the member to the allowable stress dictated by the building code. This "sizing" process is what most schools usually teach as "design." The general creative process is not easily taught; it is usually learned through experience. Of course, in the sizing process it is necessary to know the loads (forces and moments) the structural element is subjected to. In addition, building codes frequently put limits on deformations a structure may undergo. Thus, we are back to structural analysis.

As previously mentioned, most of the theoretical base of structural analysis has been laid down in mechanics of materials; consequently, a course in structural analysis is primarily technique oriented. This text will lean heavily on fully worked examples and, of course, homework problems. It is written as a text, as opposed to a reference book; therefore, only a limited number of the many techniques available for treating deformations and stress resultants will be included. The presented techniques are ones we feel are best and most fundamental for the type of structure under consideration. We feel it is better for the student to study fully a small number of widely applicable techniques rather than a large number of more specialized approaches. Understanding is in the practice and with understanding comes confidence in one's abilities—too often a neglected component in education.

Throughout the book will be reminders of the importance of calculating stresses. While most structural analysis books end their analysis with the computation of internal forces and moments, we shall frequently go the extra step and calculate, for a given structural shape, the corresponding stresses.

Many homework problems throughout the book are designed as a series for which the results are meant to be compared. These problems are primarily of the type in which design parameters are varied. By judiciously comparing and discussing the results of the problems a feeling (albeit limited) for design may be instilled. Whenever possible, notes and problems will indicate how analysis is naturally extended to design. In general, we see structural analysis not as a course in itself but as just one phase in the analysis-design continuum.

A REVIEW OF BASIC MECHANICS

These next few paragraphs are meant to be reiterations of the major features of statics and mechanics of materials that will be of use in structural analysis. These statements are not meant to be derivations or full explanations of the involved phenomena but reminders, to bring certain ideas to the fore. If any of these statements appears unfamiliar, we recommend a thorough reappraisal in a statics or mechanics of materials text.

Statics

The important quantities in statics are those of force and the moment about a point due to force.

Force, a vector quantity, is defined as that quantity which causes a body to accelerate in a translational manner. Forces on a body are due either to contact with another body or to noncontact field phenomena (gravitational, electromagnetic and electrostatic fields). On the other hand, rotational acceleration of a body about a point P is due to the moment about point P caused by a force. This moment is also a vector quantity and has magnitude equal to the magnitude of the force times the lever arm (perpendicular distance from point P to the force).

If a body is in equilibrium, the sum of all forces and the sum of all moments about any point must balance to zero. The forces on a body are best seen by drawing a free body diagram of the body. A free body diagram is a picture of the body isolated from its surroundings. Where the body had come into contact with its surroundings, forces and/or couples are shown that properly represent the type of contact. In addition, noncontact forces, mostly gravitational, are shown on the free body diagram.

Some important types of contact supports are the (frictionless) pin, (frictionless) roller and the (ideal) wall or rigid weld support (Figure 1.1).

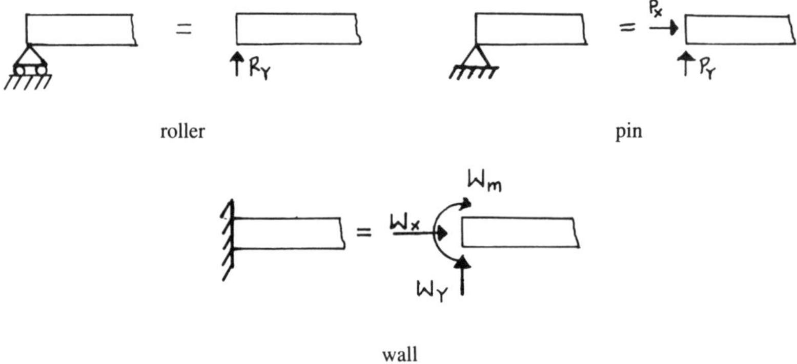

Figure 1.1

Mechanics of Materials

As previously mentioned, mechanics of materials is characterized by the internal phenomena stress and strain.

Stress is basically the internal force per unit area (with units of psi, ksi, N/cm^2, etc.). There are two distinct types of stress: normal stress (σ_x and σ_y on figure) which acts normal to a surface and shear stress (τ_{xy} on figure) which acts parallel to a surface. Shown in Figure 1.2 are planar stresses that may act on an element of a plate. The block has a horizontal, x, and vertical, y, orientation. Note rotational equilibrium requires that the shear stresses on orthogonal planes have the same magnitude.

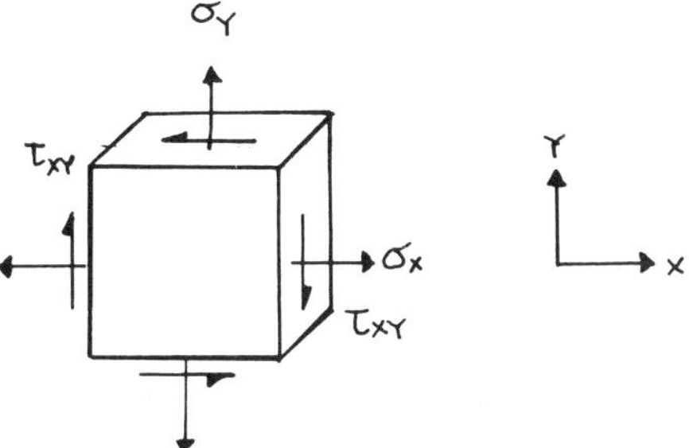

Figure 1.2

It is important to realize that the magnitude of both normal and shear stresses acting at a point are dependent upon the orientation of the cutting planes. The Mohr's circle is useful in finding the stresses acting on oblique planes once the values of the stresses acting at the *xy* orientation are known (Figure 1.3).

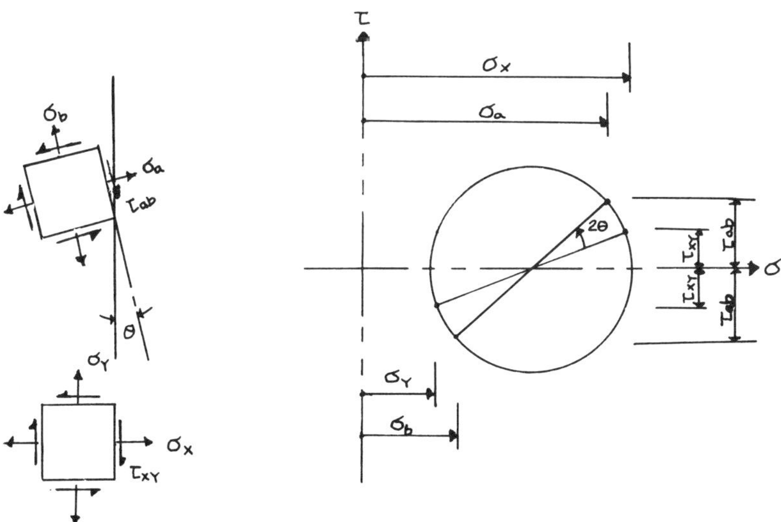

Figure 1.3

Strains, which also transform according to a Mohr's circle, are quantities that are defined in terms of deformation. Again there are two types of strain: lineal or axial strains (ε_x, ε_y) are defined as the changes in length per unit length; shear strain (γ_{xy}) is defined as the change in angle. Strains, while literally percentages, are usually measured in units of in./in., μin./in., or cm/cm.

The amount of deformation a body undergoes for certain levels of stress depends upon the properties of the materials. Most of the common structural materials obey "Hooke's law," which implies that strain is linearly related to stress. Hooke's law relates axial strains to only normal stresses and relates shear strain directly to shear stress. (Of course, here the same coordinate system is used for stress as for strain.) For a two-dimensional state of stress Hooke's law says:

$$\varepsilon_x = \frac{1}{E}(\sigma_x - \mu\sigma_y).$$

Here E and μ are Young's modulus and Poisson's ratio, respectively, which are experimentally determined. Further,

$$\varepsilon_y = \frac{1}{E}(\sigma_y - \mu\sigma_x),\ \varepsilon_z = \frac{-\mu}{E}(\sigma_x + \sigma_y),$$

$$\gamma_{xy} = \frac{\tau_{xy}}{G}\ (G = \text{shear modulus}).$$

Note in general a two-dimensional stress field causes three-dimensional deformation.

For our purposes the two most important loading situations studied in mechanics of materials are those of uniaxial tension and pure bending.

If a cylindrical member is subjected to an axial resultant force, P, acting through the centroid of the cross-section, a one-dimensional state of stress results. Here the axial normal stresses, σ_x, are uniform throughout and are equal to the resultant force, P, divided by the cross-sectional area, A (Figure 1.4).

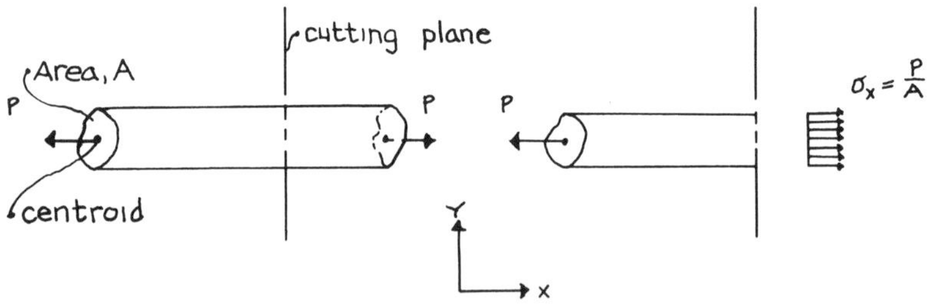

Figure 1.4

For a cylindrical member, with a cross-sectional plane of symmetry, subjected to only end couples, M, acting in the plane of symmetry, again a one-dimensional state of stress develops. Here the bending (axial, normal) stresses vary linearly through the depth of the cross-section:

$$\sigma_x = \frac{-My}{I},$$

where y is the coordinate at the point of concern measured from the centroid and I is the moment of inertia of the cross-section about the axis of bending (Figure 1.5).

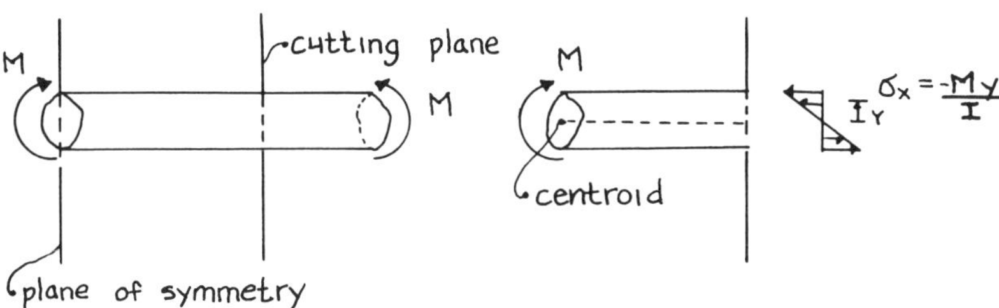

Figure 1.5

Loadings

A dominant consideration in the design of a structure is that of the loads to which the structure is subjected. These loads include the weight of the structure itself (called "dead load") and several "live loads," beginning with the weight of people, furniture and machines (collectively called "occupancy loads"). Other important live loads are the weight of snow, the forces of winds and earthquakes, the pressures of water (as for dams) and earth (as for basements) and truck and train loads (for bridges).

Many of the live load specifications are dictated by law in state and local building codes. Curiously, there is no nationally adopted code although there are several "unofficial" codes. To name two: the "American Standard Building Code," primarily for buildings, and the provisions set forth by the American Association of State Highway and Transportation Officials. Often local codes will refer to one or more of the "unofficial" codes for specifications or the local code will adopt its own rules.

Dead loads—the weight of walls, beams, roofs, floors, etc.—must be estimated before analysis and design are begun. Tables for estimating are available in the American Standard Building Code, as well as in other codes.

The live loads due to occupancy are postulated on a statistical basis. For instance, in a classroom it is impossible to reliably determine, for the life of the building, where students will sit and walk. Thus, instead of a series of questionably placed concentrated loads (the individual occupants) the codes recommend a uniformly distributed load for classrooms of 40 psf. Other cases are delineated in the American Standard Building Code.

Snow loads on roofs vary greatly with locale. Again a statistical approach is used, based on local weather bureau data. Typical snow loads vary from 80 psf in Northern Maine to zero in Southern Florida. In high mountain elevations, however, 100 psf snow loads may be specified.

Wind load specification is more complex than snow load determination because wind loads depend not only upon statistical maximum local wind speeds but on the building's shape, height, surface characteristics and overall building stiffness. In recent years a lack of understanding of wind loading has been emphasized by bothersome "failures." The "popping" of windows in the John Hancock Building in Boston in the early 1970's comes to mind. It is reasonably common these days for wind loads on large buildings to be estimated through the use of wind tunnel model testing. These models deal with the specific building and its surroundings—other large structures, mountains, etc.

Earthquake loads are generally specified as horizontal dynamic forces induced on the building's foundation. The character of these forces depends upon many not-well-understood factors like the building's stiffness, height and foundation, as well as soil stiffness and the geographical situation of the structure.

Paradoxically, structural engineers spend a great deal of time and effort calculating internal forces and moments in structures while the value of the loads that cause these internal reactions are at best arrived at through "educated guesswork." This does point up the inherent conservativeness of structural design. When in doubt the safe route must be chosen.

2

TRUSSES

A truss is a structure composed entirely of straight members arranged to form one or more triangles. The triangular configuration gives the structure stability.

In analyzing a truss, three assumptions are imposed:

Assumption 1) all joints are frictionless pin connections;

Assumption 2) all loads are applied at the truss joints;

Assumption 3) the weight of truss members can be neglected.

In reality, of course, no joint connection is truly frictionless and, in fact, in current practice high strength bolts and welds have all but replaced pin connections. Nonetheless the couples formed by friction, bolts and welds are considered to be secondary in importance. Thus, for a first analysis the frictionless pin assumption is used.

While the first assumption is made for analytical expediency, the second is followed in practice. For example, for a roof, snow loads are transferred from the roof surface itself to the truss joint through secondary structural components known as purlins (Figure 2.1). In other structures similar framing is provided to insure that all loads are transferred to the joints.

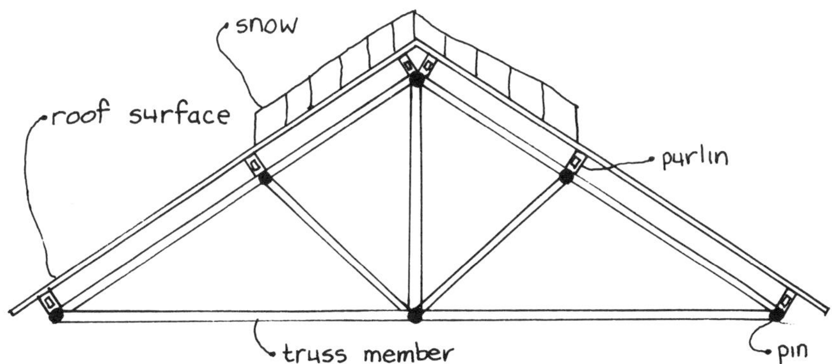

Figure 2.1

Clearly no structural member is weightless; however, usually the weight of a truss is less than 5% of the total design load. In cases where weight cannot be neglected it is assumed a member's weight will act at its two ends rather than more correctly at its centroid, thus, transferring weight to joint connections as imposed by Assumption 2.

The total result of these three assumptions is that each ideal truss member is subjected only to axial forces. That is, each member is either in a state of pure tension or pure compression with no bending. In statics these members are referred to as "two-force members." In general, a two-force member is one in which the member is subjected to only two resultant forces at only two points of contact and no couples (Figure 2.2).

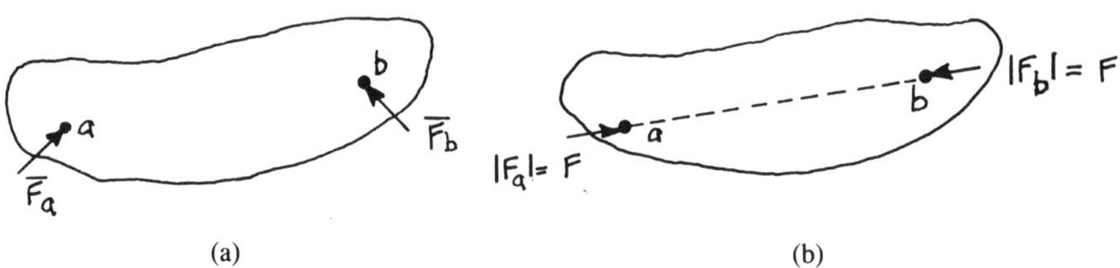

Figure 2.2

Equilibrium requires that the moments about any point must sum to zero. If in Figure (a) we take moments about point *a* this would require that $\bar{F}_b$ cause no moment about *a*. Therefore $\bar{F}_b$ must pass through point *a*; similarly, $\bar{F}_a$ must pass through point *b*. Finally, summation of forces requires $\bar{F}_a$ and $\bar{F}_b$ be equal in magnitude and opposite in direction. The result: the forces of a two-force member must be equal in magnitude, opposite in direction and collinear. For a truss member, say *ab* of Figure 2.3, only two points of contact are present, *a* and *b*. Even though many members and loads may come into joint *a* these members and external loads yield only one resultant force. Thus each ideal truss member is a two-force member.

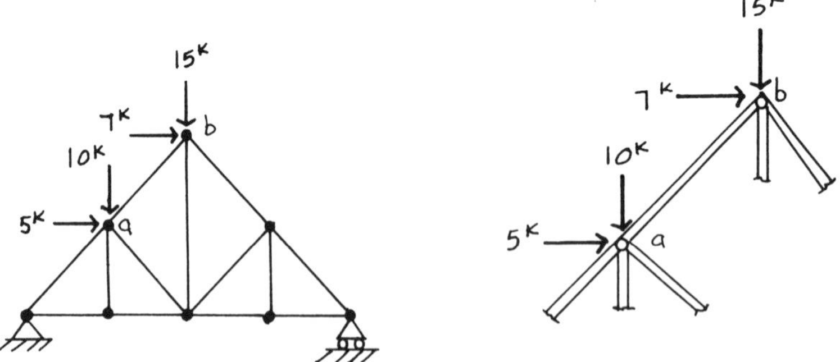

Figure 2.3

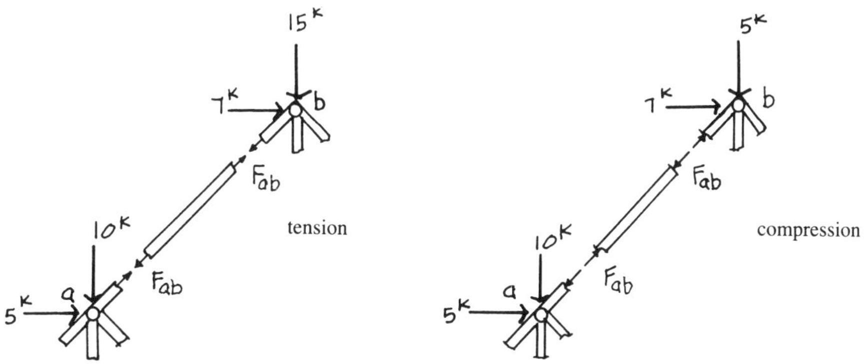

Figure 2.4

Each member is in either tension or compression (Figure 2.4).

FORCE ANALYSIS

Method of Joints

Since each member is subjected to only axial force, each joint is then a point of concurrency of all member forces and loads meeting at that joint. For each joint the number of independent coplanar equilibrium equations collapse to only two since the summation of moments about the point of concurrency yields absolutely no information. Thus the number of independent equations of equilibrium for the total truss is two times the number of joints minus three which is the number of equations used to find the external reactions for the truss. That is, for a planar truss to be in equilibrium and statically determinate:

$$2j - 3 = m, \quad \text{where } j = \text{number of joints and}$$
$$m = \text{number of truss members.}$$

If $2j - 3 > m$ then the truss is either not in equilibrium or is in an unstable equilibrium position; if $2j - 3 < m$ then there are not sufficient available equations to determine all forces in the truss, i.e., the truss is statically indeterminate.

In the method of joints one usually starts by finding the external reactions. From there it is wise to identify a joint where only two truss members meet. By drawing the free body diagram of the joint (with each truss member force along the axis of the member and all external loads, including truss support reactions) the sum of the forces in the x and y directions allows the computation of the two hitherto-unknown forces. For bookkeeping purposes, we recommend assuming all unknown forces to be tensile, then after computation, a positive sign will mean tension while a negative sign will signify compression. Once the forces at this first joint are found, move close by to another joint where only two unknown truss forces exist; draw the free body diagram, sum forces in x and y directions and solve for these two "unknowns," then on to another "two-unknown" joint, and so on. As a point of caution, keep in mind that if a member is in tension at one joint (i.e., the force is pulling away from the joint) then it must also be pulling away from the joint at its other end (Figure 2.5). Saying that the force in member *ab* is to the right is totally meaningless unless one refers to the joint, e.g., "force *ab* is to the right at joint *a*." It is less ambigious and more meaningful to say "*ab* is in tension."

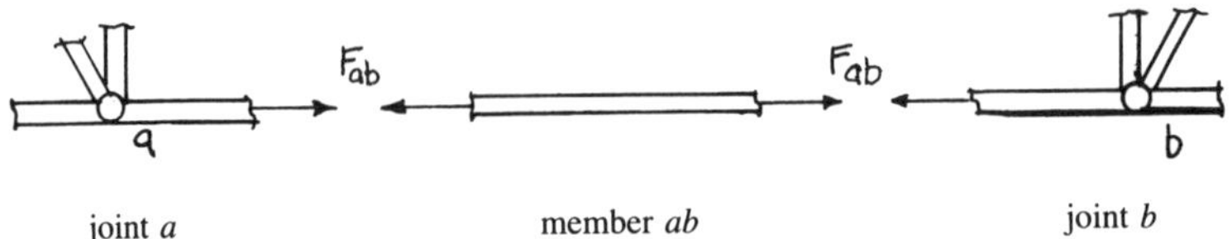

Figure 2.5

After sweeping systemically through the truss one would find at the last two joints only one unknown left but four equations of equilibrium. If all calculations up to that point were correct the excess three equations would each yield zero equals zero, so these three equations are useful as checks.

The entire process is amplified by Example 1. But first we must emphasize the use of clear free body diagrams and well organized work sheets. Force analysis for trusses is not difficult, but the amount of arithmetic can be substantial; therefore, if care is taken arithmetic errors can be avoided.

Method of Sections

Where the method of joints is a systematic way of determining all internal forces, the method of sections is best used to find only a few forces—either as a check to the method of joints or to find only the forces in members of concern.

In this method usually support forces are first determined (from the total FBD of the truss and the equations of equilibrium), then a section cut is made in order to divide the truss into two parts. When a member of a truss is "cut" the force within (along the member axis) is exposed. Since the entire truss is in equilibrium, every section of it is also in equilibrium. We thus arrive at a general coplanar system of forces in which three independent equations of equilibrium are available. Consequently by cutting through only three members we are assured of finding all three forces. Frequently, by judiciously choosing points about which moments are taken we can find one unknown with only one equation. Occasionally, if one is clever, it is possible to cut through more than three members and still determine some (but, of course, not all) of the unknowns. Example 2 will clarify the general process of sections.

Other Comments on Truss Force Analysis

A good "eye-ball" check for truss analysis comes by analogy of a truss and a beam. When a simply-supported beam is loaded with vertical forces the top fibers of the beam will be compressed and the bottom fibers will be extended (these will be in a state of tension). Similarly, the top cord members of a truss (Figure 2.6) will be in compression while the bottom cord will be in tension. A cantilever beam lends still another analogy. No general statement can be made, however, for the internal truss members.

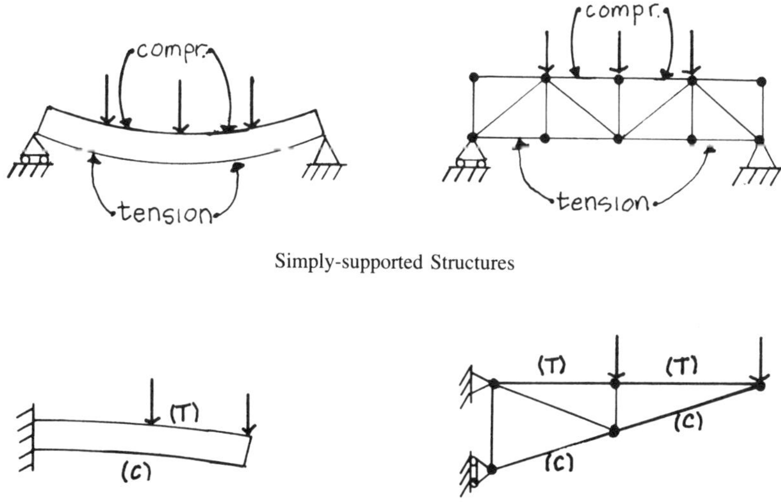

Simply-supported Structures

Cantilever Structures

Figure 2.6

Given: Truss as shown

Find: External reactions and all internal forces of truss

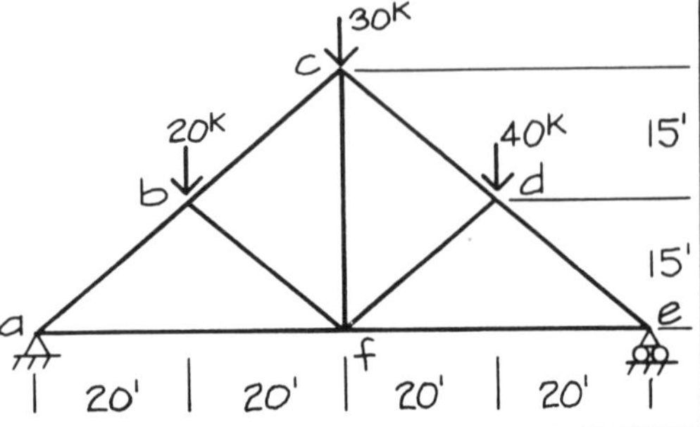

First, find the external reactions at a and e.

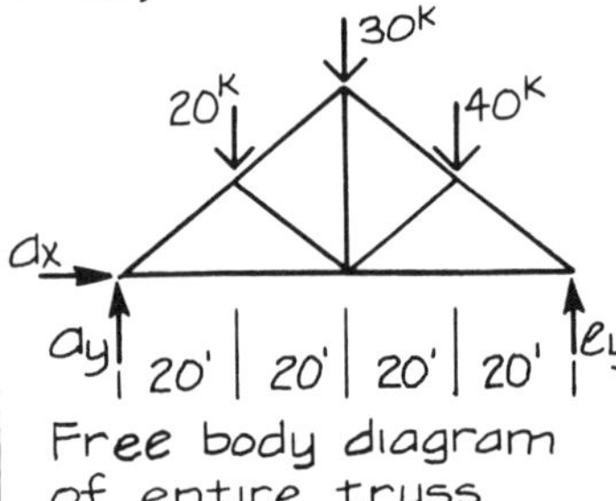

Free body diagram of entire truss

$\Sigma F_x = 0$: $a_x = 0$

$\Sigma M_a = 0$: $20(20) + 40(30) + 60(40) - 80(e_y) = 0$

$e_y = 50^K$ (↑)

$\Sigma F_y = 0$: $a_y - 20 - 30 - 40 + e_y = 0$

$a_y = 40^K$ (↑)

Start the joint analysis at a connection where only two members meet; either joint a or joint e. Choose a. Assume all members to be in tension.

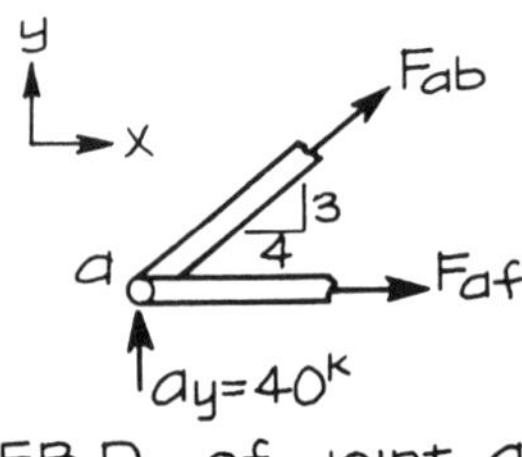

F.B.D. of joint a

$\Sigma F_y = 0$: $40 + \frac{3}{5}F_{ab} = 0$

$F_{ab} = -66.7^K$ (compr.)

$\Sigma F_x = 0$: $F_{af} + \frac{4}{5}F_{ab} = 0$

$F_{af} + (.8)(-66.7) = 0$

$F_{af} = 53.3^K$ (tension)

Next, on to joint b, being sure F_{ab} is in compression.

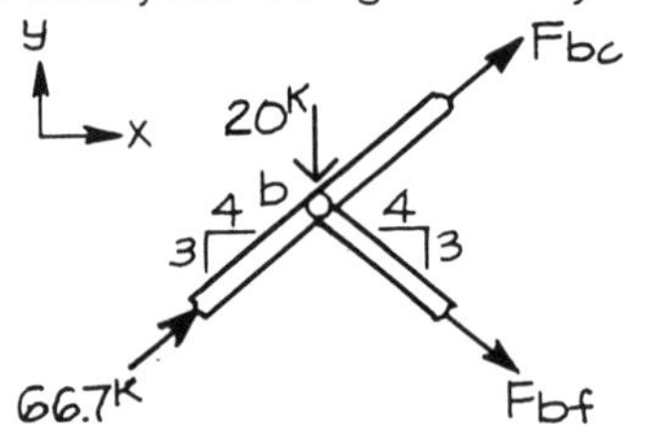

$\Sigma F_x = 0$ $\quad \frac{4}{5}F_{bc} + \frac{4}{5}F_{bf} + \frac{4}{5}(66.7) = 0$

$\Sigma F_y = 0$ $\quad \frac{3}{5}F_{bc} - \frac{3}{5}F_{bf} + \frac{3}{5}(66.7) - 20 = 0$

solving simultaneously yields

$F_{bc} = -50^K$ (C) $\qquad F_{bf} = -16.7^K$ (C)

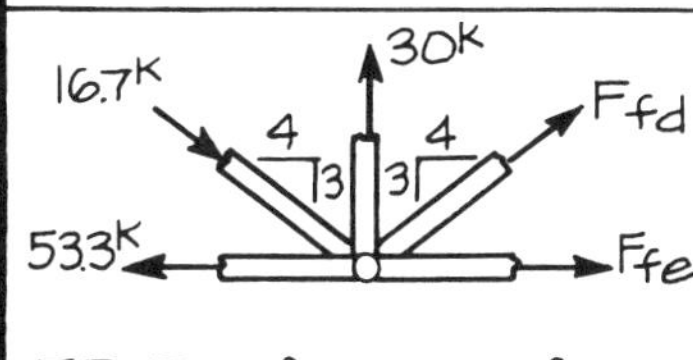

F.B.D. of joint c

$$\Sigma F_x = 0: \quad \tfrac{4}{5}F_{cd} + \tfrac{4}{5}(50) = 0$$
$$F_{cd} = -50^K (C)$$

$$\Sigma F_y = 0: \quad -F_{cf} - 30 + \tfrac{3}{5}(50) - \tfrac{3}{5}(-50) = 0$$
$$F_{cf} = 30^K (T)$$

16.7K, 30K, Ffd, 53.3K, Ffe, 4, 3, 4, 3

F.B.D. of joint f

$$\Sigma F_y = 0: \quad \tfrac{3}{5}F_{fd} - \tfrac{3}{5}(16.7) + 30 = 0$$
$$F_{fd} = -33.3^K (C)$$

$$\Sigma F_x = 0: \quad F_{fe} + \tfrac{4}{5}(-33.3) + \tfrac{4}{5}(16.7) - 53.3 = 0$$
$$F_{fe} = 66.3^K (T)$$

At joint d there is only one unfound force (F_{de}); therefore, we can use one equation as a check.

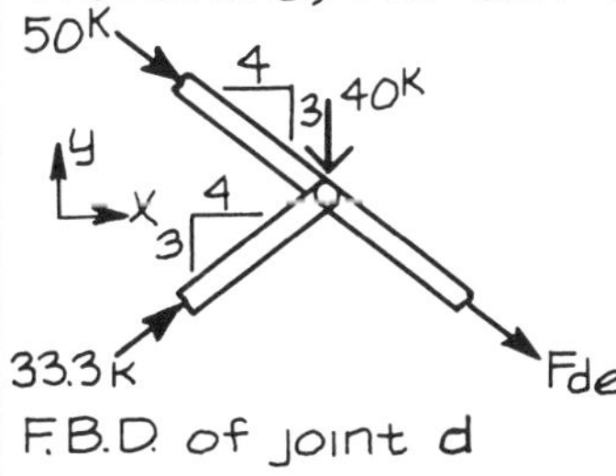

F.B.D. of joint d

$$\Sigma F_x = 0: \quad \tfrac{4}{5}F_{de} + \tfrac{4}{5}(50) + \tfrac{4}{5}(33.3) = 0$$
$$F_{de} = -83.3^K (C)$$

$$\Sigma F_y = 0: \quad -\tfrac{3}{5}F_{de} - 40 - \tfrac{3}{5}(50) + \tfrac{3}{5}(33.3) = 0$$
$$F_{de} = -83.3^K (C)$$
checks

At joint e there are no unknown forces, thus, both available equations of equilibrium will provide checks.

83.3K, 4, 3, 66.3K, 50K, y, x

F.B.D. of joint e

$$\Sigma F_y = 0: \quad -\tfrac{3}{5}(83.3) + 50 \equiv 0 \quad \text{checks}$$

$$\Sigma F_x = 0: \quad \tfrac{4}{5}(83.3) - 66.3 \equiv 0 \quad \text{checks}$$

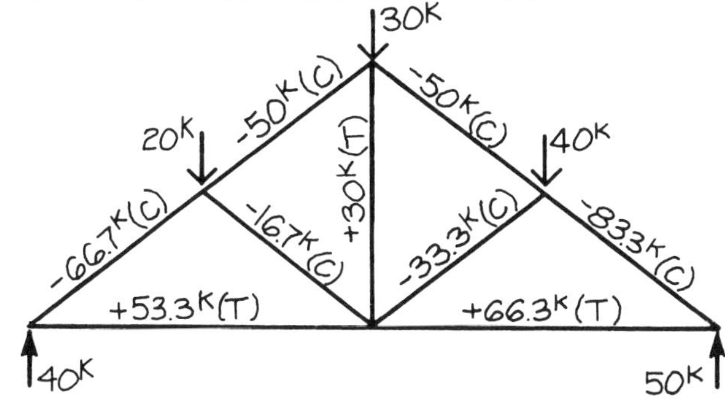

EXAMPLE #2: TRUSS FORCE ANALYSIS: METHOD OF SECTIONS

Find the forces in members af, bc, and bf of the truss in example #1.

First find the support reactions at pin a and roller e.

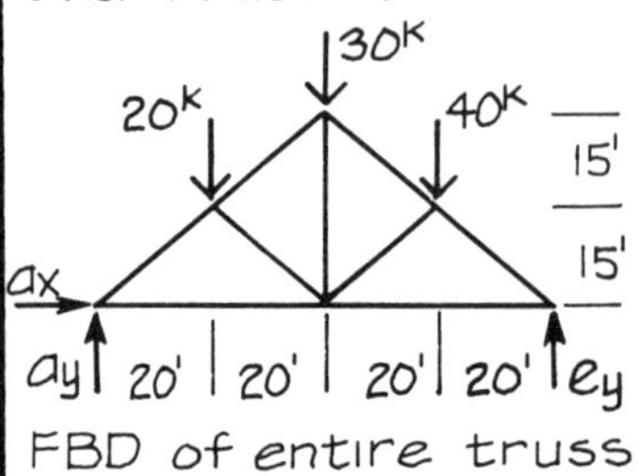

FBD of entire truss

$\Sigma F_x=0$: $a_x=0$

$\Sigma M_a=0$: $20(20)+40(30)+60(40)=80e_y$

$e_y=50^K(\uparrow)$

$\Sigma F_y=0$: $a_y-20-30-40+50=0$

$a_y=40^K(\uparrow)$

To find the desired forces we should not cut through more than three bars. The natural cut is that shown below.

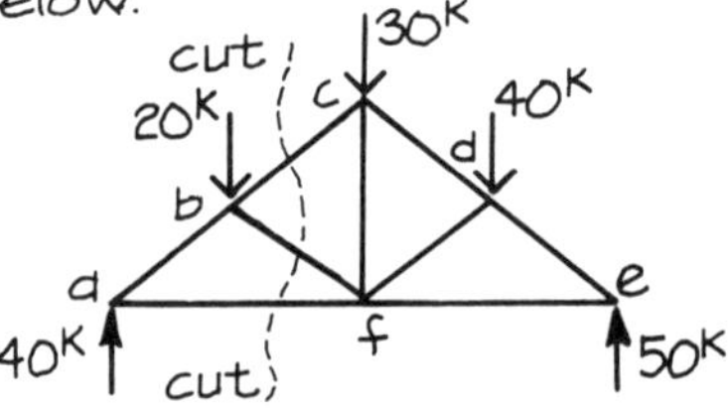

We can use either the left side or the right side as our section. Clearly, here the left side is less complicated. Since moments will be taken it is convenient to decompose F_{bc} and F_{bf} into their X & Y components as shown.

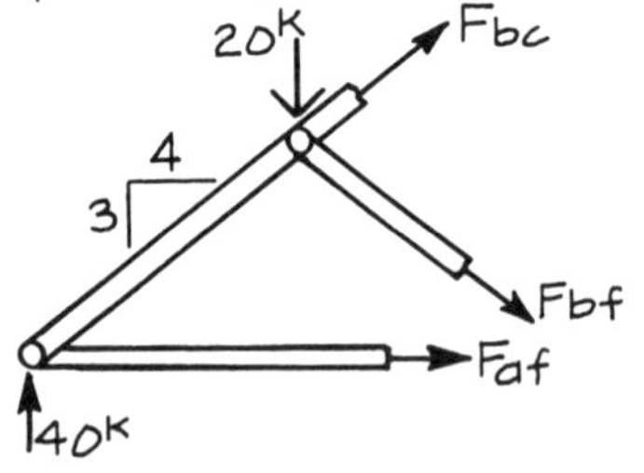

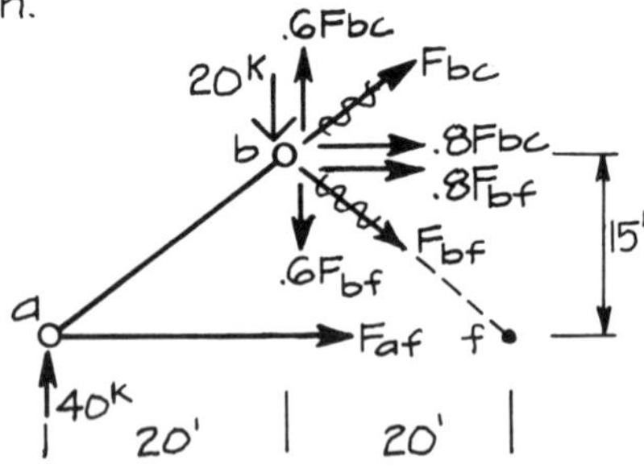

By taking moments around point f we eliminate unknowns F_{af} and F_{bf}.

$\Sigma M_f=0$: $40(40)-20(20)+15(.8F_{bc})+20(.6F_{bc})=0 \therefore F_{bc}=-50^K(C)$

Similiary, taking moments about b & a finds F_{af} & F_{bf}

$\Sigma M_b=0$: $20(40)-15(F_{af})=0 \quad \therefore F_{af}=53.3^K(T)$

$\Sigma M_a=0$: $20(20)+20(.6F_{bf})+15(.8F_{bf})=0 \therefore F_{bf}=-16.7^K(C)$

If in Example 1 the external load at joint *d* was not applied, the FBD of that joint would look like that in Figure 2.7.

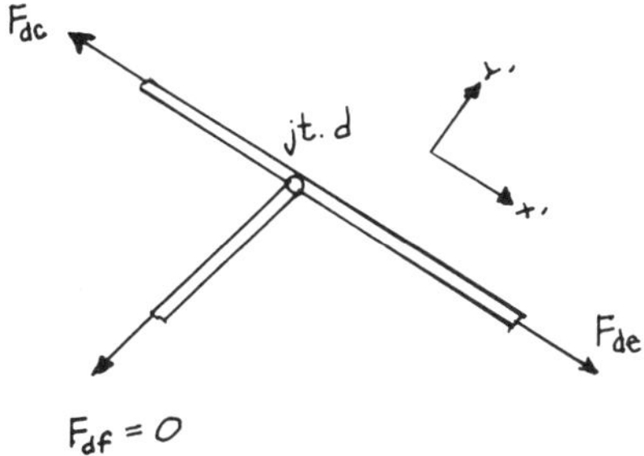

Figure 2.7

If we choose axis x' and y' as shown, then $\Sigma Fy' = 0$ implies $F_{df}\sin\theta = 0$ or $F_{df} = 0$. To generalize, any time only three forces meet at a joint and two are collinear the third one must be a zero force member. Similarly, if only two forces meet at a joint (Figure 2.8), then $\Sigma Fy = 0$ and $\Sigma Fx = 0$ implies both members are zero force members.

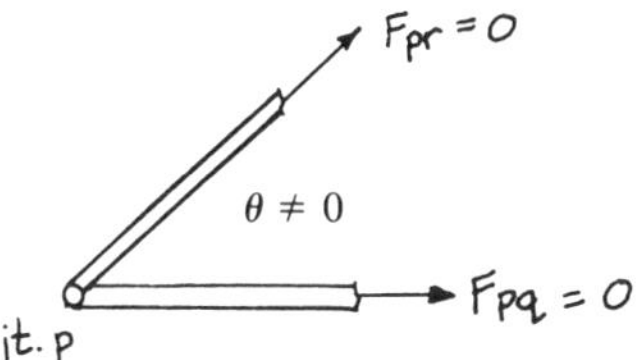

Figure 2.8

The natural question arises as to why zero force members are included in trusses. There are two practical answers: 1) usually the considered load pattern is only one of several to which the truss is subjected; with other loadings this member may not be zero in force; 2) a zero force member may be designed as a brace which prevents a long compressive member from buckling.

Stresses in Truss Members

As long as a truss member does not buckle the state of stress is simple: there exists a constant axial normal stress (along the axis of the member) equal to the axial force, P, divided by the cross-sectional area. When, however, the member is subjected to compression, the chance of buckling occurs. Before buckling, the stress is P/A but after buckling takes place, the stress rapidly changes to $P/A \pm My/I$. Thus buckling is a catastrophic phenomenon to be avoided. Before we investigate buckling we shall first generate the general deflection equation which is applicable to both buckling of truss members and bending of beams.

THE DEFLECTION EXPRESSION

When a truss member buckles it is actually bending. Thus, to determine the minimum load that causes a truss member to buckle we must look at the relationship between the geometry of bending and the loads that cause bending.

Consider a cantilever beam (Figure 2.9) subject to only an applied moment. Further consider two neighboring points a and b on the neutral axis of the beam. The radius of curvature of the neutral axis, ρ, sweeps out the arc $a'b'$, i.e., $a'b' = (\Delta\theta)\rho$.

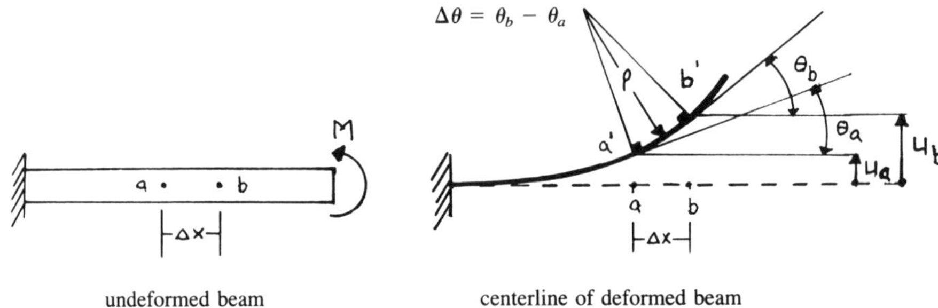

Figure 2.9

But since a and b are on the neutral axis, $a'b'$ is the same length as $ab = \Delta x$. Then

$$\frac{1}{\rho} = \frac{\Delta\theta}{\Delta x}.$$

in the limit

$$\frac{1}{\rho} = \lim_{\Delta x \to 0} \frac{\Delta\theta}{\Delta x} = \frac{d\theta}{dx},$$

also in the limit

$$\theta \equiv \theta_a = \lim_{\Delta x \to 0} \frac{u_b - u_a}{\Delta x} = \lim_{\Delta x \to 0} \frac{\Delta u}{\Delta x} = \frac{du}{dx},$$

where u is the displacement (deflection) of a point on the neutral axis. Hence the curvature of the neutral axis is

$$\frac{1}{\rho} = \frac{d^2u}{dx^2}.$$

Now look at points c and d, y units off the neutral axis (Figure 2.10).

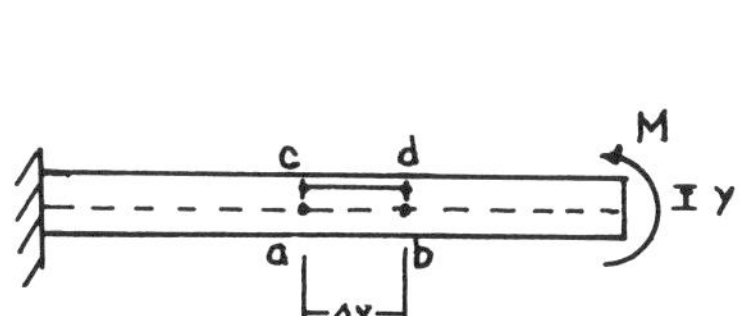

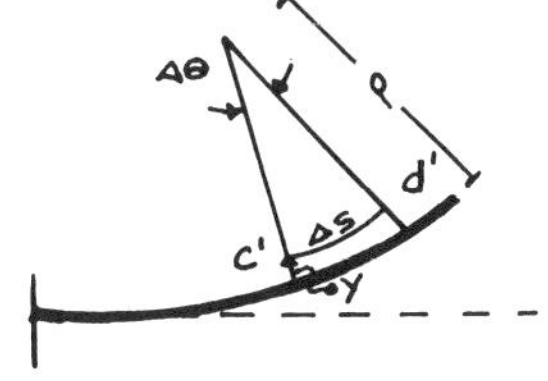

Figure 2.10

The distance between c' and d', Δs, is

$$c'd' \equiv \Delta s = (\rho - y)\,\Delta\theta.$$

Therefore the strain of line cd

$$\varepsilon_{cd} \equiv \lim_{\Delta x \to 0} \frac{\Delta s - \Delta x}{\Delta x} = \frac{(\rho - y)\,\Delta\theta - \rho\,\Delta\theta}{\rho\,\Delta\theta}$$

$$\varepsilon_{cd} = \frac{-y}{\rho} \equiv -y\,\frac{d^2u}{dx^2}. \tag{1}$$

Recall the axial stress at cd is

$$\sigma_{cd} = \frac{-My}{I}.$$

Since we have a one-dimensional stress picture, Hooke's law implies

$$\varepsilon_{cd} = \frac{\sigma_{cd}}{E} = \frac{-My}{EI}. \tag{2}$$

Equating the two statements for ε_{cd} Equations 1 and 2,

$$\varepsilon_{cd} = -y\,\frac{d^2u}{dx^2} = \frac{-My}{EI}, \text{ or}$$

$$\frac{d^2u}{dx^2} = \frac{M}{EI}. \tag{3}$$

This relation says the curvature at a point, which in itself is the second derivative of the deflection with respect to the axial variable x, is proportional to the bending moment at that point. The proportionality constant is the inverse of Young's modulus times the moment of inertia (about the plane of bending).

Note: A positive bending moment means a positive curvature, i.e., the deformed beam "smiles" for positive bending and thus "frowns" for negative bending.

BUCKLING

Now consider a pinned end member subjected to a resultant axial compressive load that acts through the centroid of the cross-section (Figure 2.11). We seek the value of the load that causes the member to buckle. In our investigation there are two assumptions:

1) the member has buckled,
2) the stress is still in the linear elastic range.

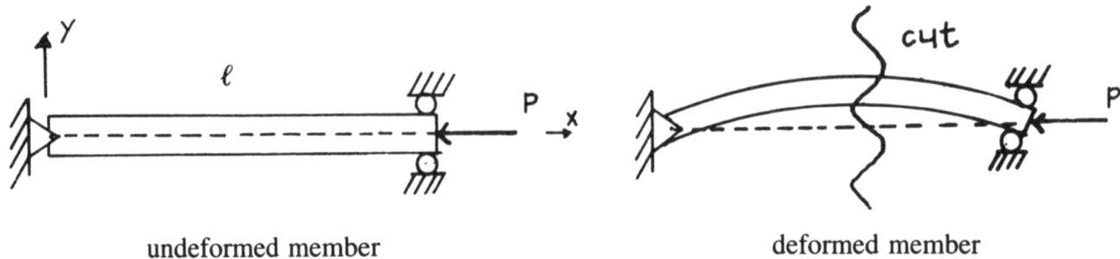

undeformed member

deformed member

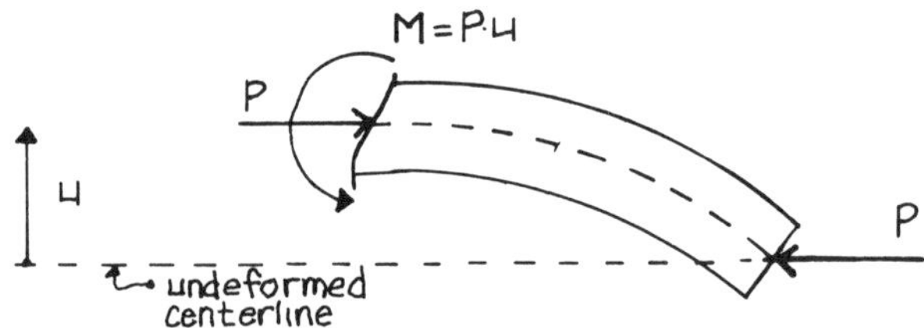

F.B.D. of right section

Figure 2.11

For equilibrium we see that a bending moment is necessary. Further, the magnitude of this moment equals the load, P, times the value of the deflection, u, at the point of concern. Here we have negative bending; thus Equation 3, which is restricted to elastic bending, states:

$$\frac{d^2u}{dx^2} = -\frac{M}{EI} \equiv \frac{-(Pu)}{EI}, \text{ or}$$

$$\frac{d^2u}{dx^2} + \frac{Pu}{EI} = 0. \tag{4}$$

This second order ordinary differential equation has the general solution:

$$u = C_1\sin\left(\sqrt{\frac{P}{EI}}\,x\right) + C_2\cos\left(\sqrt{\frac{P}{EI}}\,x\right), \tag{5}$$

where C_1 and C_2 are constants.

For those unfamiliar with differential equations it is easily proved that Equation 5 is the solution to Equation 4; take two derivatives of Equation 5, i.e.,

$$\frac{du}{dx} = \sqrt{\frac{P}{EI}}\, C_1\cos\left(\sqrt{\frac{P}{EI}}\, x\right) - \sqrt{\frac{P}{EI}}\, C_2\sin\left(\sqrt{\frac{P}{EI}}\, x\right)$$

and

$$\frac{d^2u}{dx^2} = -\left(\frac{P}{EI}\right)C_1\sin\left(\sqrt{\frac{P}{EI}}\, x\right) - \frac{P}{EI}\, C_2\cos\sqrt{\frac{P}{EI}}\, x$$

$$\equiv -\frac{P}{EI}\left(C_1\sin\sqrt{\frac{P}{EI}}\, x + C_2\cos\sqrt{\frac{P}{EI}}\, x\right).$$

By Equation 5 the team in brackets is exactly u. Therefore

$$\frac{d^2u}{dx^2} = -\frac{P}{EI}\, u \qquad \text{or} \qquad \frac{d^2u}{dx^2} + \frac{Pu}{EI} = 0,$$

which is precisely Equation 4. Thus solution 5 satisfies Equation 4.

Applying boundary conditions to general solution 5 will give a specific solution. We know from the physical nature of the problem that there is no deflection at each end; thus

Boundary Condition 1, at $x = 0$, $u = 0$ and

Boundary Condition 2, at $x = \ell$, $u = 0$.

Putting Boundary Condition 1 in Equation 5 yields

$$0 = C_1\sin 0 + C_2\cos 0, \quad \text{or } C_2 = 0.$$

Putting Boundary Condition 2 in Equation 5 (with $C_2 = 0$) yields

$$0 = C_1\sin\sqrt{\frac{P}{EI}}\, \ell.$$

This means either $C_1 = 0$ or $\sin\sqrt{P/EI}\, \ell = 0$. If $C_1 = 0$ then no deflections exist at all, thereby violating our Assumption 1. Thus

$$\sin\sqrt{\frac{P}{EI}}\, \ell = 0.$$

This means $\sqrt{P/EI}\, \ell = 0, \pm \pi, \pm 2\pi$, etc.

Again the zero value means no deflection. Thus the minimum value to cause buckling, predicted by L. Euler in 1759, is

$$P_{\text{Euler}} = \frac{\pi^2 EI}{\ell^2}. \tag{6}$$

The corresponding stress at the onset of buckling is

$$\sigma_{\text{Euler}} = \frac{P_{\text{Euler}}}{A} = \frac{\pi^2 EI/A}{\ell^2}. \tag{7}$$

Recalling that the radius of gyration is defined as

$$r = \sqrt{I/A},$$

we can write (7) as

$$\sigma_{\text{Euler}} = \frac{\pi^2 E}{(\ell/r)^2}. \tag{8}$$

The ratio ℓ/r is referred to as the slenderness ratio. Plotting the Euler buckling stress, σ_{Euler}, versus the slenderness ratio we get a parabolic curve, Figure 2.12.

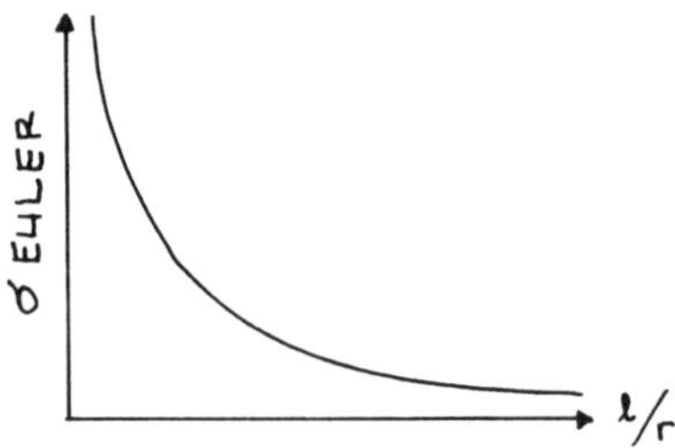

Figure 2.12

As with any theory, it is important to test the conclusions. In this case it is seen that for low ℓ/r values the predicted buckling stress doesn't conform with test results. The reason, on reflection, is simple. For lower values of ℓ/r (for steel around 100) Euler stresses are found to be above the proportional limit stresses and, of course, Euler's theory is only applicable for the linearly elastic range (Assumption 2).

Similarly, a brittle material might crush, rather than buckle, in this ℓ/r range. Therefore, in the low ℓ/r range where Euler's theory doesn't account for plastic buckling and crushing, empirical data is used to predict failure, Figure 2.13.

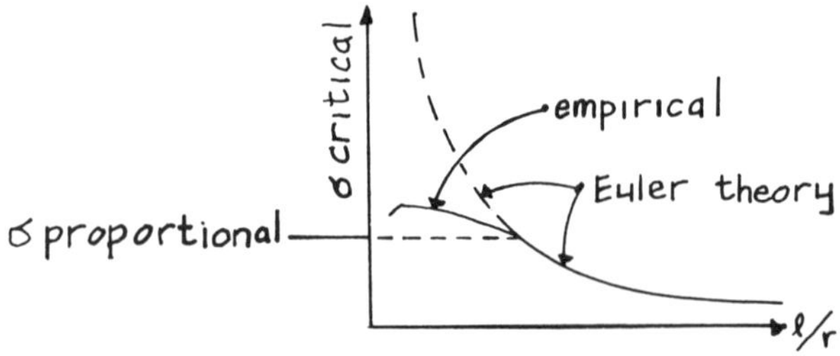

Figure 2.13

A similar approach is taken for other support conditions. The sketches (Figure 2.14) show the results. In general, the more constraint, the larger the required buckling load.

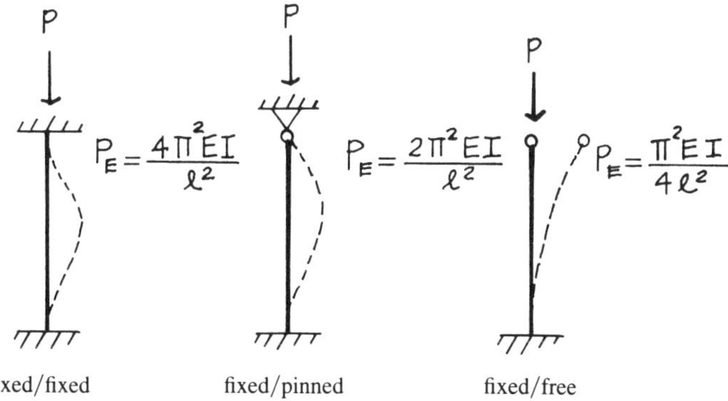

Figure 2.14

To illustrate how these buckling results are used in analysis and design, we shall look at a few example problems, but before buckling examples are cited, a few general words about moments of inertia are called for.

As we recall from previous courses (Statics and/or Strength of Materials), the moment of inertia of area about an axis is defined as a double integral ($I_x = \iint y^2\, dA$ and $I_y = \iint x^2\, dA$). The resulting number (units: length to fourth power) gives an indication of how the area of concern is clustered about a particular axis. The more the area is spread away from the axis, the higher the value of the moment of inertia about that axis.

Of course, a different orientation of the axis will yield a different value for the moment of inertia. This new value can be found directly by integration. By using a change of variables, we can relate the "new value" to "old values," namely Figure 2.15,

$$I_x' = \frac{I_x + I_y}{2} + \left(\frac{I_x - I_y}{2}\right)\cos 2\theta - I_{xy} \sin 2\theta,$$

where I_x and I_y are moments of inertia about x and y axes and I_{xy} is the product of inertia ($I_{xy} = \iint xy\, dA$) about the x and y axes. I_x' is the moment of inertia about the oblique x'-axis. The product of inertia, I_{xy}, can have positive, negative or zero values while I_x, I_y, I_x', etc. can only have positive values. In particular, I_{xy} is zero if either the x- or y-axis is an axis of symmetry. When I_{xy} is zero, the corresponding I_x and I_y

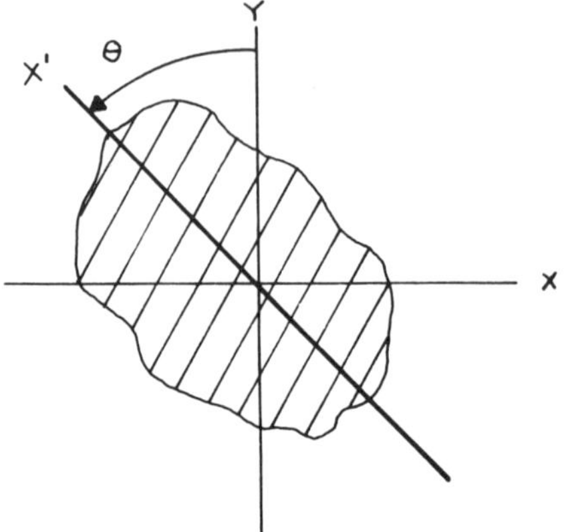

Figure 2.15

values are maximum and minimum moments of inertia through the point of concern (they are often called principle moments of inertia).

The parallel axis theorem (for moments of inertia) deserves some review. This theorem says that the difference in magnitude between moments of inertia about two parallel axes—one being a *centroidal* axis—is equal to the area of concern times the distance between the axes squared; the centroidal moment of inertia is the lesser value (Figure 2.16a). For example, knowing the moment of inertia of a rectangle about its centroid (Figure 2.16b) is $bh^3/12$, we can compute the moment of inertia about the base axis:

$$I_{\text{base}} = I_{\text{centroid}} + Ad^2$$

$$= \frac{bh^3}{12} + (bh)\left(\frac{h}{2}\right)^2 = \frac{bh^3}{3}.$$

Caution: The transfer must be done from the *centroidal* axis.

$$I_a = I_{ac} + Ad^2.$$

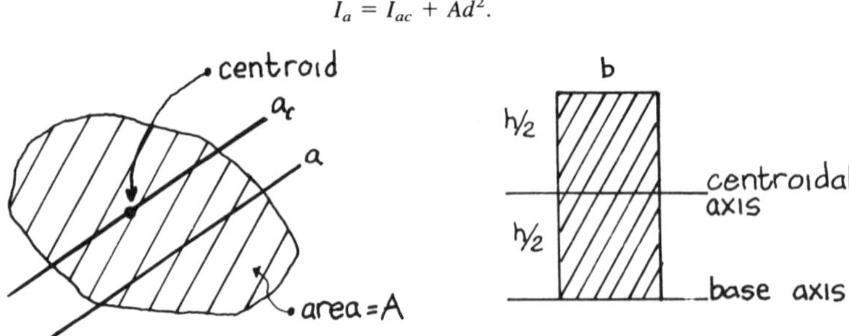

Figure 2.16a

Figure 2.16b

EXAMPLE #3: BUCKLING

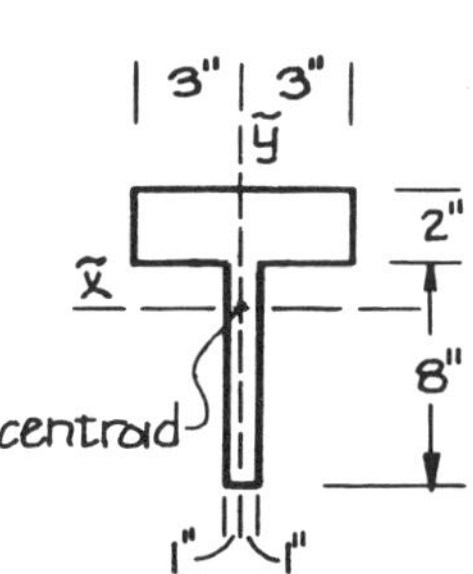

Find the buckling load and buckling stress for the column whose cross-section is shown.

The column is steel ($E = 29{,}000$ ksi), 20' long, load through its centroid and supported at both ends by ball and socket connections.

We first note that the area is symmetric about the $\tilde{y}$ axis (see figure above), thus the product of inertia $I_{\tilde{x}\tilde{y}}$ is zero. This leaves the $\tilde{x}$ and $\tilde{y}$ axes as principal axes of inertia which means the column will buckle about either of these two axes.

Find centroid: We will use the base as our reference axis and break area into two rectangles.

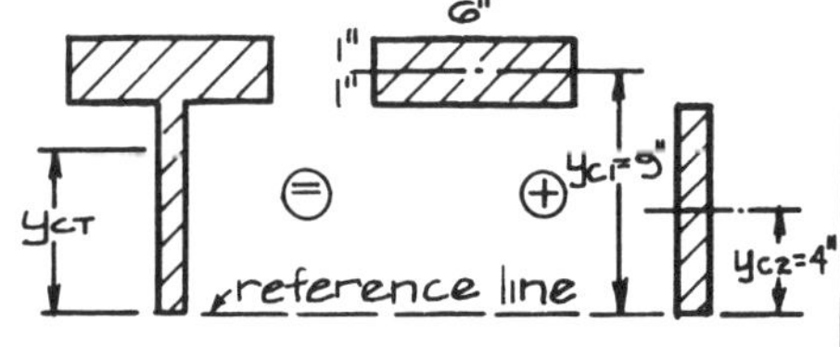

$$y_{CT} = \frac{\Sigma A y}{\Sigma A} = \frac{A_1 y_{c1} + A_2 y_{c2}}{A_1 + A_2}$$

$$= \frac{(12)(9)+(8)(4)}{12+8}$$

$y_{CT} = 7"$ from base

Find $I_{\tilde{x}}$: Using the parallel axis theorem twice.

$$I_{\tilde{x}} = I_{xc1} + Ad_1^2 + I_{xc2} + Ad_2^2$$

$$= \frac{(6)(2)^3}{12} + (12)(2)^2 + \frac{(1)(8)^3}{12} + (8)(3)^2$$

$$= 166.7 \text{ in}^4$$

Find $I_{\tilde{y}}$: No transfer is necessary since $\tilde{y}$ coincides with both y_{c1} and y_{c2} axes.

$$I_{\tilde{y}} = I_{y1} + I_{y2} = \frac{(2)(6)^3}{12} + \frac{(8)(1)^3}{12} = 36.7 \text{ in}^4$$

Clearly then buckling will initiate about $\tilde{y}$ axis since the resistance to rotation, as measured by the moment of inertia, is least about this axis.

Buckling Load: $P_{euler} = \dfrac{\pi^2 E I_{\tilde{y}}}{\ell^2} = \dfrac{\pi^2(29{,}000)(36.7)}{(20 \times 12)^2} = 182.3$ Kips

Buckling Stress: $\sigma = P/A = 182.3/20 = 9.12$ Ksi

$$\text{or} \quad \sigma = \frac{\pi^2 E}{(\ell/r_{\tilde{y}})^2} = \frac{\pi^2 (29{,}000)}{\left(240/\sqrt{36.7/20}\right)^2} = 9.12 \text{ Ksi}$$

EXAMPLE #4: BUCKLING

Given: A 20' pin-connected column which is subjected to an axial compressive load of 200K. We wish to make two designs. Both designs are based on 36 ksi yield stress steel. (E=29,000 ksi)

Design #1: Select a solid circular column.

Design #2: Select a tubular circular section with ½" wall thickness.

In both cases use the following design conditions.

a) Allowable compressive load is to be no more than .52 of the Euler critical load. (This means the safety factor against buckling is 1/.52 = 1.92)

b) To insure against plastic deformation the actual stress must be less than .52 of the yield stress (i.e. (.52)(36) = 18.7 ksi). Here a 1.92 safety factor is being used against plastic deformation.

Design #1

$P_{euler} = \frac{\pi^2 EI}{\ell^2}$; therefore the required moment of inertia is:

$$I_{Req} = \frac{1.92\,(P_{actual})\,\ell^2}{\pi^2 E} = \frac{1.92(200)(240)^2}{\pi^2(29{,}000)} = 77.28 \text{ in}^4$$

For a circular section I, of course, is the same for every axis passing through the centroid. This value for I is $\pi R^4/4$.

Therefore: $\pi R^4/4 = 77.28 \text{ in}^4$

$$R = 3.15''$$

Let's round this off (upward) to a nominal radius of R = 3.25"

With R = 3.25" check the second criteria.

$$\sigma_{actual} = \frac{P_{actual}}{A_{actual}} = \frac{200}{\pi(3.25)^2} = 6.03 \text{ ksi} < 18.7 \text{ ksi ok}$$

Thus, use a solid shaft of 3¼" radius

Design #2

Here the same required moment of inertia is found. $I_{required} = 77.28 \text{ in}^4$

Subtracting the "hole" from the larger circle with radius R_o, we find:

$$I = \pi/4\,[R_o^4 - (R_o - .5)^4]$$

thus: $\pi/4[R_o^4 - (R_o - .5)^4] = 77.28 \text{ in}^4$

solving we find $R_o = 3.91''$; round off.

Try $R_o = 4''$ (inside radius of 3.5")

Check actual stress:

$$\sigma_{actual} = \frac{P_{act}}{A_{act}} = \frac{200}{\pi(4)^2 - \pi(3.5)^2} = 16.98 \text{ ksi} < 18.7 \text{ ksi ok}$$

Use hollow shaft with 4" outside radius and 3.5" inside radius.

Note: For the hollow shaft design, if we used a thinner wall the possibility of local buckling of the wall may occur (similar to wall buckling when one steps on an upright empty beer can). This is a different type of buckling than the previously discussed buckling of the entire column.

Note: clearly the hollow column is a more efficient section compared to the solid one, for we get a much smaller cross-sectional area for the same value of the moment of inertia from the hollow section. In general, a hollow tube's area is less clustered to the centroidal axis than is a solid section's area.

PROBLEMS

The following truss problems are designed in sets of four in which all trusses in a set support the same total load and have the same span. The differences between trusses in a set are due to variations in truss depth and internal configuration.

It is recommended that one or more problems in a set be assigned per student so that all trusses in a set are covered by the class. It is also recommended that the effect of these design variations be discussed in class.

Problems 1–12

a) Find the force in each member of the truss by the method of joints; and
b) check three members in the truss by the method of sections.

Problems 13–24

For the truss shown, find the least weight of the truss using standard (steel) pipe (Figure 2.17) and the following design criteria:

For tension members:

The allowable (design) stress cannot exceed .6 of the yield stress of 36 ksi, i.e.,

$$\sigma_{\text{allowable}} \leq .6\sigma_{\text{yield}} = 21.6 \text{ ksi}$$

For compression members (two conditions):

1) The allowable (design) load cannot exceed .52 of the Euler critical buckling load, and
2) the actual stress cannot exceed .52 of the yield stress, i.e., $\sigma_{\text{actual}} \leq .52\sigma_{\text{yield}} =$ 18.7 ksi

COLUMNS Standard steel pipe								
Nominal Dia. (in.)	12	10	8	6	5	4	3.5	3
Wall Thickness (in.)	0.375	0.365	0.322	0.280	0.258	0.237	0.226	0.216
Weight per Foot (#)	49.56	40.48	28.55	18.97	14.62	10.79	9.11	7.58
	Properties							
Area A (in.^{2})	14.6	11.9	8.40	5.58	4.30	3.17	2.68	2.23
I (in.^{4})	279.	161.	72.5	28.1	15.2	7.23	4.79	3.02
r (in.)	4.38	3.67	2.94	2.25	1.88	1.51	1.34	1.16

Figure 2.17 From the *Manual of Steel Construction*, 8th edition, and used by permission of American Institute of Steel Construction, Inc., the copyright owner.

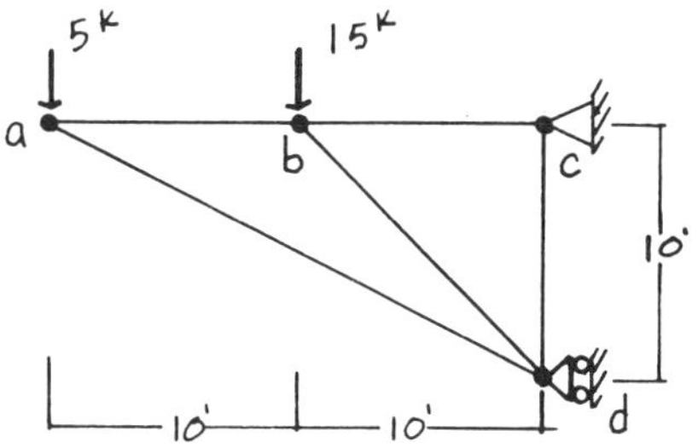

1 & 13

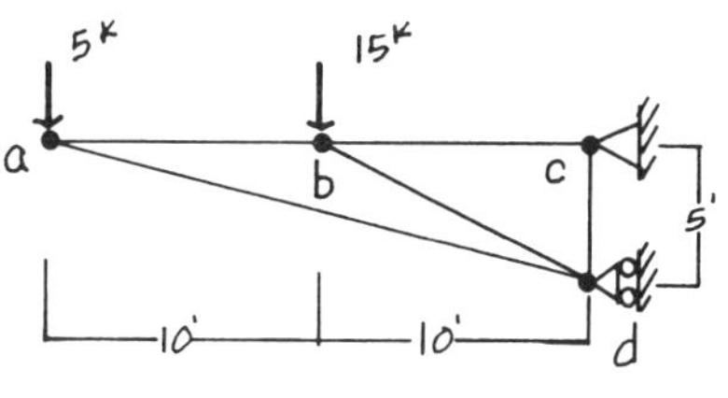

2 & 14

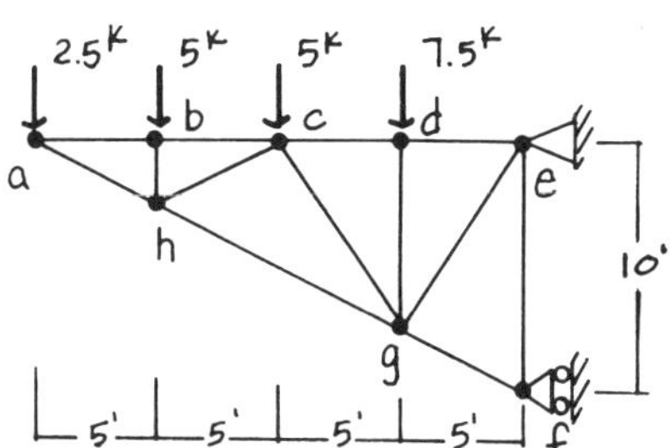

3 & 15

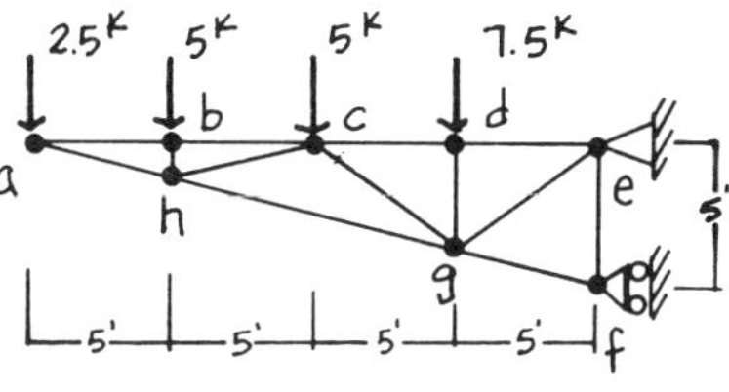

4 & 16

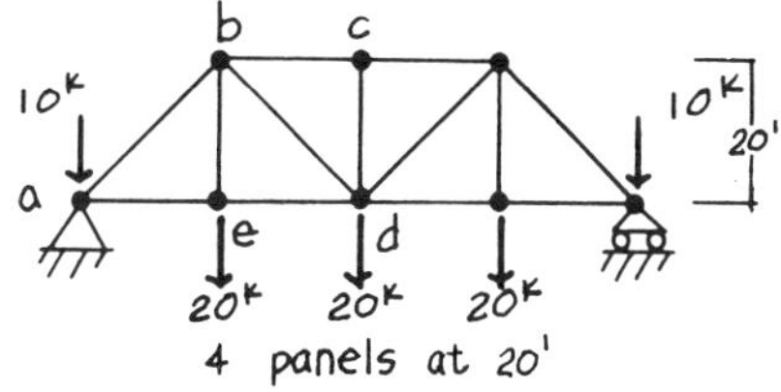

5 & 17

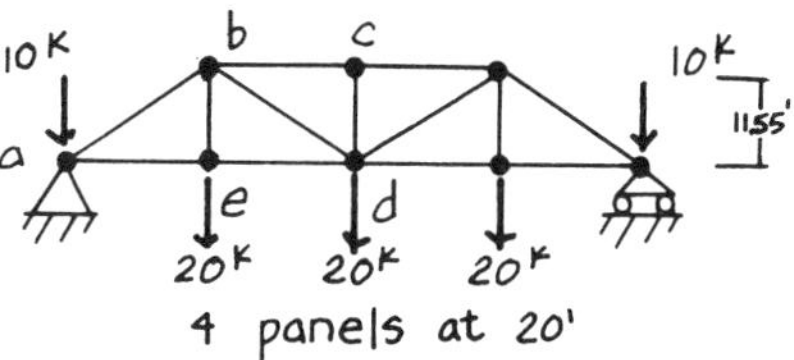

6 & 18

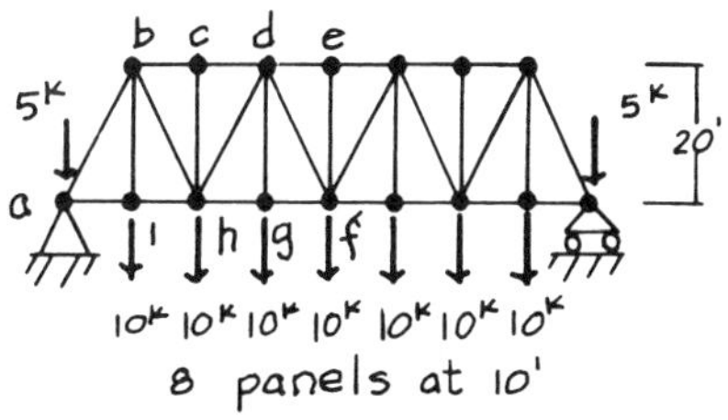

7 & 19

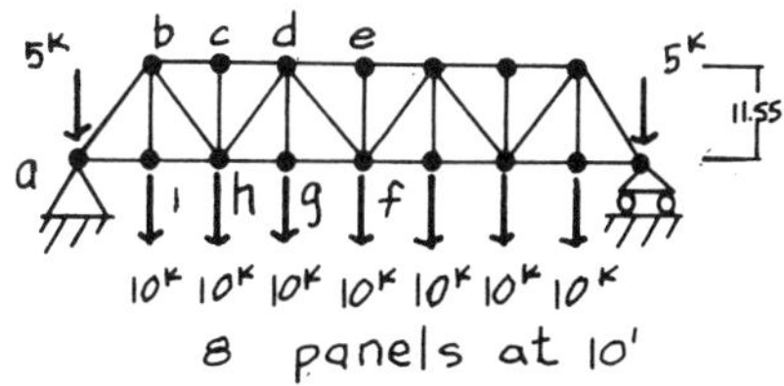

8 & 20

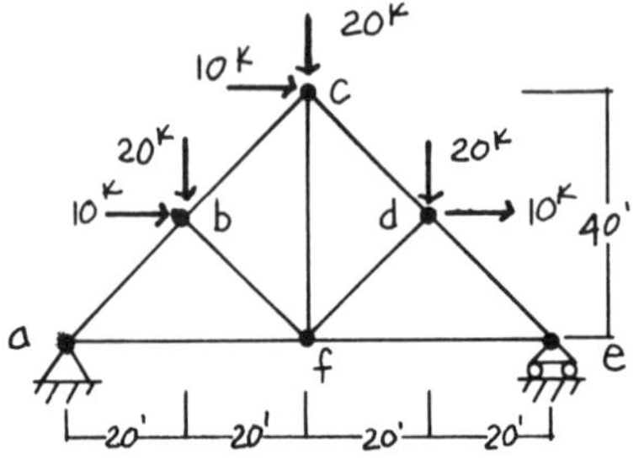

9 & 21

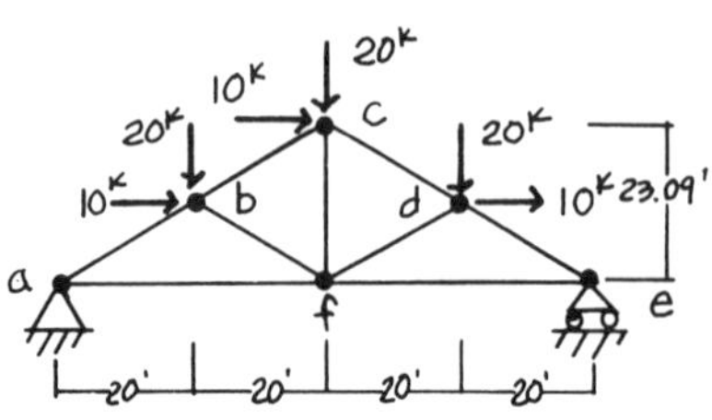

10 & 22

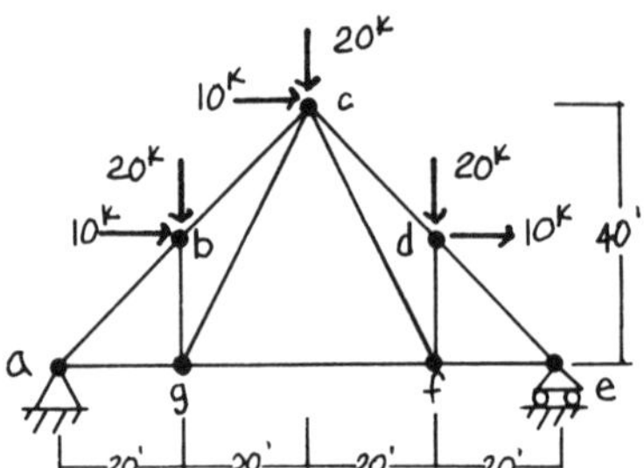

11 & 23

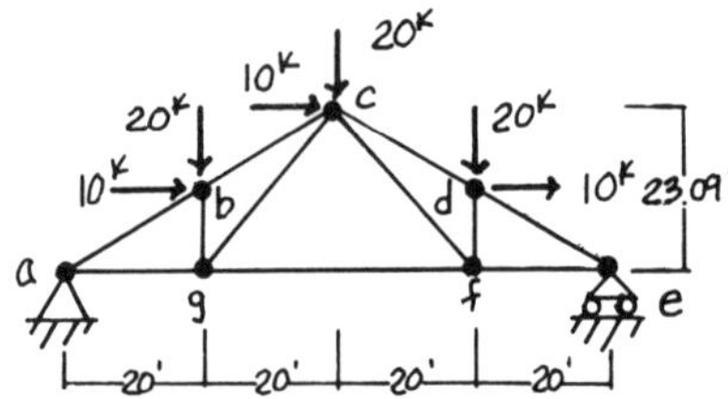

12 & 24

3

BEAMS

INTRODUCTION

Beams are the most common structural members. They are designed to support transverse, usually-gravity type, loads. The design of beams, like the design of most structural elements, is usually based on some stress criteria. In addition some codes restrict, generally for reasons other than structural integrity, the maximum beam deflection.

Both beam stresses and beam deflections will be dealt with in this text. Beam stresses will be encountered in this chapter; beam deflections will be investigated first in this chapter and then extended in Chapter 5.

In general, beams are subjected internally to three stresses: bending (axial, normal) stresses, shear stresses and normal stresses due to axial loads. The latter stresses are usually negligible. Bending and shear stresses in beams may have been adequately dealt with in a Strength of Materials course. Nonetheless, since the subjects are so important, a review of these stress developments may be found later in this chapter. Of major importance at this stage are the realizations that the bending stress at a point is directly related to the internal bending moment, M, at that point and that the shear stress at a point is directly related to the internal shear force, V, at that point.

The first part of this chapter deals with calculating shear forces and bending moments for beams. Two methods will be generated: first, the direct method of using free body diagrams and the equations of equilibrium; second, a semi-graphical approach. The chapter progresses with the relationships between stresses (shear and bending) and shear forces and bending moments, and ends with beam deflections by direct integration.

SHEAR FORCES AND BENDING MOMENTS

Let us take as an illustration a statically determinate beam subjected to gravity loads. (Figure 3.1a). If an imaginary cut is made through point p, the beam is severed into two parts (Figure 3.1b). On the cut surfaces resultant internal shear forces and bending moments appear. These resultants are, of course, necessary in order to maintain equilibrium of both severed parts. The resultants on the left part are generated by contact with the right part and vice versa. The left side of the beam is restrained from rotating and translating, in the vertical direction, due to its contact with the right portion of the beam. On each part of the severed beam (Figure 3.1b) we have drawn what we consider positive shear forces and positive bending moments. Positive shear forces cause clockwise moments about a point inside the beam part; positive bending moments cause the beam to deflect concavely upward in a smiling fashion.

Note: V and M are shown in the positive sense on both FBDs.

Figure 3.1a **Figure 3.1b**

The magnitudes of the shear force V and the bending moment M can be directly found by applying two equations of equilibrium to either the left or right part of the severed beam.

From a design point of view the largest values of V and M are of prime concern. Thus, we will find V and M at every point and then graph these values to find their maximums. In so doing the question arises as to how many cuts are needed to find V and M at every point in the beam. The number of cuts depends upon the loading situation. In general, a cut is taken wherever the loading pattern is altered. For example, cuts are taken when possible to the left and to the right of each concentrated load and reaction (Figure 3.2a). For uniformly distributed loading an additional cut is taken through the loading (Figure 3.2b).

Figure 3.2a **Figure 3.2b**

Examples 1–3 illustrate in detail the direct method for finding shear force and bending moment expressions.

EXAMPLE #1: V & M EXPRESSIONS

Given: Beam shown

Find: V & M expressions for the entire beam. (with graphs)

10ᴷ, 5ᴷ; 5' | 7' | 4'

Given beam

The FBD at right shows the beam is subjected to four concentrated forces: two loads and two support reactions. Cuts are taken (where possible) to the left and to the right of all forces. Three cuts are necessary. The reference for the distance variable, X, will be taken as the left end.

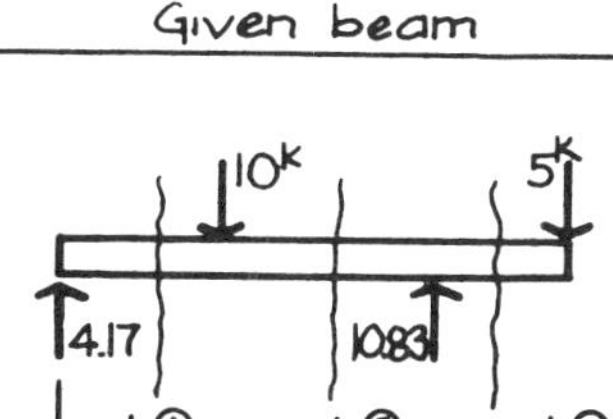

FBD of given beam

Cut ① is taken anywhere between the left end and the 10ᴷ load. We take a FBD of everything to the left of cut ①. (We could, instead, take everything to the right of the cut). Seen on the FBD are the left support reaction, the internal bending moment, and the internal shearforce. Statics allows the determination of V and M at any point within the first 5'.

cut ①: FBD (left side)

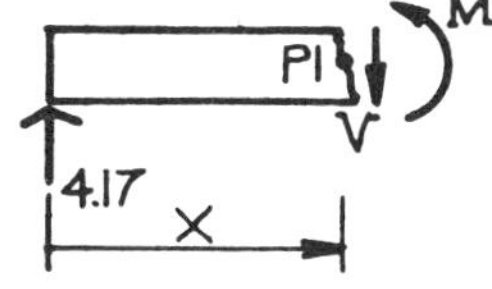

$\Sigma F_Y = 0 \quad V = +4.17 \quad (a)$

$\Sigma M_{P1} = 0 \quad M = 4.17X \quad (b)$

$(0 \leq X \leq 5')$

Cut ②: Again we take the left portion of the beam. Note that the variable, X, is taken from the <u>left end</u> to the <u>cut</u>. Also note V & M are shown in the positive sense.

cut ②: FBD (left portion)

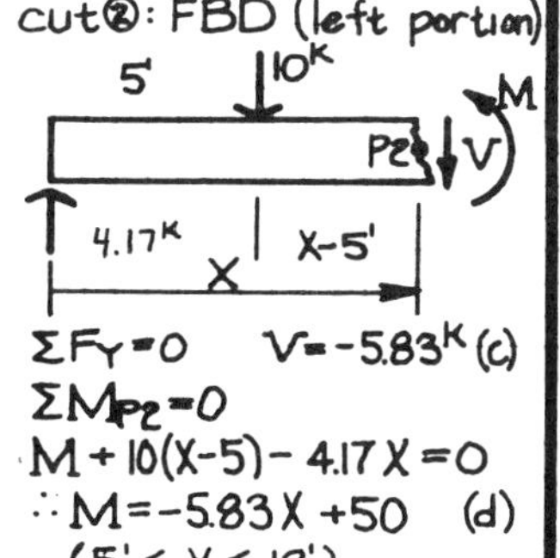

$\Sigma F_Y = 0 \quad V = -5.83^K \quad (c)$

$\Sigma M_{P2} = 0$

$M + 10(X-5) - 4.17X = 0$

$\therefore M = -5.83X + 50 \quad (d)$

$(5' \leq X \leq 12')$

Cut ③: Here it is easier to consider the right portion than the left since the right has only one force acting on it as opposed to three for the left. Our FBD is from the cut to the right end. Note V & M are shown in positive sense. Also, X is still taken from the left end.

Cut ③: FBD (right side)

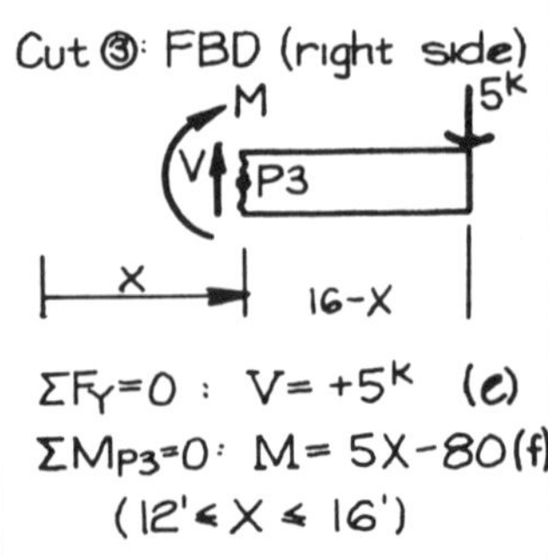

$\Sigma F_Y = 0$: $V = +5^K$ (e)

$\Sigma M_{P3} = 0$: $M = 5X - 80$ (f)

$(12' \leq X \leq 16')$

Equations a-f give complete descriptions of V & M. In graphing it is important to realize the range of each equation.

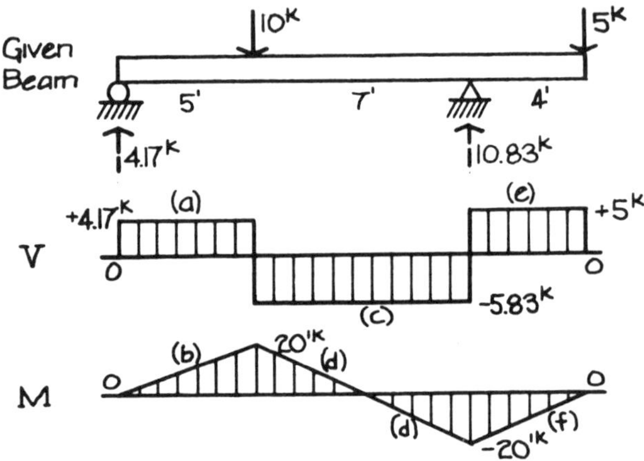

Notes:

1) Jumps in the shear diagram occur when concentrated forces exist; the magnitude and the direction of these jumps correspond to the magnitude and direction of the concentrated forces. The value of the shear force at a jump is not clear, however, keep in mind that a point load is a fiction; loads are always applied over some finite distance.
2) The moment diagram is continuous since no concentrated moments are applied; equations b and d give consistant values, allowing for round-off, at 5' as do d and f at 12'.
3) The moment at the roller is zero; an end roller cannot support a moment. The moment at the pin is not zero since the pin is not at an end.

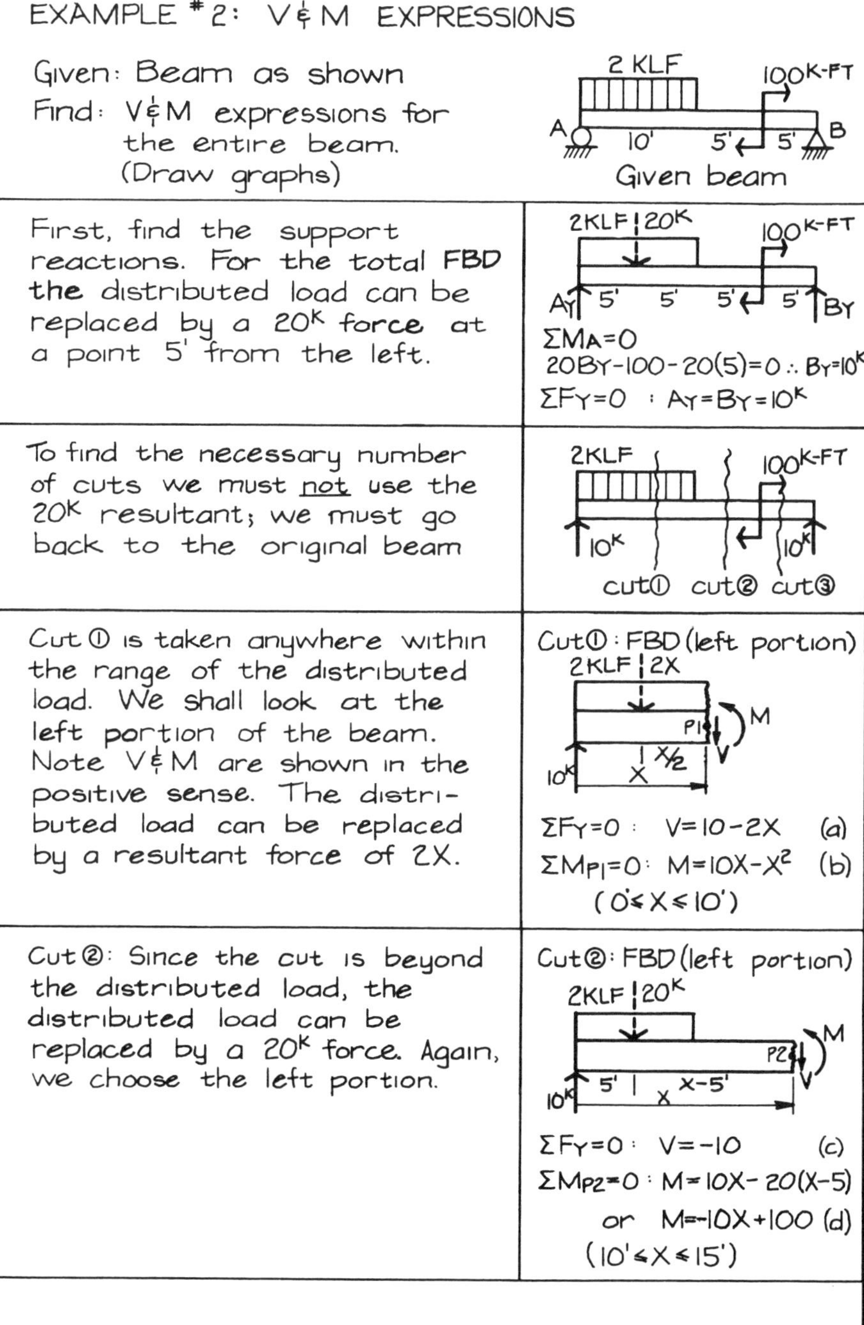

EXAMPLE #2: V & M EXPRESSIONS

Given: Beam as shown

Find: V & M expressions for the entire beam. (Draw graphs)

Given beam

First, find the support reactions. For the total FBD the distributed load can be replaced by a 20^K force at a point 5' from the left.

$\Sigma M_A = 0$

$20B_Y - 100 - 20(5) = 0 \therefore B_Y = 10^K$

$\Sigma F_Y = 0 \; : \; A_Y = B_Y = 10^K$

To find the necessary number of cuts we must not use the 20^K resultant; we must go back to the original beam

Cut ① is taken anywhere within the range of the distributed load. We shall look at the left portion of the beam. Note V & M are shown in the positive sense. The distributed load can be replaced by a resultant force of 2X.

Cut ①: FBD (left portion)

$\Sigma F_Y = 0 : \quad V = 10 - 2X \qquad (a)$

$\Sigma M_{P1} = 0 : \quad M = 10X - X^2 \qquad (b)$

$(0' \leq X \leq 10')$

Cut ②: Since the cut is beyond the distributed load, the distributed load can be replaced by a 20^K force. Again, we choose the left portion.

Cut ②: FBD (left portion)

$\Sigma F_Y = 0 : \quad V = -10 \qquad (c)$

$\Sigma M_{P2} = 0 : \quad M = 10X - 20(X-5)$

$\text{or} \quad M = -10X + 100 \quad (d)$

$(10' \leq X \leq 15')$

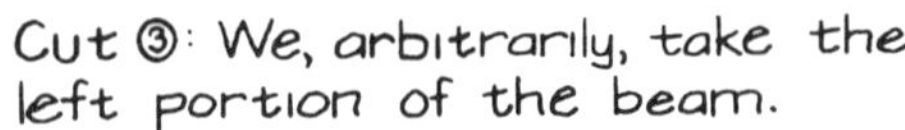

Cut ③: We, arbitrarily, take the left portion of the beam.

Cut③: FBD (left portion)

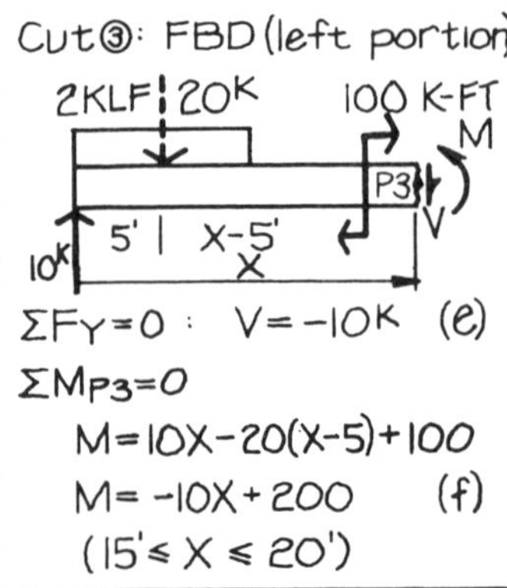

$\Sigma F_Y = 0$: $V = -10^K$ (e)

$\Sigma M_{P3} = 0$

$M = 10X - 20(X-5) + 100$

$M = -10X + 200$ (f)

$(15' \leq X \leq 20')$

Equations a-f give complete descriptions of V & M. Keep in mind the range of the equations.

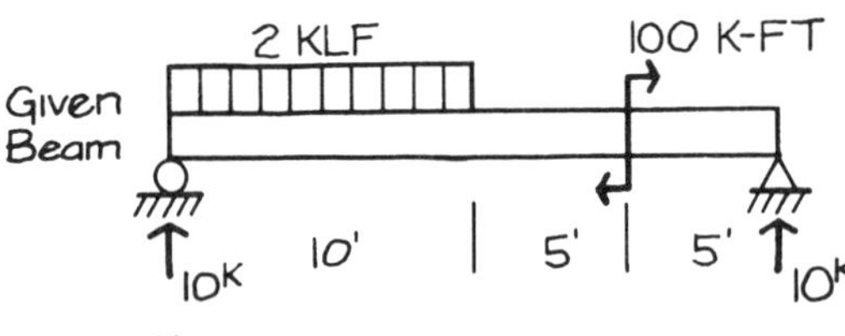

From eq.(a) we find V=0 at X=5'

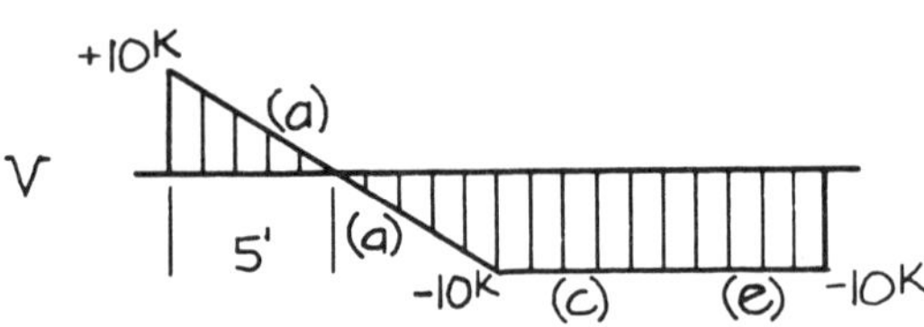

From eq(b) we find M=+25 at X=5'. Note from eqs. (b) and (d) we find that both equations give the slope at X=10' to be −10.

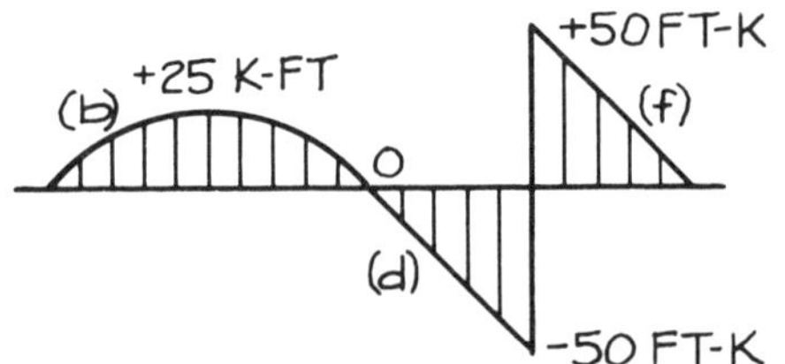

Notes:

1) There are no jumps (except at ends) for the V diagram but there is a jump in the M diagram due to the concentrated couple.
2) We reiterate, cuts are taken to the left and to the right of changes in loading.
3) Caution must be exercised in dealing with distributed loads; resultant force can be used but only to replace that portion of the loading that appears on the FBD.

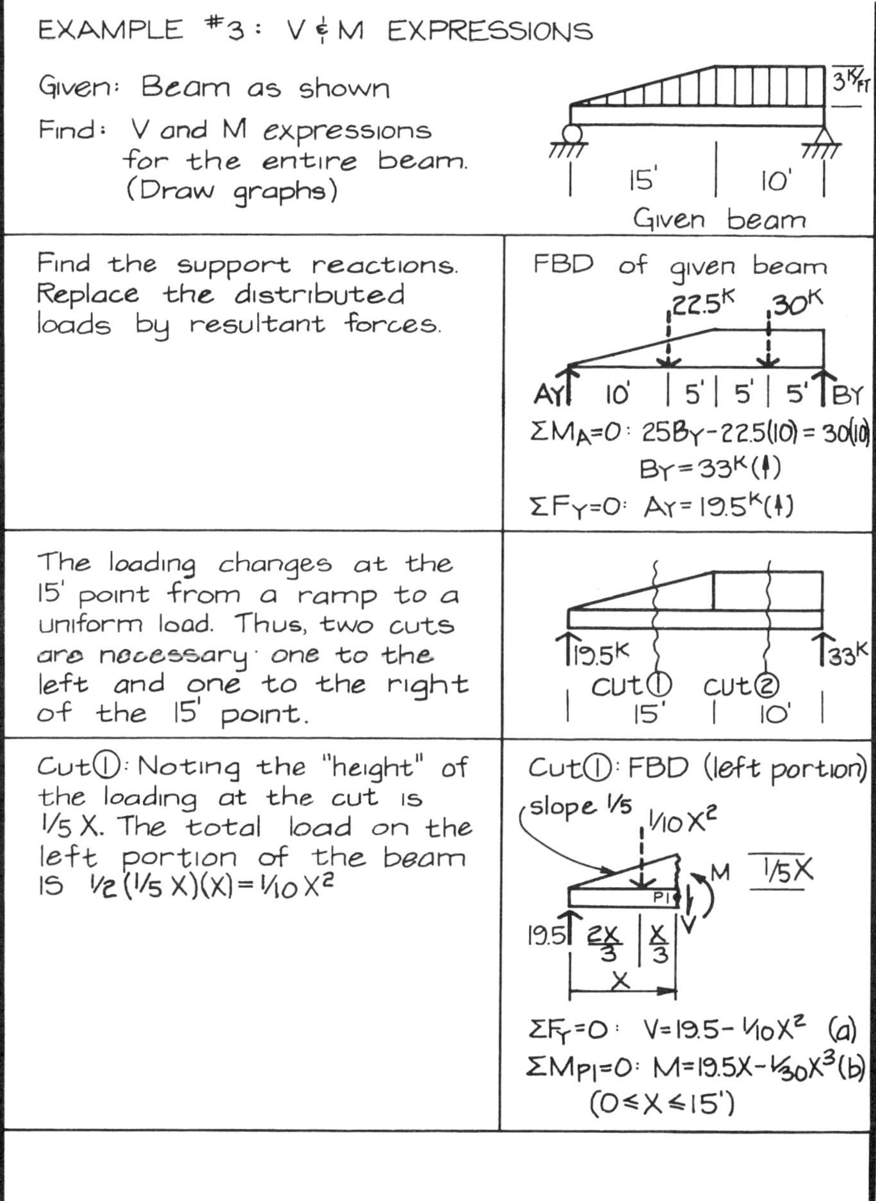

EXAMPLE #3: V & M EXPRESSIONS

Given: Beam as shown

Find: V and M expressions for the entire beam. (Draw graphs)

Find the support reactions. Replace the distributed loads by resultant forces.

FBD of given beam

$\Sigma M_A = 0$: $25B_Y - 22.5(10) = 30(10)$

$B_Y = 33^K (\uparrow)$

$\Sigma F_Y = 0$: $A_Y = 19.5^K (\uparrow)$

The loading changes at the 15' point from a ramp to a uniform load. Thus, two cuts are necessary: one to the left and one to the right of the 15' point.

Cut①: Noting the "height" of the loading at the cut is $1/5\,X$. The total load on the left portion of the beam is $1/2(1/5\,X)(X) = 1/10\,X^2$

Cut①: FBD (left portion)

$\Sigma F_Y = 0$: $V = 19.5 - 1/10\,X^2$ (a)

$\Sigma M_{P1} = 0$: $M = 19.5X - 1/30\,X^3$ (b)

$(0 \le X \le 15')$

Cut②: For the left portion of the beam the ramp is seen in its entirety. The uniform load is only partially seen.

Cut②: FBD (left portion)

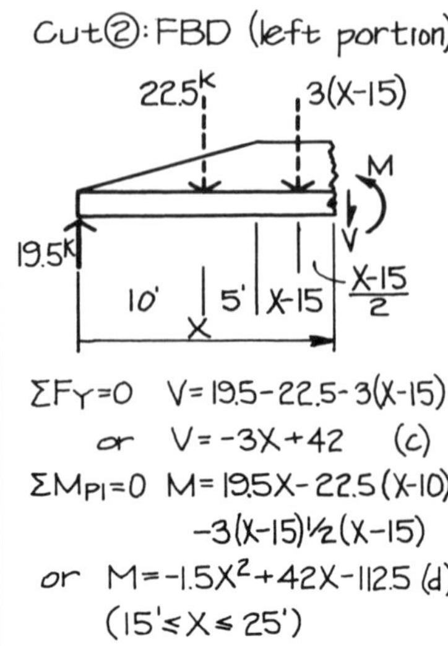

$\Sigma F_Y = 0 \quad V = 19.5 - 22.5 - 3(X-15)$

$\text{or} \quad V = -3X + 42 \quad (c)$

$\Sigma M_{P1} = 0 \quad M = 19.5X - 22.5(X-10) - 3(X-15)\tfrac{1}{2}(X-15)$

$\text{or} \quad M = -1.5X^2 + 42X - 112.5 \quad (d)$

$(15' \leq X \leq 25')$

Equations a–d give complete descriptions of V & M

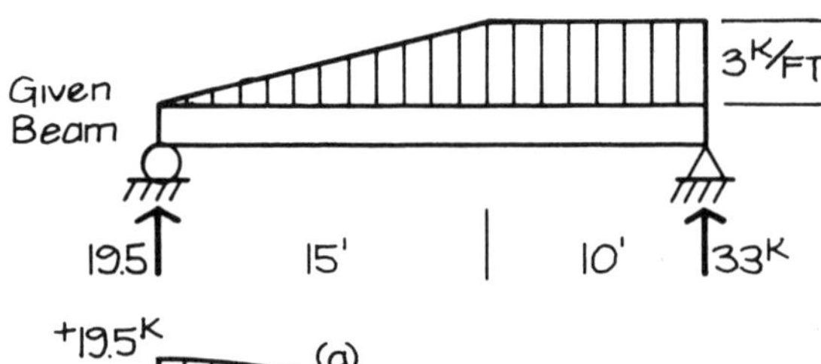

From (a) we find V=0 at 13.96' from the left.

From (b) at X=13.96' we find M=181.5 K-FT
From (b) and (d) we find M=180 K-FT at X=15'

V

+19.5K
(a)
13.96'
-3K
(c)
-33K

M

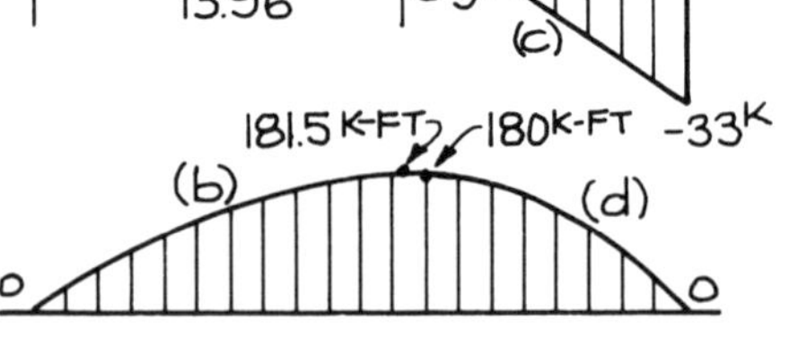

Shear Force and Bending Moment Diagrams: Semi-graphical Approach

As we saw in Examples 1–3, the direct approach of making a series of cuts and then writing the equilibrium equations is easy and reliable but time consuming. The more cuts needed, the more time is consumed by this process. So we'll introduce a short-cut approach to take us directly from the loaded beam to the shear force, V, and bending moment, M, diagrams. (The V and M equations are generally not found by this approach, although they can be).

We start with a generic beam (Figure 3.3a) having a distributed load which can vary along the beam. We call the load intensity function w. It is measured in force per unit length and is taken as positive when the loads are in the gravity direction. Looking at a slice (Figure 3.3b) formed by two closely spaced cuts, Δx apart, we see shear forces and bending moments on both surfaces of the slice. In general, the shear forces on the two surfaces are slightly different in magnitude. Following the calculus approach, ΔV is labeled as this difference. Similarly, ΔM designates the difference in bending moments on the two surfaces. (All shear force and bending moments are shown, using the sign convention of the last section, in the positive sense.)

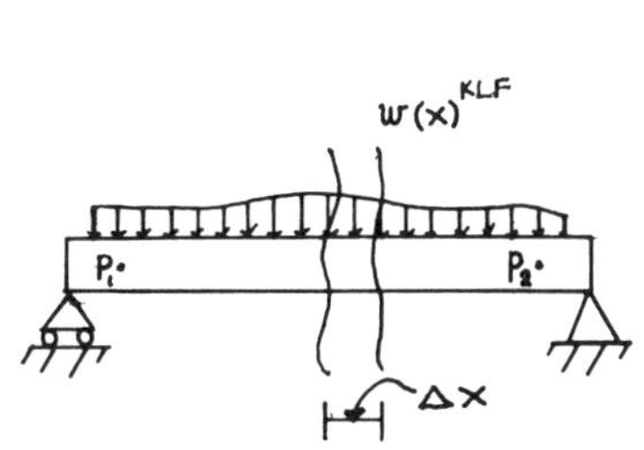

Figure 3.3a

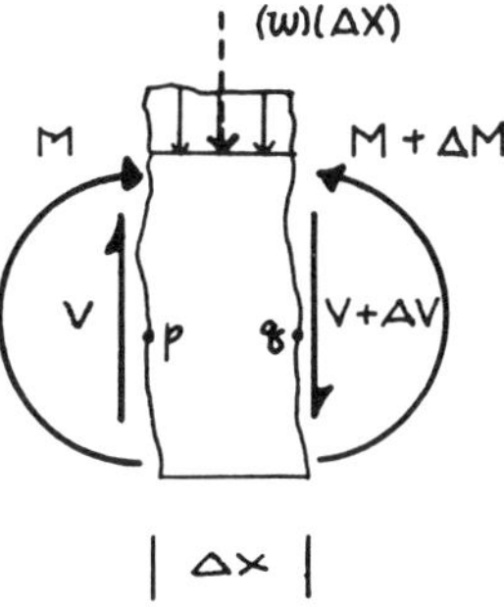

Figure 3.3b

Considering the FBD of the slice (Figure 3.3b) we sum forces in the vertical direction, taking upward as positive. (Note $w\ \Delta x$ is the *force* due to the load).

$$\sum F_y = 0 : V - (V + \Delta V) - (w)(\Delta x) = 0 \tag{1}$$

or

$$\frac{\Delta V}{\Delta x} = -w.$$

Going to the limit as Δx approaches zero, we get

$$\frac{dV}{dx} = -w. \tag{2}$$

If we integrate Equation 2 between two arbitrary points p_1 (left) and p_2 (right) we get

$$\int_{p_1}^{p_2} \frac{dV}{dx}\, dx = -\int_{p_1}^{p_2} w\, dx, \quad \text{or}$$

$$V(p_2) - V(p_1) = -\int_{p_1}^{p_2} w\, dx. \tag{3}$$

Taking moments about point p of the slice and arbitrarily taking clockwise moments as positive, we get

$$\sum M_p = 0 : M - (M + \Delta M) + (\Delta x)(V + \Delta V) + (w\, \Delta x)\left(\frac{\Delta x}{2}\right) = 0$$

$$\text{or} \quad -\Delta M + V\, \Delta x + (\Delta V)(\Delta x) + \frac{w}{2}(\Delta x)^2 = 0.$$

Dividing by Δx:

$$-\frac{\Delta M}{\Delta x} + V + \Delta V + \frac{w}{2}(\Delta x) = 0.$$

When Δx goes to zero, we note from Equation 1 that ΔV also goes to zero; thus

$$\frac{dM}{dx} = V. \tag{4}$$

Upon integrating from left (p_1) to right (p_2), we get

$$M(p_2) - M(p_1) = \int_{p_1}^{p_2} V\, dx. \tag{5}$$

Equations 2 and 4 tell the shape of the V and M diagrams, respectively, while Equations 3 and 5 allow calculations of the shear values and moment values, respectively.

In words, Equation 2 says the *slope* of the shear diagram at any point equals the negative value of the load intensity at that point.

Equation 3 says the *difference* in shear values between points p_2 and p_1 equals the negative of the area, between p_2 and p_1, under the load curve.

Equation 4 says the *slope* of the moment diagram at any point equals the shear value at that point.

Equation 5 says the *difference* in moment values between points p_2 and p_1 equals the area, between p_2 and p_1, under the shear curve.

Equations 2 and 3 provide then a means of using the loading curve to establish the shape and magnitudes of the shear force curve; Equations 4 and 5 use the shear force curve as a basis in forming the bending moment curve.

Since Equations 3 and 5 involve differences in values, it is necessary to have "starting" values. For this purpose, support reactions, found via an FBD of the entire beam, are used.

We note, from Equation 2, the degree of the shear curve is one degree higher than the loading curve and Equation 4 implies the moment curve is one degree higher than the shear curve. For instance, a uniformly distributed load (a zero order curve) yields a straight sloping shear curve (first degree curve) and a second degree parabolic moment curve.

On occasions when an area under a curve is difficult to calculate, making a shear or moment value hard to find, it is often advantageous to use a cut and an FBD to obtain the desired value. That is, use the direct approach to resolve an impasse that the short-cut method presents.

Examples 4–6 illustrate the semi-graphical method.

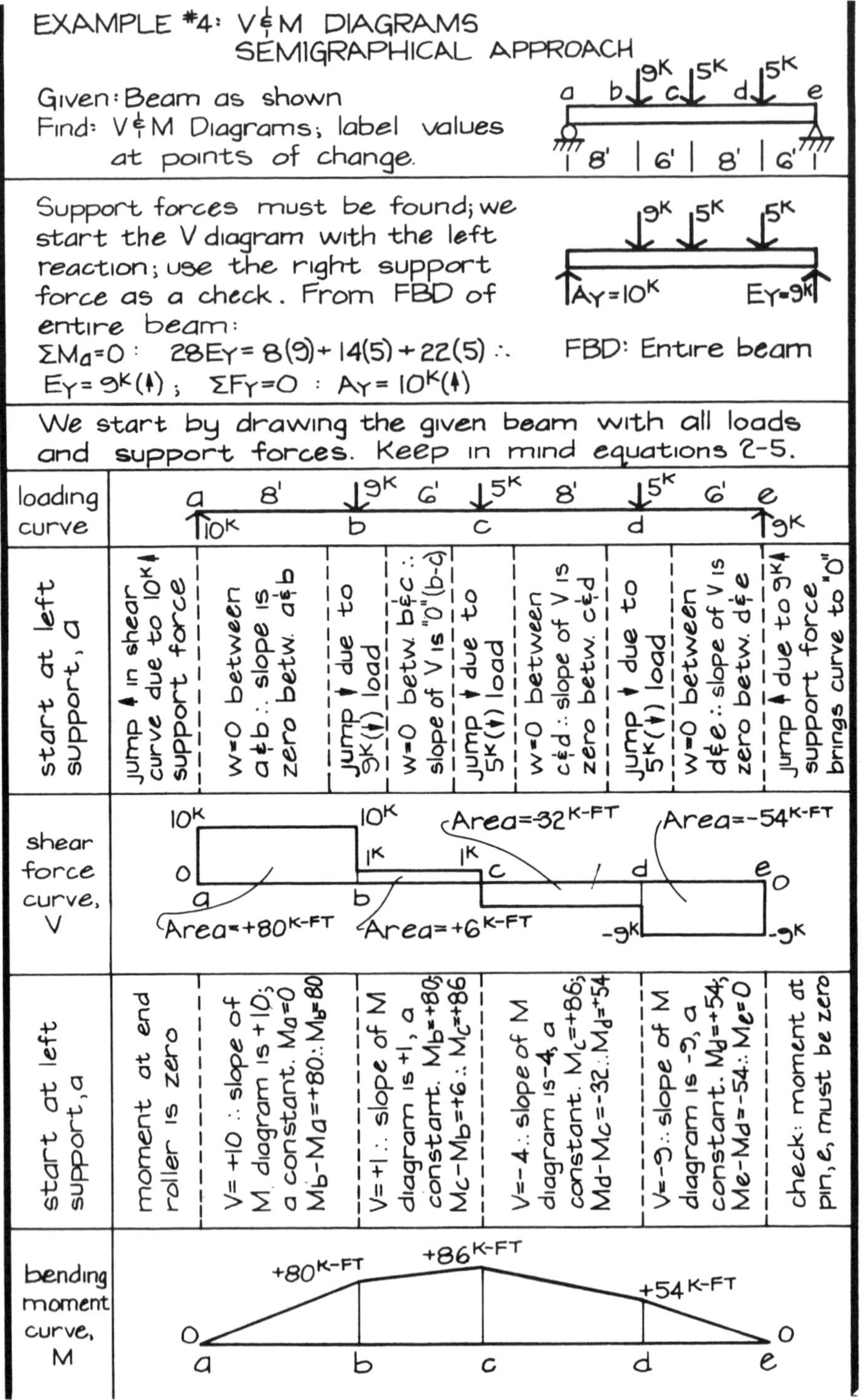
EXAMPLE #4: V&M DIAGRAMS
SEMIGRAPHICAL APPROACH
Given: Beam as shown
Find: V&M Diagrams; label values at points of change.
a b c d e
9K 5K 5K
8' 6' 8' 6'
Support forces must be found; we start the V diagram with the left reaction; use the right support force as a check. From FBD of entire beam:
ΣMa=0 : 28EY = 8(9)+14(5)+22(5) ∴
EY = 9K(↑) ; ΣFY=0 : AY = 10K(↑)
AY=10K
EY=9K
FBD: Entire beam
We start by drawing the given beam with all loads and support forces. Keep in mind equations 2-5.
loading curve
shear force curve, V
bending moment curve, M
start at left support, a
jump ↑ in shear curve due to 10K↑ support force
w=0 between a&b ∴ slope is zero betw. a&b
jump ↓ due to 9K(↓) load
w=0 betw. b&c ∴ slope of V is "0" (b-c)
jump ↓ due to 5K(↓) load
w=0 between c&d ∴ slope of V is zero betw. c&d
jump ↓ due to 5K(↓) load
w=0 between d&e ∴ slope of V is zero betw. d&e
jump ↑ due to 9K↑ support force brings curve to "0"
Area=+80K-FT
Area=+6K-FT
Area=-32K-FT
Area=-54K-FT
10K
1K
-9K
moment at end roller is zero
V=+10 ∴ slope of M diagram is +10; a constant. Ma=0 Mb-Ma=+80 ∴ Mb=80
V=+1 ∴ slope of M diagram is +1, a constant. Mb=+80; Mc-Mb=+6 ∴ Mc=+86
V=-4 ∴ slope of M diagram is -4, a constant. Mc=+86; Md-Mc=-32 ∴ Md=+54
V=-9 ∴ slope of M diagram is -9, a constant. Md=+54; Me-Md=-54 ∴ Me=0
check: moment at pin, e, must be zero
+80K-FT
+86K-FT
+54K-FT
0

Once the concepts are understood on how the load diagram establishes the shear diagram and how the shear diagram gives the moment diagram this procedure can be put in table form. Remember concentrated loads or reactions cause "jumps" in the shear and moment diagrams. Concentrated forces cause "jumps" in shear; concentrated couples cause "jumps" in moment. Also keep in mind that these diagrams are <u>statics</u> based.

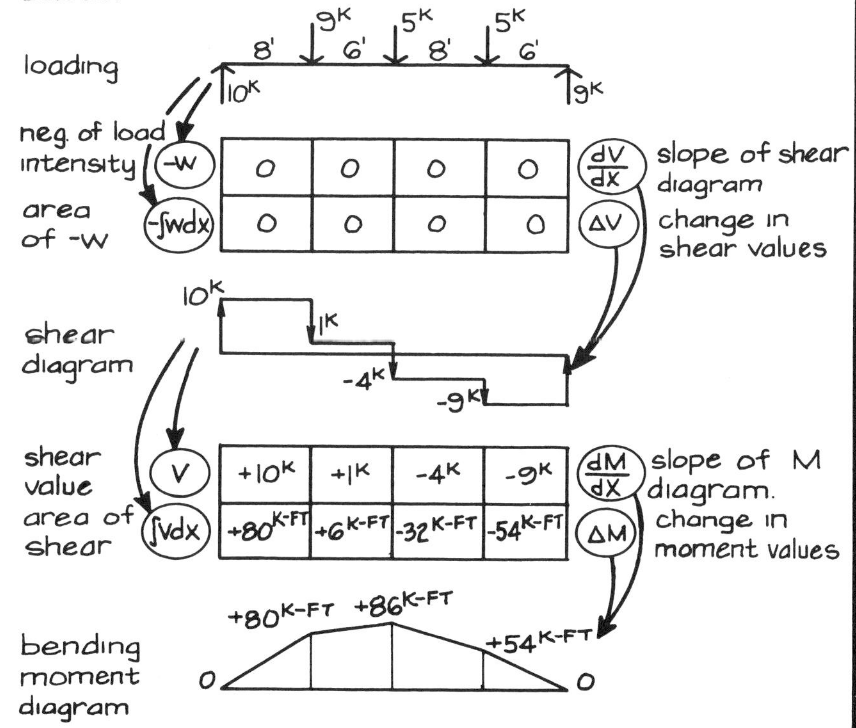

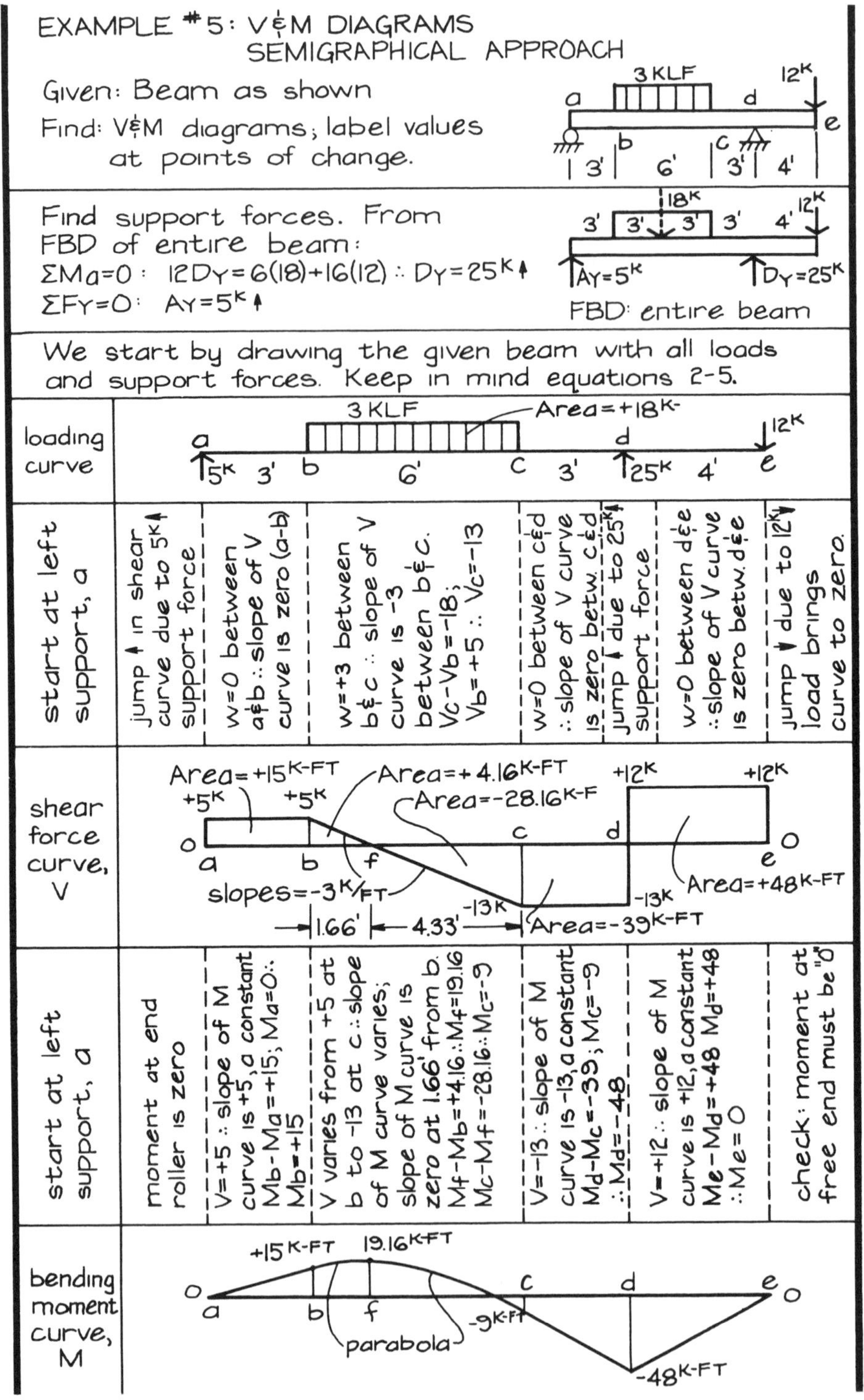

EXAMPLE #5: V&M DIAGRAMS
SEMIGRAPHICAL APPROACH
Given: Beam as shown
Find: V&M diagrams; label values at points of change.
3 KLF
12K
a b c d e
3' 6' 3' 4'
Find support forces. From FBD of entire beam:
ΣMa=0: 12DY=6(18)+16(12) ∴ DY=25K↑
ΣFY=0: AY=5K↑
18K
AY=5K
DY=25K
FBD: entire beam
We start by drawing the given beam with all loads and support forces. Keep in mind equations 2-5.
loading curve
Area=+18K
5K
25K
start at left support, a
jump ↑ in shear curve due to 5K↑ support force
w=0 between a&b ∴ slope of V curve is zero (a-b)
w=+3 between b&c ∴ slope of V curve is -3 between b&c. Vc-Vb=-18; Vb=+5 ∴ Vc=-13
w=0 between c&d ∴ slope of V curve is zero betw. c&d
jump ↑ due to 25K↑ support force
w=0 between d&e ∴ slope of V curve is zero betw. d&e
jump ↓ due to 12K↓ load brings curve to zero.
shear force curve, V
Area=+15K-FT
Area=+4.16K-FT
Area=-28.16K-F
+5K
+12K
-13K
slopes=-3K/FT
1.66'
4.33'
Area=-39K-FT
Area=+48K-FT
start at left support, a
moment at end roller is zero
V=+5 ∴ slope of M curve is +5, a constant Mb-Ma=+15; Ma=0 ∴ Mb=+15
V varies from +5 at b to -13 at c ∴ slope of M curve varies; slope of M curve is zero at 1.66' from b. Mf-Mb=+4.16 ∴ Mf=19.16 Mc-Mf=-28.16 ∴ Mc=-9
V=-13 ∴ slope of M curve is -13, a constant Md-Mc=-39; Mc=-9 ∴ Md=-48
V=+12 ∴ slope of M curve is +12, a constant Me-Md=+48 Md=+48 ∴ Me=0
check: moment at free end must be "0"
bending moment curve, M
+15K-FT
19.16K-FT
-9K-FT
-48K-FT
parabola

Or using table form

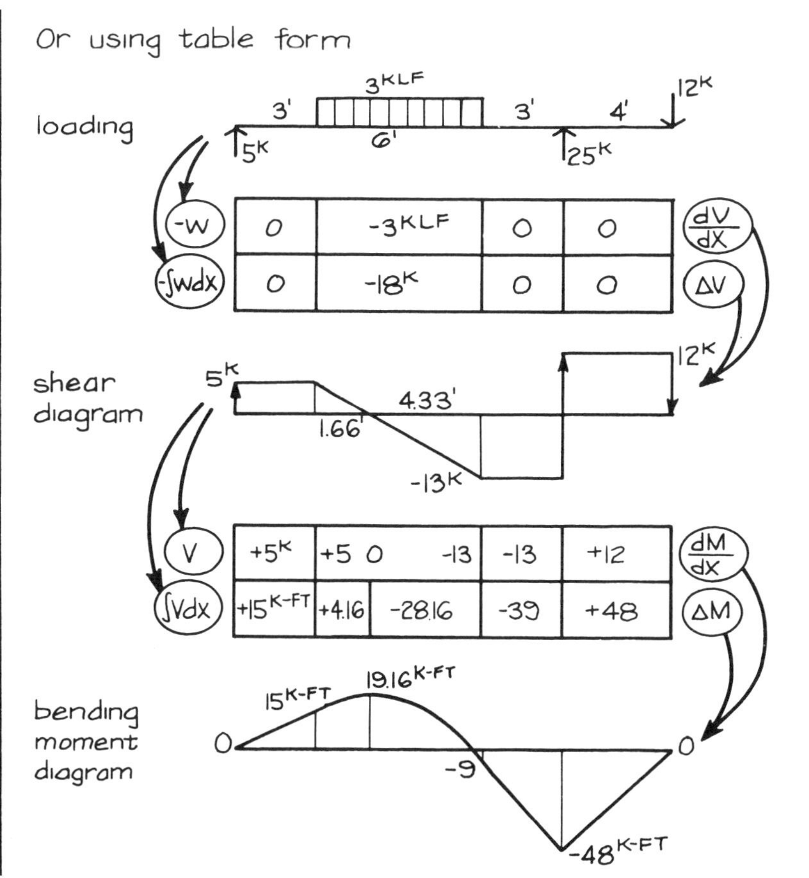

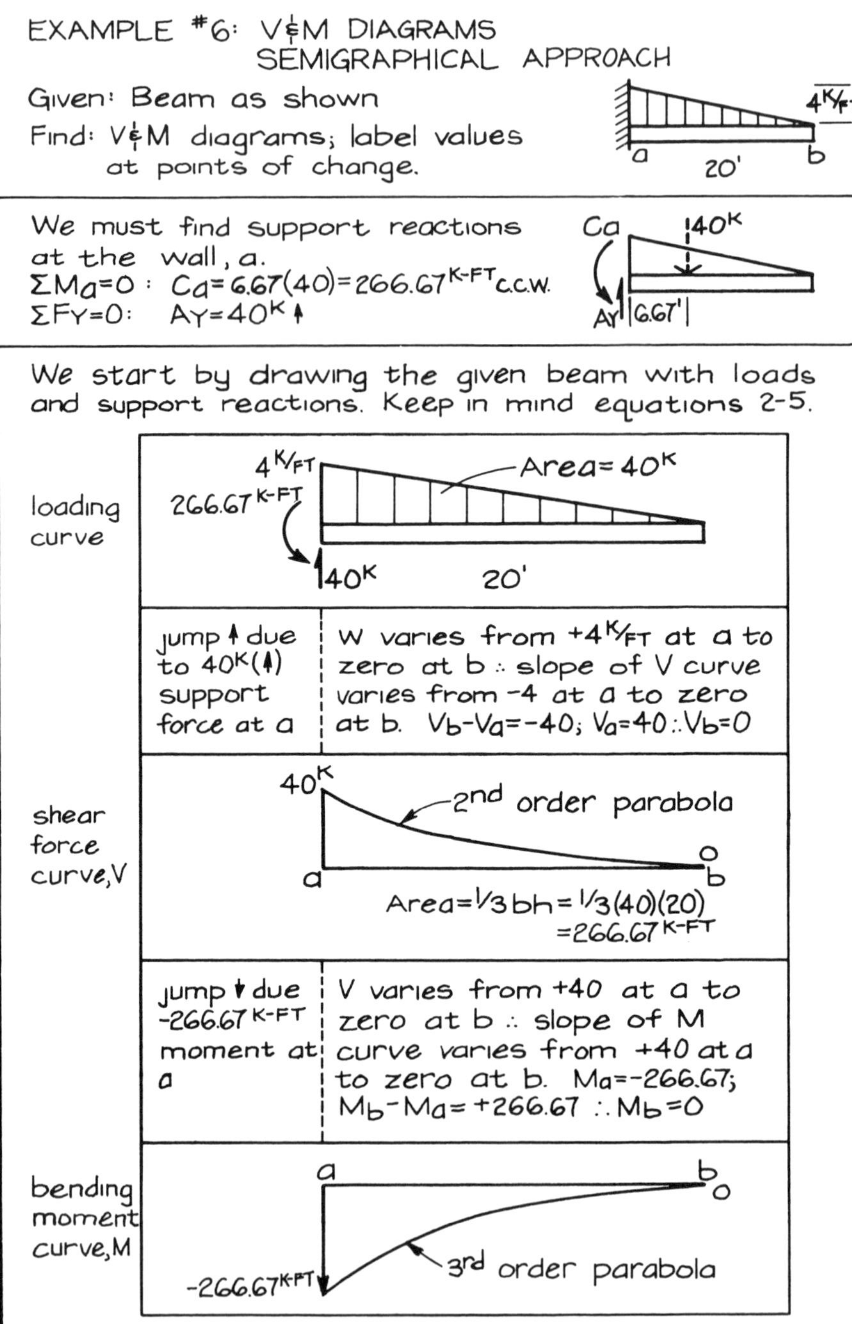
EXAMPLE #6: V&M DIAGRAMS
SEMIGRAPHICAL APPROACH
Given: Beam as shown
Find: V&M diagrams; label values at points of change.
4K/FT
a
20'
b
We must find support reactions at the wall, a.
ΣMa=0 : Ca=6.67(40)=266.67K-FT C.C.W.
ΣFY=0: AY=40K ↑
Ca
40K
AY
6.67'
We start by drawing the given beam with loads and support reactions. Keep in mind equations 2-5.
loading curve
4K/FT
Area=40K
266.67K-FT
40K
20'
jump ↑ due to 40K(↑) support force at a
W varies from +4K/FT at a to zero at b ∴ slope of V curve varies from -4 at a to zero at b. Vb-Va=-40; Va=40 ∴ Vb=0
shear force curve, V
40K
2nd order parabola
a
0
b
Area=1/3bh=1/3(40)(20)
=266.67K-FT
jump ↓ due -266.67K-FT moment at a
V varies from +40 at a to zero at b ∴ slope of M curve varies from +40 at a to zero at b. Ma=-266.67; Mb-Ma=+266.67 ∴ Mb=0
bending moment curve, M
a
b
0
-266.67K-FT
3rd order parabola

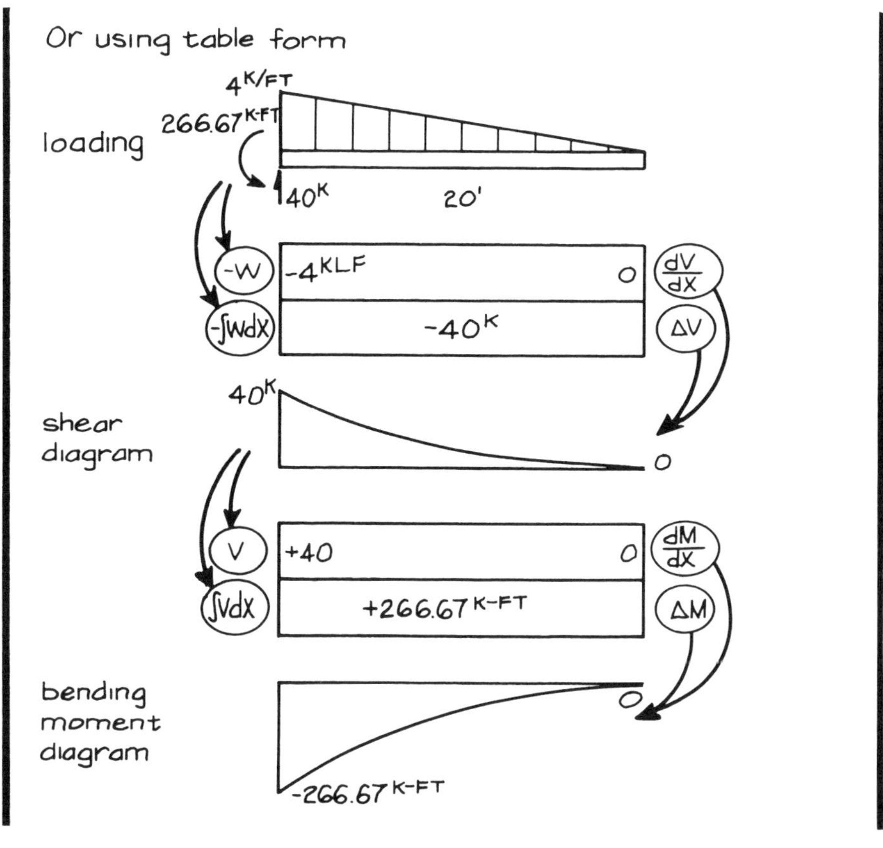
Or using table form
4 K/FT
266.67 K-FT
loading
40K
20'
-W
-4 KLF
0
dV/dX
-∫wdX
-40K
ΔV
40K
shear diagram
0
V
+40
0
dM/dX
∫VdX
+266.67 K-FT
ΔM
bending moment diagram
0
-266.67 K-FT

BEAM STRESSES

The basis for developing the stress picture in beams is strength of materials. We will use the strength of materials results for pure bending and then extend these results to more general beams.

If a beam is loaded only with end couples, M, the pure bending case, then only bending (normal, axial) stresses occur (see introduction): $\sigma_x = (-M/I)y$. Here "y" is the vertical coordinate, measured from the centroid of the cross-section, and I is the moment of inertia of the cross-section taken about the centroidal axis (Figure 3.4).

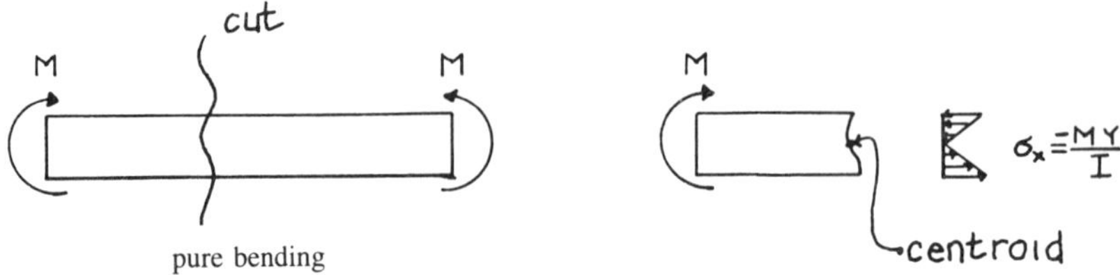

Figure 3.4a

Figure 3.4b

The same bending stress relation is used, as an approximation, when gravity-type loads are encountered on beams. In such problems, called "Engineering Bending," shear forces and thus shear stresses also occur. In engineering bending, the moments vary with the axial coordinate, x. In previous sections we investigated how M varies. Thus $\sigma_x = [-M(x)/I]y$.

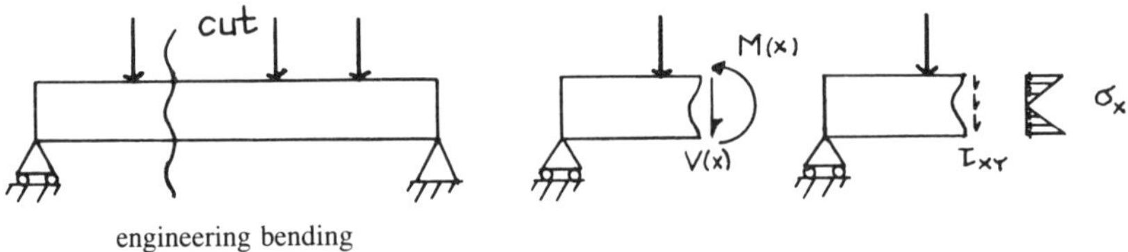

Figure 3.5a

Figure 3.5b

Figure 3.5c

To find the shear stress, τ_{xy}, we take two closely spaced cuts from a beam (the cross-section or the beam must be symmetric about the vertical axis: we take an I-beam). Since the set of stresses (Figure 3.6) on the two faces differs in magnitude we again use the differential idea. On the left face we see σ_x; it varies linearly from zero at the centroid to maximum compression and tension values at the top and bottom, respectively. On the right face is $\sigma_x + \Delta\sigma_x$. The shear stresses on the two faces also differ but shear differentials would only make the issue more confused.

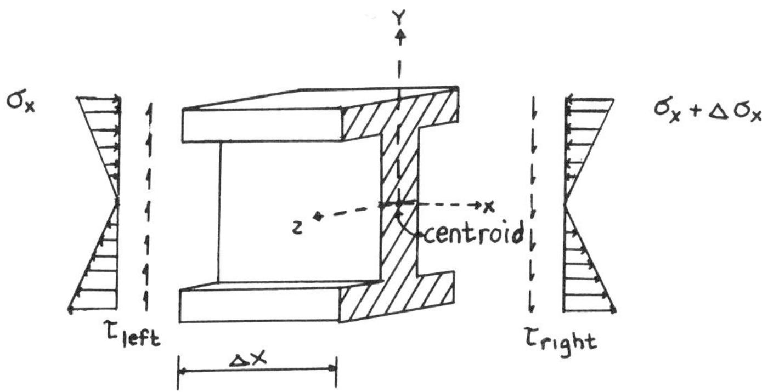

Figure 3.6 (Note: no stress variation in z-direction)

Now on the slice, Figure 3.6, we'll take a horizontal cut through the arbitrary point p. Our mission is to find the shear stress at point p, τ_p. Note that the shear stresses at a point must have the same magnitude on the horizontal and vertical planes (Figure 3.7b).

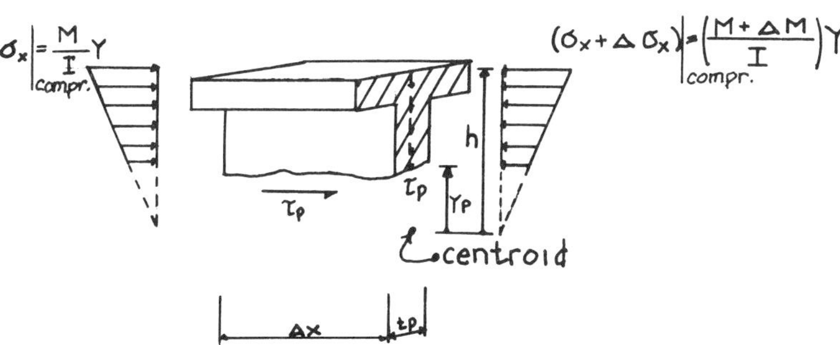

Figure 3.7a

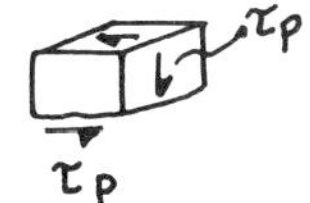

Figure 3.7b

Summing forces in the x-direction (note that the average shear stress on the horizontal surface acts on an area of t_p times Δx),

$$\sum Fx = 0 : + \int_{y_p}^{h} \sigma_x \, dA - \int_{y_p}^{h} (\sigma_x + \Delta\sigma_x) \, dA + (\tau_p)(t_p \, \Delta x) = 0.$$

Here the integration is over the top portion of the section, from point p to the top where $y = h$. Since $\sigma_x = My/I$ then $\Delta\sigma_x = (\Delta M/I)y$ thus; $\Sigma F_x = 0$ yields:

$$-\int_{y_p}^{h} \left(\frac{\Delta M}{I}\right)(y)\, dA + \tau_p(t_p\, \Delta x) = 0.$$

Both ΔM and I can be taken outside the integral: ΔM is only a function of the axial coordinate, x, and I is a number. Therefore:

$$\tau_p = \left(\frac{\Delta M}{\Delta x}\right) \frac{\displaystyle\int_{y_p}^{h} y\, dA}{It_p}.$$

In the limit as $\Delta x \to 0$, $\Delta M/\Delta x$ becomes dM/dx which equals the shear force, V. Thus we arrive at

$$\tau_p = \frac{V}{It_p} \int_{y_p}^{h} y\, dA = \frac{VQ}{It_p}.$$

We call the integral, $\int y\, d\Delta$, Q.

Q is the "moment of area," above (or below) p, taken about the centroid;

I is the moment of inertia, of the cross-section, about the centroidal axis;

t_p is the cross-sectional thickness at point p;

V is the shear force, along the beam, at point p.

Q and t_p depend upon the actual location in the cross-section of point p; V depends upon the location of p along the beam and I is a constant for most beams.

Example 7 illustrates the finding of both bending and shear stresses. Example 8 lends insight to the design of bends.

EXAMPLE # 7: SHEAR AND BENDING STRESSES

Given: Beam in EX #1 with symmetrical "tee" cross-section. (shown)

Find: Maximum tension and compression bending stresses; maximum shear stress.

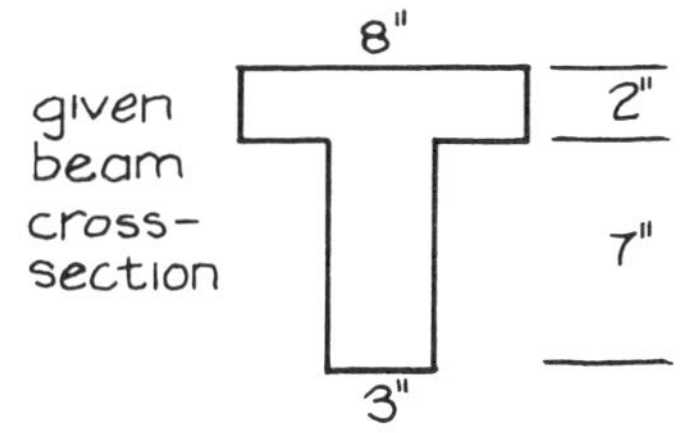

The stress equations are: $\tau = \frac{VQ}{It}$ and $\sigma_x = \frac{-M}{I} Y$.
Here, V and M are found in example #1.

First locate centroid of cross-section. $Y = \frac{\Sigma A Y}{\Sigma A}$

$$Y_C = \frac{(21in^2)(3.5'') + (16in^2)(8'')}{21in^2 + 16in^2} = 5.45''$$ from bottom

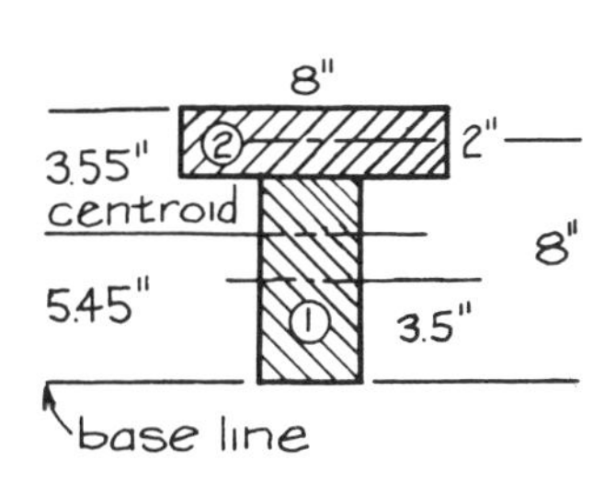

Calculate moment of inertia of cross-section about centroidal axis.

$$I = \left\{\frac{(3)(7)^3}{12} + 21[5.45 - 3.5]^2\right\} + \left\{\frac{(8)(2)^3}{12} + 16[5.45 - 8]^2\right\} = 274.97 in^4$$

Calculate maximum bending stresses. Looking at moment diagram we see positive moment on left meaning compression on top and tension on bottom. Negative bending on the right means tension on top and compression on bottom. Taking both peak values we calculate:

at 5' point

$$\sigma_{\text{tension bottom}} = \frac{(20 \times 12)(5.45)}{274.97} = 4.76^{KSI}(T)$$

$$\sigma_{\text{compr. top}} = \frac{(20 \times 12)(3.55)}{274.97} = 3.03^{KSI}(C)$$

at right support

$$\sigma_{\text{tension top}} = \frac{(20 \times 12)(3.55)}{274.97} = 3.03^{KSI}(T)$$

$$\sigma_{\text{compr. bottom}} = \frac{(20 \times 12)(5.45)}{274.97} = 4.76^{KSI}(C)$$

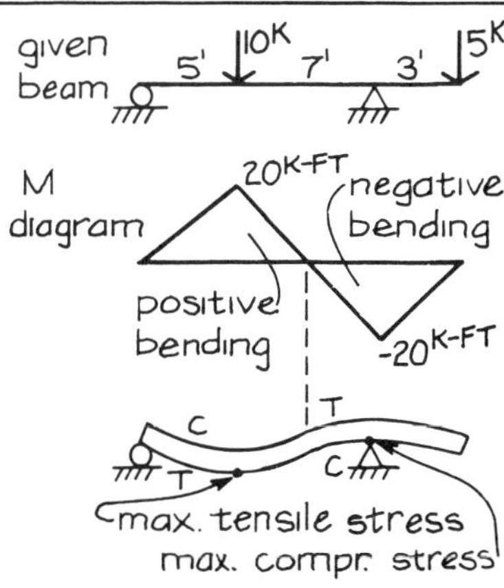

Thus max. tens. stress is 4.76^{KSI} at 5' point (bott. of beam); max compr. stress is also 4.76^{KSI} at right support (bottom of beam).

Calculate Q. Since there is an abrupt geometrical change at the web/flange interface Q must calculated in two parts. First for a point in the web:

Area = 3"[5.45" − Y]

Lever arm = Y + 1/2[5.45" − Y]

$= \frac{5.45 + Y}{2}$

$Q_{web} = 3"[5.45 - Y]\left[\frac{5.45 + Y}{2}\right]$

$Q_{web} = 1.5[29.70 - Y^2]\ in^3$

$-5.45" \leq Y \leq 1.55"$

For every point in the web t = 3"

$\left.\frac{Q}{t}\right|_{web} = 1/2[29.70 - Y^2]$

$-5.45" \leq Y \leq 1.55"$

For a point in the flange

Area = 8"[3.55 − Y]

Lever arm = Y + 1/2(3.55 − Y]

$= \frac{3.55 + Y}{2}$

$Q_{Fl.} = 4[3.55 - Y][3.55 + Y]$

$Q_{Fl.} = 4[12.60 - Y^2]\ in^3$

$1.55" \leq Y \leq 3.55"$

For every point in the flange t = 8": $\left.\frac{Q}{t}\right|_{Fl.} = \frac{12.60 - Y^2}{2}$

$1.55" \leq Y \leq 3.55"$

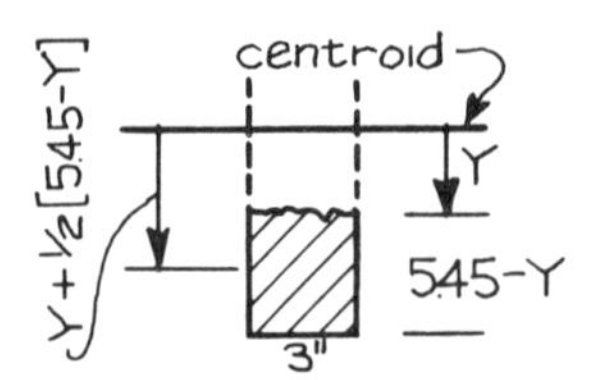

For a point in the web

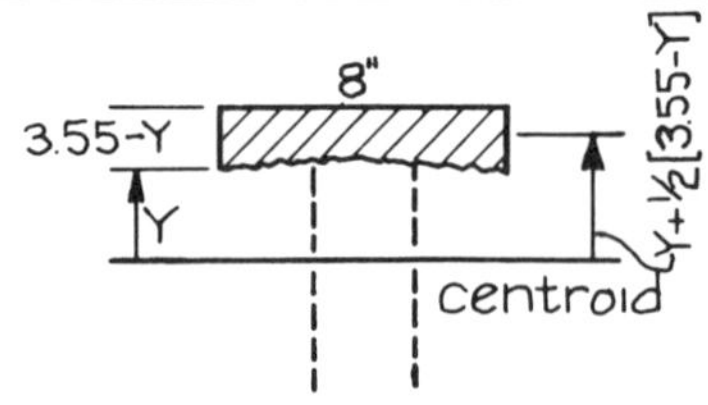

For a point in the flange

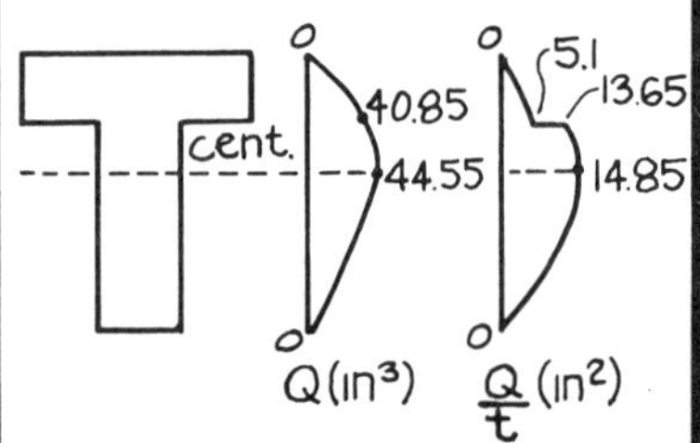

Note: Q is always max. at centroid; jump in Q/t diagram due to change of t from 8" to 3".

The sign of the shear, V, value can be ignored. Beam response is basically the same for positive and negative shears (not so, however for tensile and compressive normal stresses). Thus V max. is 5.83^K between 5' point to right support.

$$\tau_{max} = \frac{VQ}{It} = \frac{[5.83^K][44.55 in^3]}{[274.97 in^4][3"]} = .315^{KSI} = 315^{PSI}$$

This stress occurs at the centroid for all points between 5' spot and right support.

EXAMPLE #8: SHEAR AND BENDING STRESSES - DESIGN

Given: Beam as shown
Allowable bending stress, 22^{KSI}
Allowable shear stress, 20^{KSI}

Find: Depth, h, of the section.

2 KLF
12'
given beam

2"
h
cross-section

Find shear force and bending moment diagrams.

Maximum shears occur at support, $\pm 12^{K}$.

Maximum bending moment occurs at midspan, $36^{K\text{-}FT}$

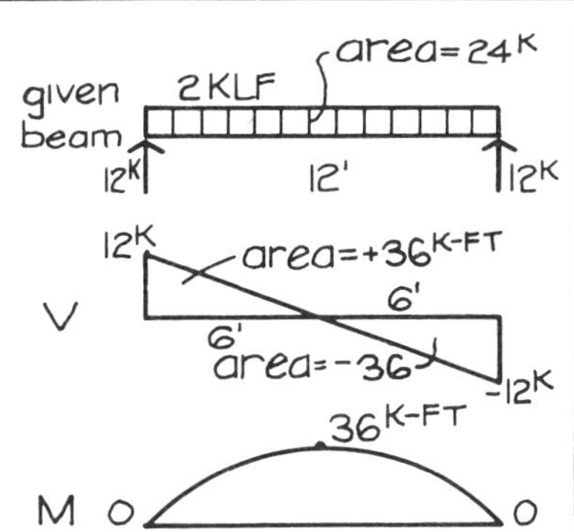

The primary design concern for beams is that of bending stress: set the bending stress equal to the allowable bending stress (since the section is symmetrical, tension and compressive values will be equal).

$$22^{KSI} = \frac{M Y_{max}}{I} \quad \frac{[(36^{K\text{-}FT})(12^{in}/_{FT})]\, Y_{max}}{I}$$

The ratio I/Y_{max} is usually called section modulus, S

$$S = \frac{I}{Y_{max}} = \frac{(b)(h^3)}{12} \Big/ \left(\frac{h}{2}\right) = \frac{bh^2}{6} = \frac{h^2}{3}$$

$$22^{KSI} = \frac{(36 \times 12)}{h^2/3} \Longrightarrow \quad h = \mathbf{7.67}'', \text{ round off to } 8''$$

Thus with h=8" the actual bending stress is less than the allowable of 22^{KSI}. Now check the shear stress. (Note: Q is max. at the centroid). Use h=8"

$$\tau_{max} = \frac{VQ_{max}}{It} = \frac{[12][(2)(4)(2)]}{\left[\frac{(2)(8)^3}{12}\right][2]} = 1.13^{KSI} < 20^{KSI} \quad \text{OK}$$

Thus with h=8" the actual shear stress is much less than the allowable shear stress of 20^{KSI}. Use h=8"

BEAM DEFLECTIONS BY DOUBLE INTEGRATION

Back in Chapter 2 the basic beam deflection equation was developed. In our development we neglected the generally miniscule effect of shear force on deflection. The resulting equation (3) was $d^2u/dx^2 = M/EI$.

Here u is the deflection (up as positive) of the neutral axis of the beam, M is the local internal bending moment, E is Young's modulus, I is the beam's principal moment of inertia about its centroidal axis of the cross-section, and x is the independent (running) variable taking from the left end of the beam and running to the right. We note that d^2u/dx^2 is the local curvature of the deflected beam.

The deflection can be found by twice integrating the above expression. The first integral, du/dx, we recognize as the slope of the deflected beam. Thus

$$\frac{du}{dx} \equiv \theta = \text{slope} = \int \frac{M}{EI}\,dx + C_1 \quad \text{and}$$

$$u = \text{deflection} = \iint \frac{M}{EI}\,ds\,dx + C_1x + C_2\,,$$

where C_1 and C_2 are constants of integration.

This procedure seems straightforward—integrate twice and find the constants of integration from the boundary (usually support) conditions. The procedure is easy as long as the beam is described by only one moment expression as in Example 9. In cases where two moment expressions are called for, four boundary conditions are necessary. Example 10 shows how matched conditions develop at the moment transition point resulting in solving three simultaneous equations. With, say, only two concentrated loads, we would get three moment expressions, six boundary conditions, and five simultaneous equations to solve! Wow! As you can see, this method, while conceptually easy, quickly results in a great struggle of algebra. So an easier, quicker method will be sought in Chapter 5. This method, called moment-area, is based on double integration but is graphical in nature.

EXAMPLE #9: BEAM DEFLECTION BY DOUBLE INTEGRATION

Given: Beam as shown

Find: The deflection at any point of the beam.

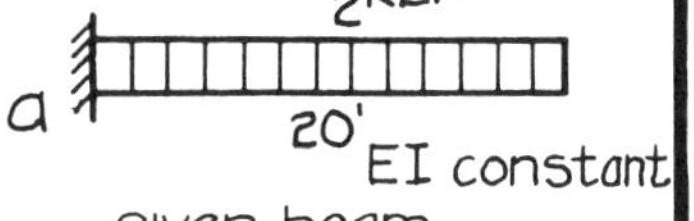

a — 2^{KLF} — 20' — EI constant

given beam

First, find the support reactions from the total FBD and $\Sigma F_Y = 0$ and $\Sigma M_a = 0$	$400^{K\text{-}FT}$, 2^{KLF}, 20', 40^K total FBD
Next, take a cut (at any arbitrary point along the beam) and find the bending moment, M(X). Two notes: 1) M(X) is always shown in the positive sense; ie, so that it causes the beam to smile. 2) The portion of the distributed load shown on the "cut" beam can be resolved into a single concentrated load acting through the centroid of the portion of the load of the "cut" beam.	$400^{K\text{-}FT}$, $12X$, 2^{KLF}, M(X), q, V(X), 40^K, $\frac{X}{2}$, X FBD / cut beam $\Sigma M_{pt.\,q} = 0 \; (+\circlearrowleft)$ $M(X) + 400 - 40X + (2X)\left(\frac{X}{2}\right) = 0$ $M(X) = -X^2 + 40X - 400$
Proceed with the integration. (C_1 and C_2 are constants of integration). Note: It is convenient to take the "EI" to the left side.	$\frac{d^2U}{dX^2} = \frac{M(X)}{EI} = -X^2 + 40X - 400$ $EI\theta = EI\frac{dU}{dX} = -\frac{X^3}{3} + 20X^2 - 400X + C_1 \quad (1)$ $EIU = -\frac{X^4}{12} + \frac{20X^3}{3} - 200X^2 + C_1X + C_2 \quad (2)$
Boundary conditions: need two (since two constants of integration) Note: From the given beam that at the wall no rotation nor deflection can occur. Therefore: B.C. #1 $X = 0$ $\theta = \frac{dU}{dX} = 0$ B.C. #2 $X = 0$ $U = 0$	B.C. #1 $X = 0$, $\theta = \frac{dU}{dX} = 0$ goes into Eq.(1) above thus $0 = 0 + 0 + 0 + C_1$ or $C_1 = 0$ B.C. #2 $X = 0$, $U = 0$ goes into Eq(2) above thus $0 = 0 + 0 + 0 + 0 + C_2$ or $C_2 = 0$
With the constants of integration found, both the slope and deflection equations are known.	slope equation $EI\theta = \frac{dU}{dX} = -\frac{X^3}{3} + 20X^2 - 400X$ deflection equation $EIU = -\frac{X^4}{12} + \frac{20X^3}{3} - 200X^2$

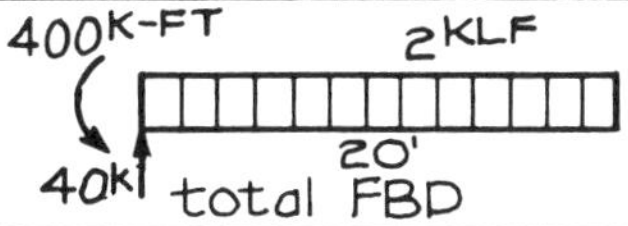

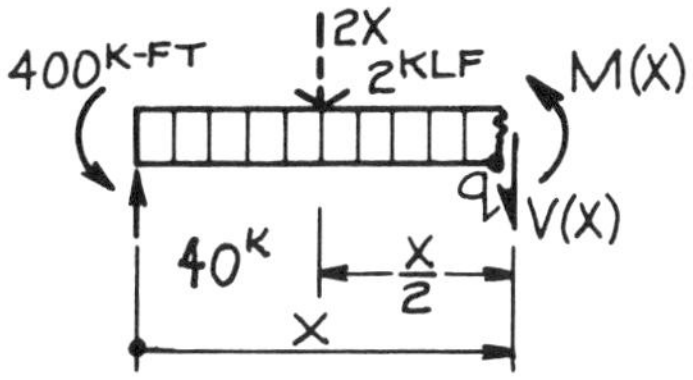

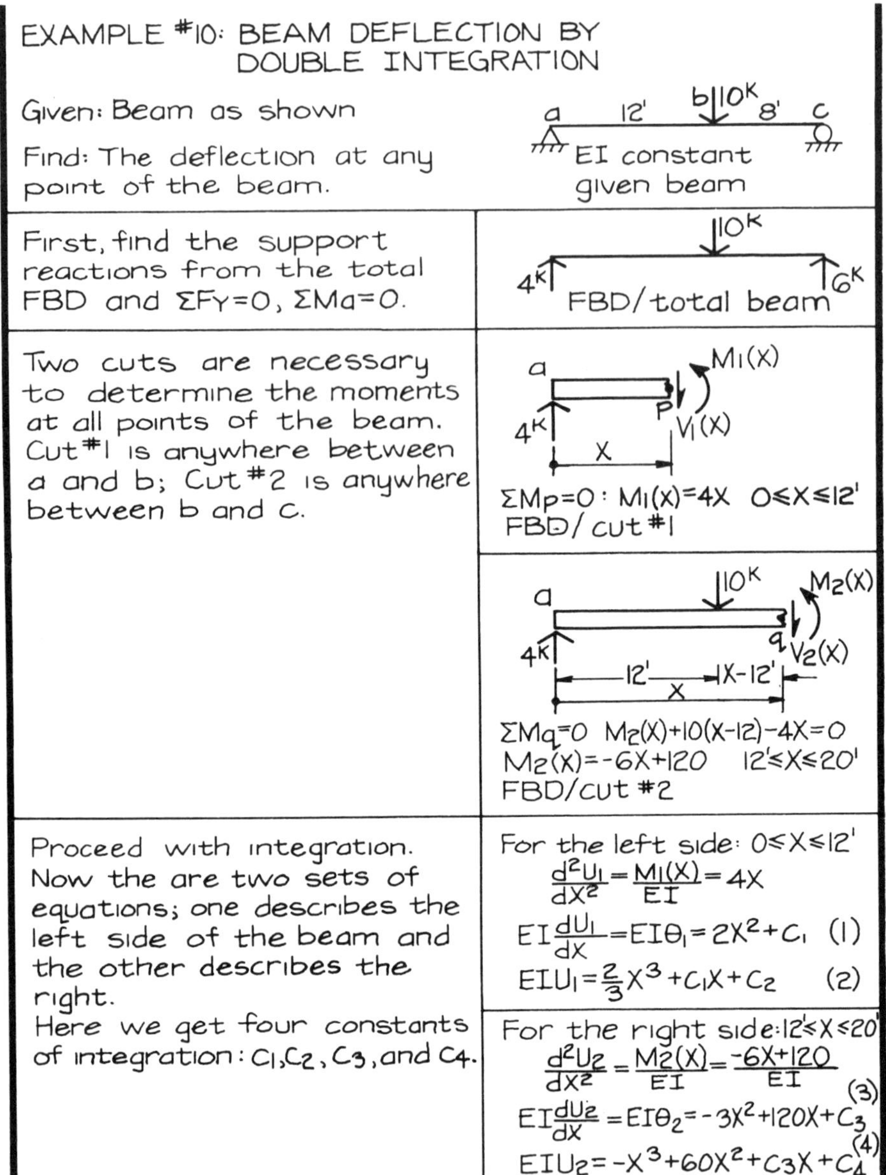

EXAMPLE #10: BEAM DEFLECTION BY DOUBLE INTEGRATION

Given: Beam as shown

Find: The deflection at any point of the beam.

given beam

First, find the support reactions from the total FBD and $\Sigma F_Y=0$, $\Sigma M_a=0$.

FBD/total beam

Two cuts are necessary to determine the moments at all points of the beam. Cut #1 is anywhere between a and b; Cut #2 is anywhere between b and c.

$\Sigma M_p=0$: $M_1(X)=4X \quad 0\leq X\leq 12'$

FBD/cut #1

$\Sigma M_q=0 \quad M_2(X)+10(X-12)-4X=0$

$M_2(X)=-6X+120 \quad 12'\leq X\leq 20'$

FBD/cut #2

Proceed with integration. Now the are two sets of equations; one describes the left side of the beam and the other describes the right.

Here we get four constants of integration: C_1, C_2, C_3, and C_4.

For the left side: $0\leq X\leq 12'$

$$\frac{d^2U_1}{dX^2}=\frac{M_1(X)}{EI}=4X$$

$$EI\frac{dU_1}{dX}=EI\theta_1=2X^2+C_1 \quad (1)$$

$$EIU_1=\frac{2}{3}X^3+C_1X+C_2 \quad (2)$$

For the right side: $12'\leq X\leq 20'$

$$\frac{d^2U_2}{dX^2}=\frac{M_2(X)}{EI}=\frac{-6X+120}{EI}$$

$$EI\frac{dU_2}{dX}=EI\theta_2=-3X^2+120X+C_3 \quad (3)$$

$$EIU_2=-X^3+60X^2+C_3X+C_4 \quad (4)$$

Boundary conditions: At the supports a and c no deflection can take place ∴

BC#1 $X=0$, $U_1=0$

BC#2 $X=20'$, $U_2=0$

The 3RD and 4TH boundary conditions are not as obvious. They come from the physical condition that when the beam deflects the deflection is smooth. This means that at point b the slopes described by eqs. (1) and (3) must be the same. Also at point b the deflections given by eqs. (2) and (4) must be equal. ∴

BC#3 $X=12'$ $\theta_1=\theta_2$

BC#4 $X=12'$ $U_1=U_2$

BC#1 $X=0$, $U_1=0$ goes into eq. (2)

$0=0+0+C_2 \therefore C_2=0$

BC#2 $X=20'$, $U_2=0$ goes into eq (4)

$0=-20^3+60(20)^2+20C_3+C_4$

or $20C_3+C_4=-16{,}000$ (5)

BC#3 $X=12'$ $\theta_1=\theta_2$

both eqs. (1) & (3) are used

$2(12)^2+C_1=-3(12)^2+120(12)+C_3$

or $C_1-C_3=720$ (6)

BC#4 $X=12'$ $U_1=U_2$

both eqs. (2) & (4) are used

$2/3(12)^3+C_1(12)+C_2=-(12)^3+60(12)^2+C_3(12)+C_4$

or $12C_1+C_2-12C_3-C_4=5760$ (7)

C_2 is easily found ($C_2=0$) but the constants of integration C_1, C_3, C_4 are found after laboriously solving simultaneous equations (5), (6) and (7). We finally get (the algebra is not shown)

$$C_1=-224, \quad C_3=-944, \quad C_4=+2880$$

thus, for the left side ($0 \le X \le 12'$)

$$EI\theta_1=2X^2-224 \quad \text{and} \quad EIU_1=\tfrac{2}{3}X^3-224X$$

and for the right side ($12' \le X \le 20'$)

$$EI\theta_2=-3X^2+120X-944 \quad \text{and} \quad EIU_2=-X^3+60X^2-944X+2880$$

In particular the deflection under the load, point b, can be evaluated using either side, ie,

$$U(X=12')=\frac{1}{EI}\left[\frac{2}{3}(12)^3-(224)(12)\right]=\frac{1}{EI}\left[-(12)^3+60(12)^2-944(12)+2880\right]$$

$$U(X=12')=\frac{1}{EI}[-1536]$$, the negative sign means downward.

PROBLEMS

Problems 1–10

Find the shear force and bending moment expressions (for beams 1–10) by using the direct approach.

Problems 11–20

Find the shear force and bending moment diagrams for beams 1–10 by using the semi-graphical approach.

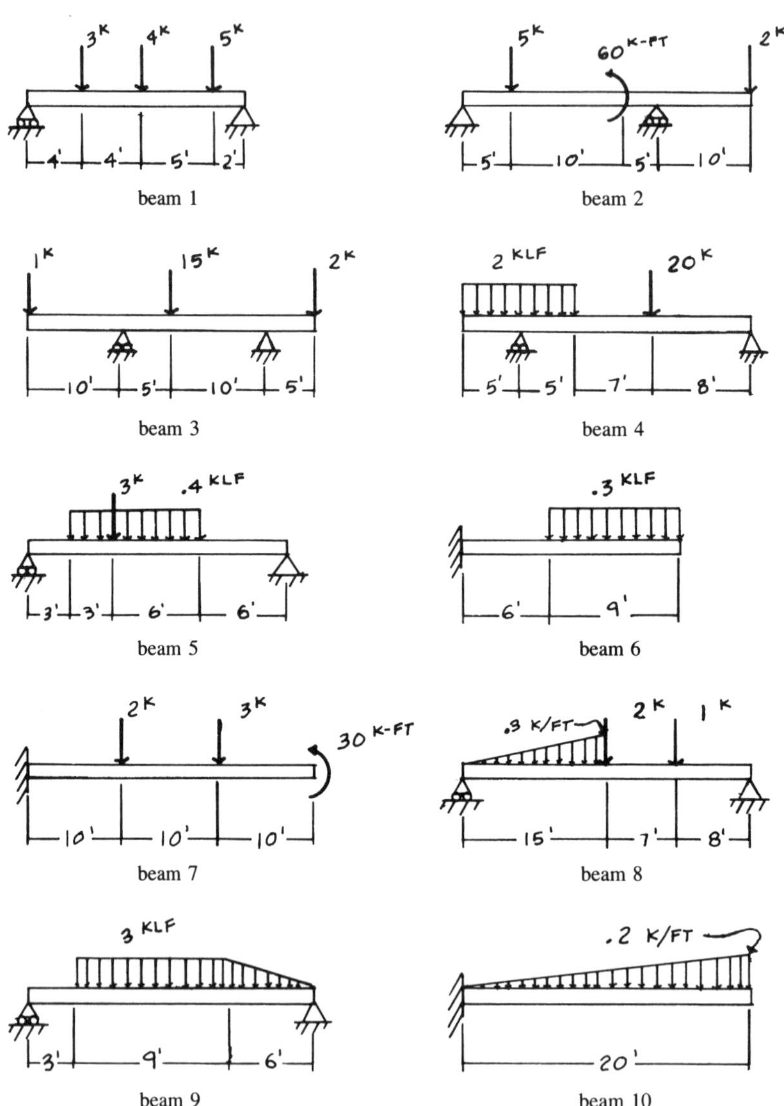

beam 1

beam 2

beam 3

beam 4

beam 5

beam 6

beam 7

beam 8

beam 9

beam 10

Problems 21(a–d)–30(a–d)

For beams 1–10 find the maximum tension and compression stresses and the maximum shear stress if the beams have cross-sections as shown.

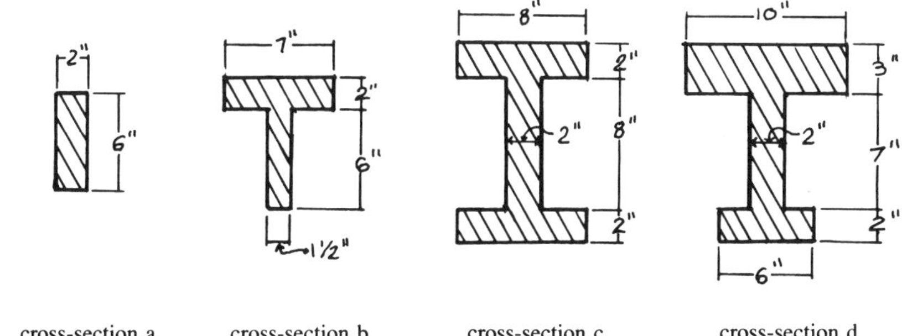

cross-section a　　cross-section b　　cross-section c　　cross-section d

Problems 31(a, b)–35(a, b)

For the beams 1–5 find the cross-sectional dimension ℓ if the beams have the cross-sections shown and the allowable bending stress (tension and compression) is 22 ksi and the allowable shear stress is 20 ksi.

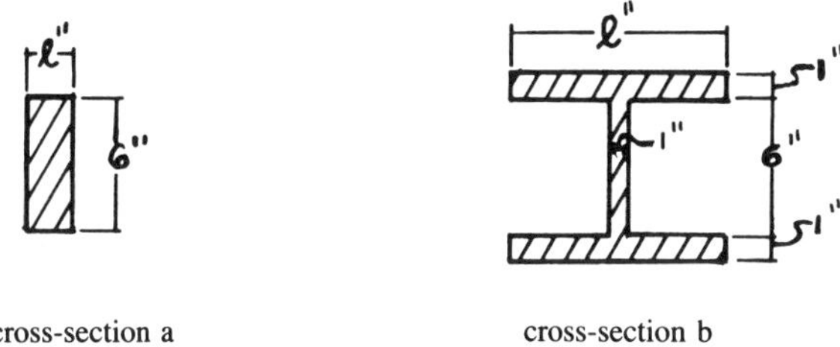

cross-section a　　cross-section b

4

FRAMES

INTRODUCTION

This chapter is concerned with finding internal resultants in determinate frames. In addition to shear forces and bending moments, frames usually have internal axial forces, P.

The basis for the determination of internal resultants is, again, the equilibrium equations. Statical analysis for frames is usually more involved than for beams, due to the number of members and the assorted connections for frames.

Internal Resultants

As a reminder we note, as we turn a corner, say at an internal pin shown in Figure 4.1, the internal forces act as equal and opposite pairs. Further, we see that the vertical force acts as an axial force on the column and as a shear force on the beam. To resolve this ambiguity we note the vertical force as ab_y on both members, instead of V and P. A similar approach is taken for the horizontal force.

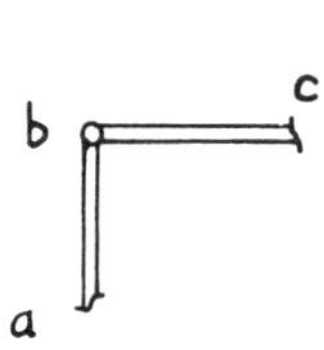

Figure 4.1a

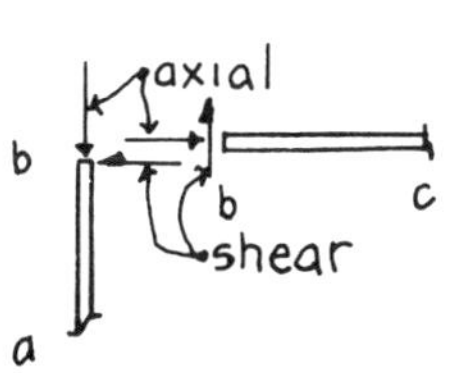

Figure 4.1b

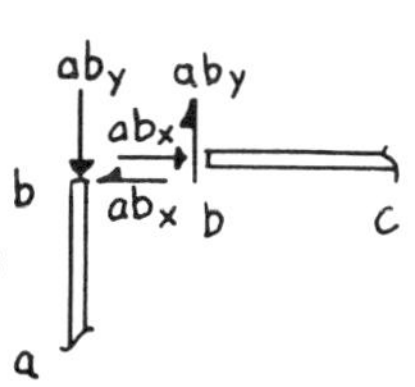

Figure 4.1c

In using the short cut approach to finding shear forces and bending moments (Equations 2–5 in the last chapter) it is necessary to integrate from left to right; thus a reference system is important. For one-bay frames it is usual to go clockwise around the frame, thus the inside of the frame becomes the bottom of each member (Figure 4.2a). In so doing the positive directions for shear forces and bending moments follow the convention for beams. In addition, a tensile force is taken as positive for *P* (Figure 4.2b). In Examples 1 and 2 this convention is illustrated further; the *V*, *M* and *P* diagrams in these examples are oriented in accordance with the position of the "viewer," as shown.

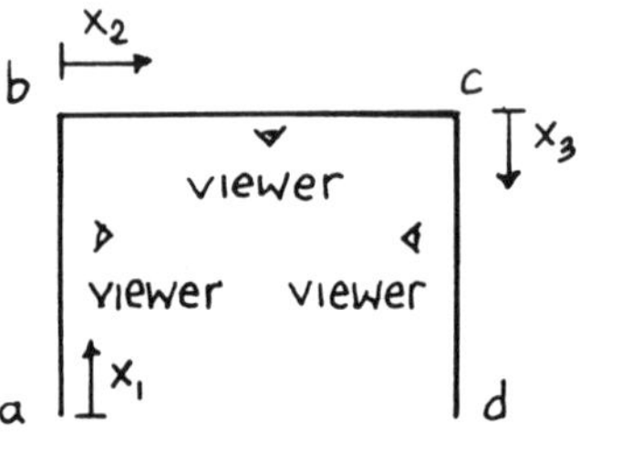

Figure 4.2a

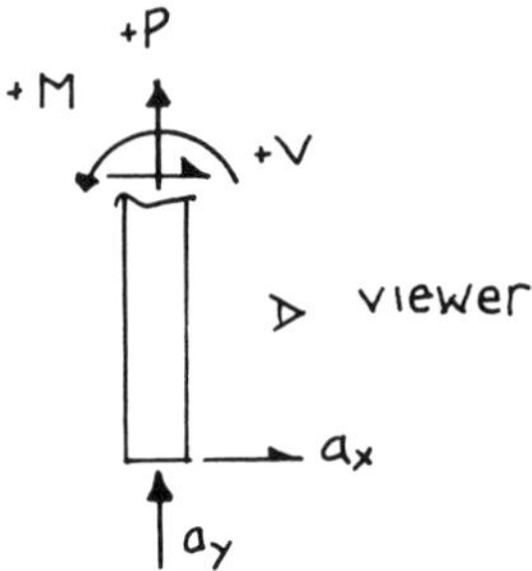

Figure 4.2b FBD: cut through *ab*

It is common for codes to specify snow roof loads in terms of the horizontal projection of the roof. When the roof is pitched, it is necessary to resolve the distributed load into normal and shear loads in order to find the internal resultants. Such a resolution is illustrated in Figure 4.3.

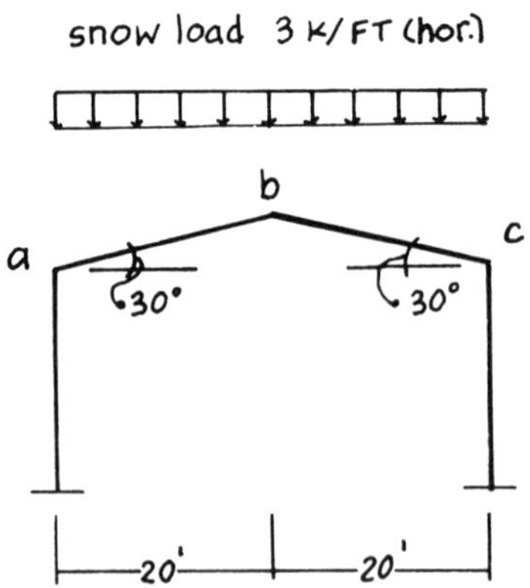

Figure 4.3a given loading

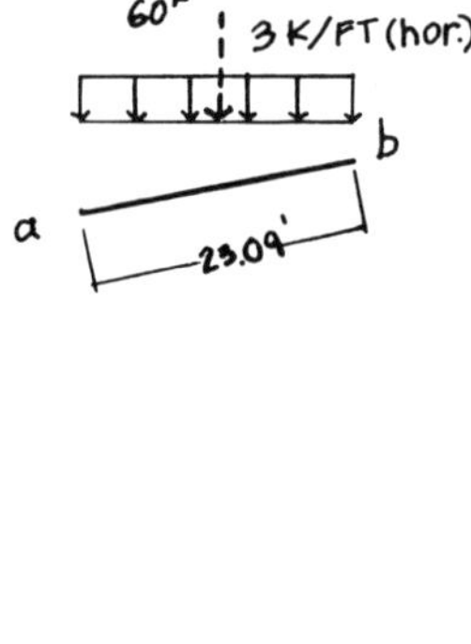

Figure 4.3b distributed load and vertical resultant on member *ab*

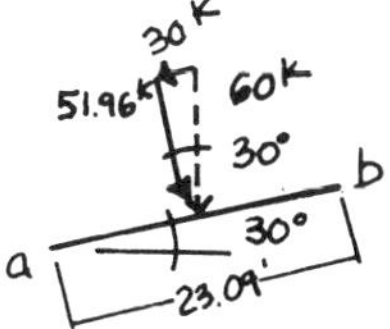

Figure 4.3c vertical resultant resolved into normal and parallel loads

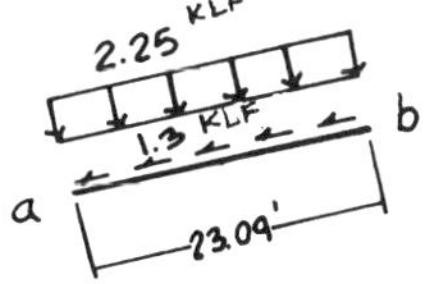

Figure 4.3d distributed normal and parallel loads

DETERMINACY AND STABILITY

It would be convenient to say that a frame is statically determinate if the available equations of equilibrium equal the restraints (unknown reactions). While this is usually true, there are tricky cases where it is not.

The determinacy issue is straightforward for one-story frames: establish all useful free body diagrams, count unknown reactions and available independent equations of equilibrium and compare the two sets. If there are more unknowns (restraints) than equations, the frame is statically indeterminate; if unknowns (restraints) equal equations, we have statical determinacy; if unknowns (restraints) are less than equations, we have instability (collapse). The following examples will illustrate these points.

In Figures 4.4a and 4.4b only one FBD is used (any internal "cuts" will expose 3 internal reactions: axial force, shear force and bending moment which cancels the additional 3 equations of equilibrium picked up by an additional FBD). In Figure 4.4c,d the internal pins allow the use of additional FBDs.

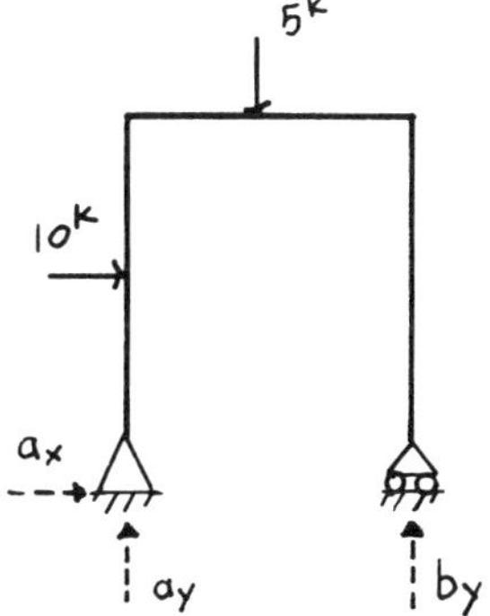

Figure 4.4a 3 reactions (a_x, a_y, b_y)
3 eqs. of equil.
0: statically determinate

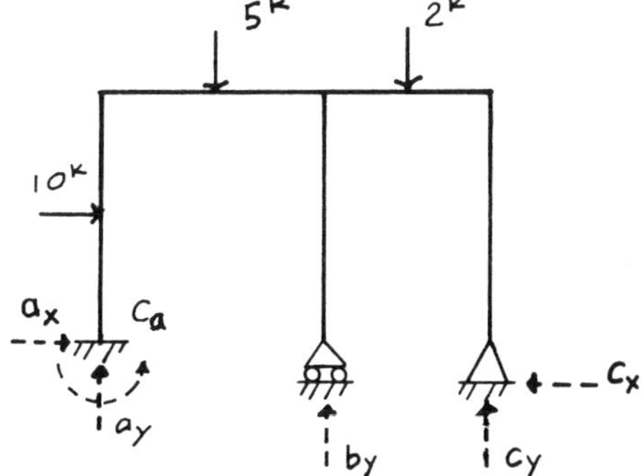

Figure 4.4b 6 reactions
3 eqs. of equil.
3° statically indeterminate

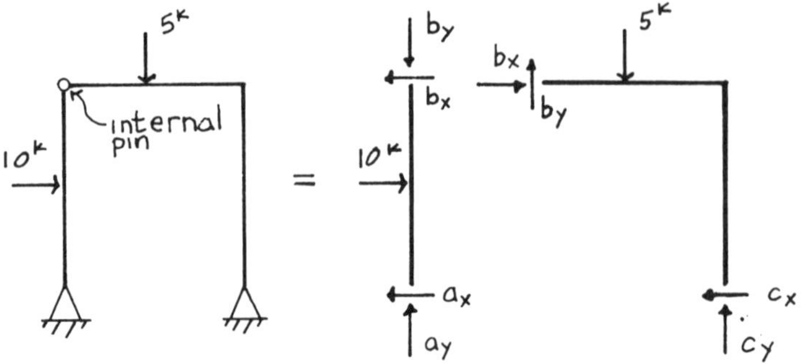

Figure 4.4c 6 reactions (a_x, a_y, b_x, b_y, c_x, c_y)
6 eqs. of equil. (2 FBDs)

0: statically determinate

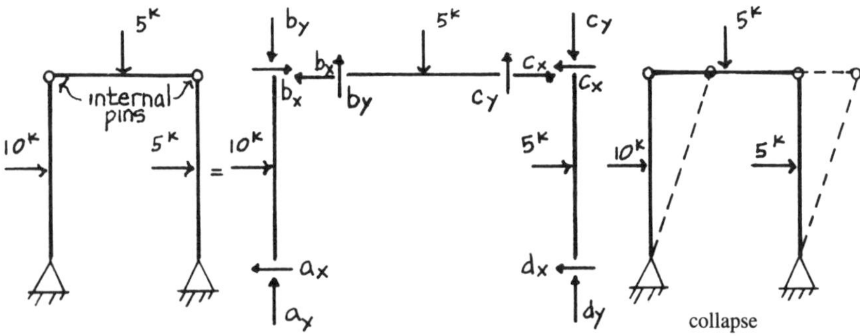

Figure 4.4d 8 reactions (a_x, a_y, b_x, b_y, c_x, c_y, d_x, d_y)
9 eqs. of equil. (3 FBDs)

−1: unstable

For multistory frames, the available FBD issue becomes involved. It is easier to cut the frame into "trees" and look at the conditions necessary to restore stability. In Figure 4.5 we see a total of 12 internal unknown reactions. If we consider these internal reactions as loads and look at the left tree as if it were free standing (which it is not), it would fall down since there is no rotational resistance at the support—a couple is necessary there for stability of the left tree. If a wall support were in fact there, the *total* frame would be statically indeterminate to 12th degree (due to internal unknown)—the pin at the left tree reduces the degree of statical indeterminacy to 11.

In Figure 4.6 rotational stability is supplied to the "trees" by the addition of knee braces, each of which reduces the degree of indeterminacy by one degree.

Note: external loads not shown for clarity.

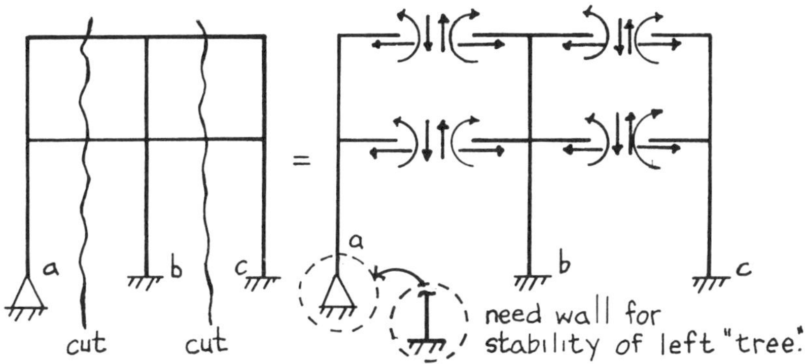

Figure 4.5 12 internal reactions
 1 needed support for stability
 11° statically indeterminate

Note: external loads not shown.

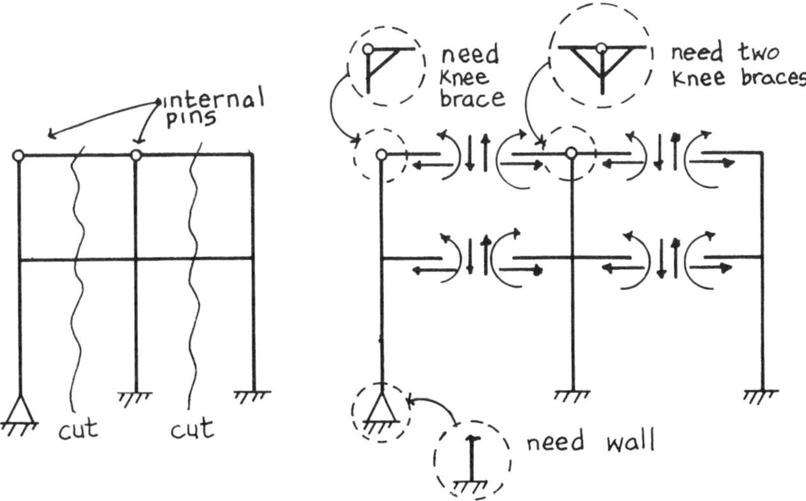

Figure 4.6 12 internal reactions
 4 need supports for stability
 8° statically indeterminate

The frame in Figure 4.7 is the frame in Figure 4.4d but with an additional pin-connected floor. The frame will still collapse. The additional floor provides no lateral restraint. This is the "tricky" case mentioned at the beginning of this section since the "numbers" imply this frame is stable (6 internal reactions and 6 needed supports for tree stability: two walls and 4 knee braces). The frame in Figure 4.7b is stable because of the lateral restraint provided by the inclined member and the wall support makes Figure 4.7c stable and statically indeterminate.

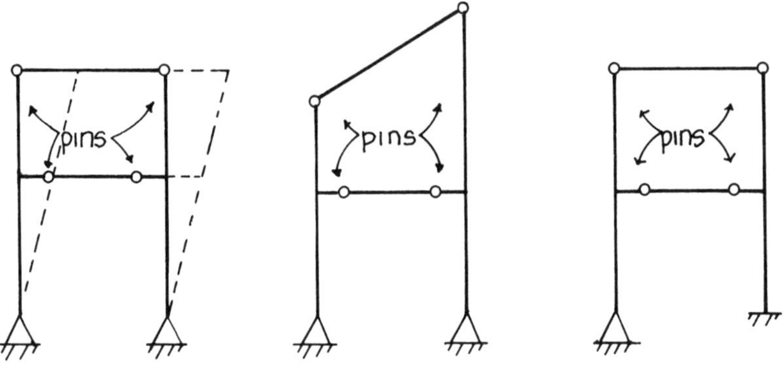

Figure 4.7a unstable

Figure 4.7b stable and determinate

Figure 4.7c stable and 1° indeterminate

STRESSES

Three stress terms occur at each point in a frame: shear stress, $\tau = VQ/It$; bending (axial) stress, $\sigma = My/I$, and direct (axial) stress, $\sigma = P/A$. When the direct (axial) stress is compressive, which is usual, we have interaction of buckling and bending; design codes reflect this interaction, which is quite a complex phenomenon.

EXAMPLE #1: INTERNAL RESULTANTS - FRAME

Given: Frame as shown

Find: V and M diagrams; axial forces, P.

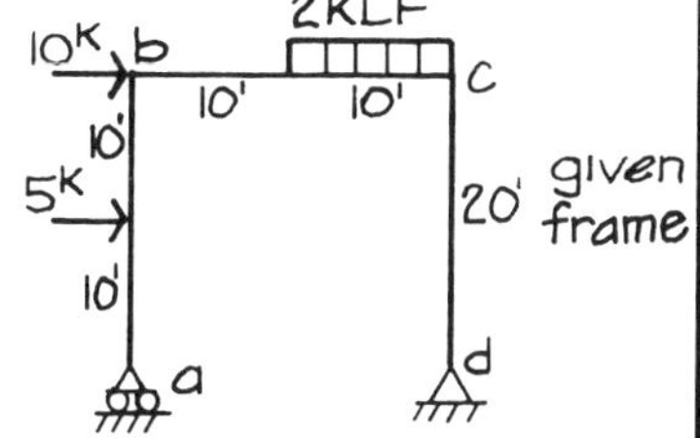

Total FBD:

$\Sigma F_X = 0$: $d_X = -15^K (\leftarrow)$

$\Sigma M_{pt\,a} = 0$:

$20 d_Y = [10(5) + 20(10) + 15(20)]$

$\therefore d_Y = 27.5^K (\uparrow)$

$\Sigma F_Y = 0$: $a_Y = -7.5^K (\downarrow)$

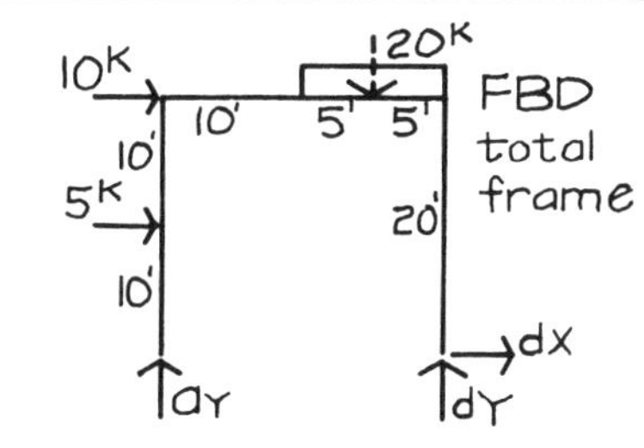

In order to find the internal joint reactions, make cuts at joints b and c and form three FBDs (of ab, bc and cd). The question arises concerning, "On which FBD does the 10^K load at joint b belong?" It can be put on ab or bc or part on ab and part on bc. The internal horizontal force will balance the transfer. What is important, though, is that when FBD ab and FBD bc are put together only a 10^K external force at b appears. In our solution we will put the 10^K on FBD bc.

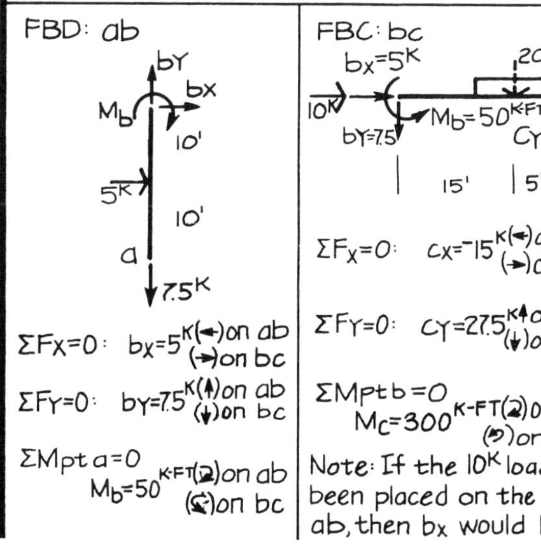

FBD: ab

$\Sigma F_X = 0$: $b_X = 5^K (\leftarrow)$ on ab, $(\rightarrow)$ on bc

$\Sigma F_Y = 0$: $b_Y = 7.5^K (\uparrow)$ on ab, $(\downarrow)$ on bc

$\Sigma M_{pt\,a} = 0$

$M_b = 50^{K\text{-}FT}$ (↻) on ab, (↺) on bc

FBC: bc

$\Sigma F_X = 0$: $c_X = -15^K (\leftarrow)$ on bc, $(\rightarrow)$ on cd

$\Sigma F_Y = 0$: $c_Y = 27.5^K (\uparrow)$ on bc, $(\downarrow)$ on cd

$\Sigma M_{pt\,b} = 0$

$M_c = 300^{K\text{-}FT}$ (↻) on bc, (↺) on cd

Note: If the 10^K load had been placed on the column, ab, then b_X would be 15^K.

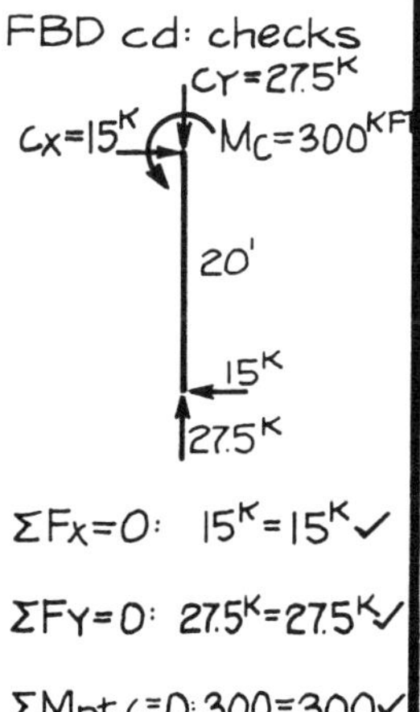

FBD cd: checks

$\Sigma F_X = 0$: $15^K = 15^K$ ✓

$\Sigma F_Y = 0$: $27.5^K = 27.5^K$ ✓

$\Sigma M_{pt\,c} = 0$: $300 = 300$ ✓

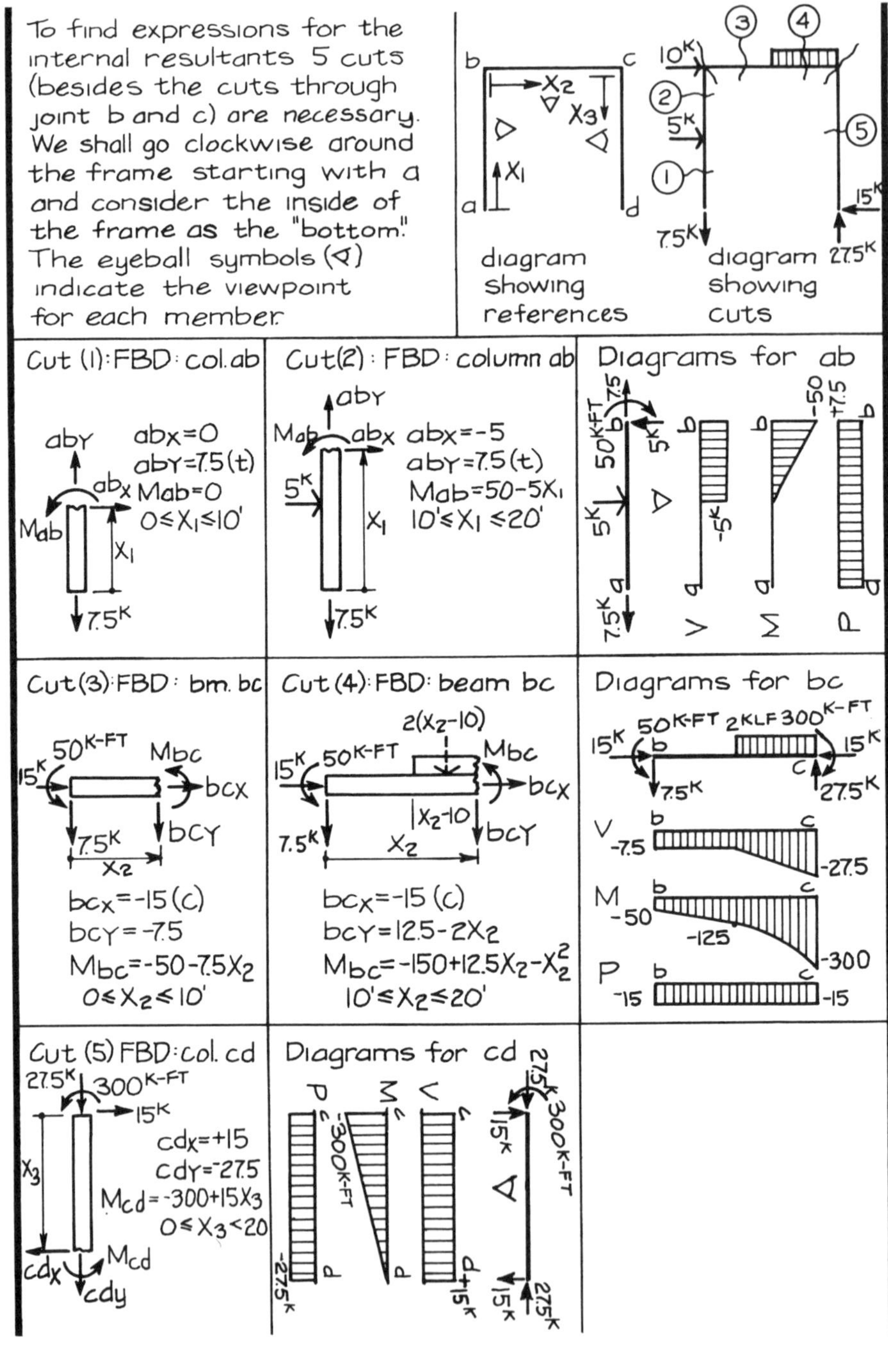

To find expressions for the internal resultants 5 cuts (besides the cuts through joint b and c) are necessary. We shall go clockwise around the frame starting with a and consider the inside of the frame as the "bottom." The eyeball symbols (◁) indicate the viewpoint for each member.
diagram showing references
diagram showing cuts
10K
5K
7.5K
15K
27.5K
Cut (1): FBD: col. ab
abx=0
abY=7.5 (t)
Mab=0
0≤X1≤10'
Cut(2): FBD: column ab
abx=-5
abY=7.5 (t)
Mab=50-5X1
10'≤X1≤20'
Diagrams for ab
Cut(3): FBD: bm. bc
bcx=-15 (c)
bcY=-7.5
Mbc=-50-7.5X2
0≤X2≤10'
Cut (4): FBD: beam bc
2(X2-10)
bcx=-15 (c)
bcY=12.5-2X2
Mbc=-150+12.5X2-X2²
10'≤X2≤20'
Diagrams for bc
50K-FT 2KLF 300K-FT
-7.5
-27.5
-50
-125
-300
-15
Cut (5) FBD: col. cd
27.5K 300K-FT
15K
cdx=+15
cdY=-27.5
Mcd=-300+15X3
0≤X3<20
Diagrams for cd
-300K-FT
-27.5K
+15K

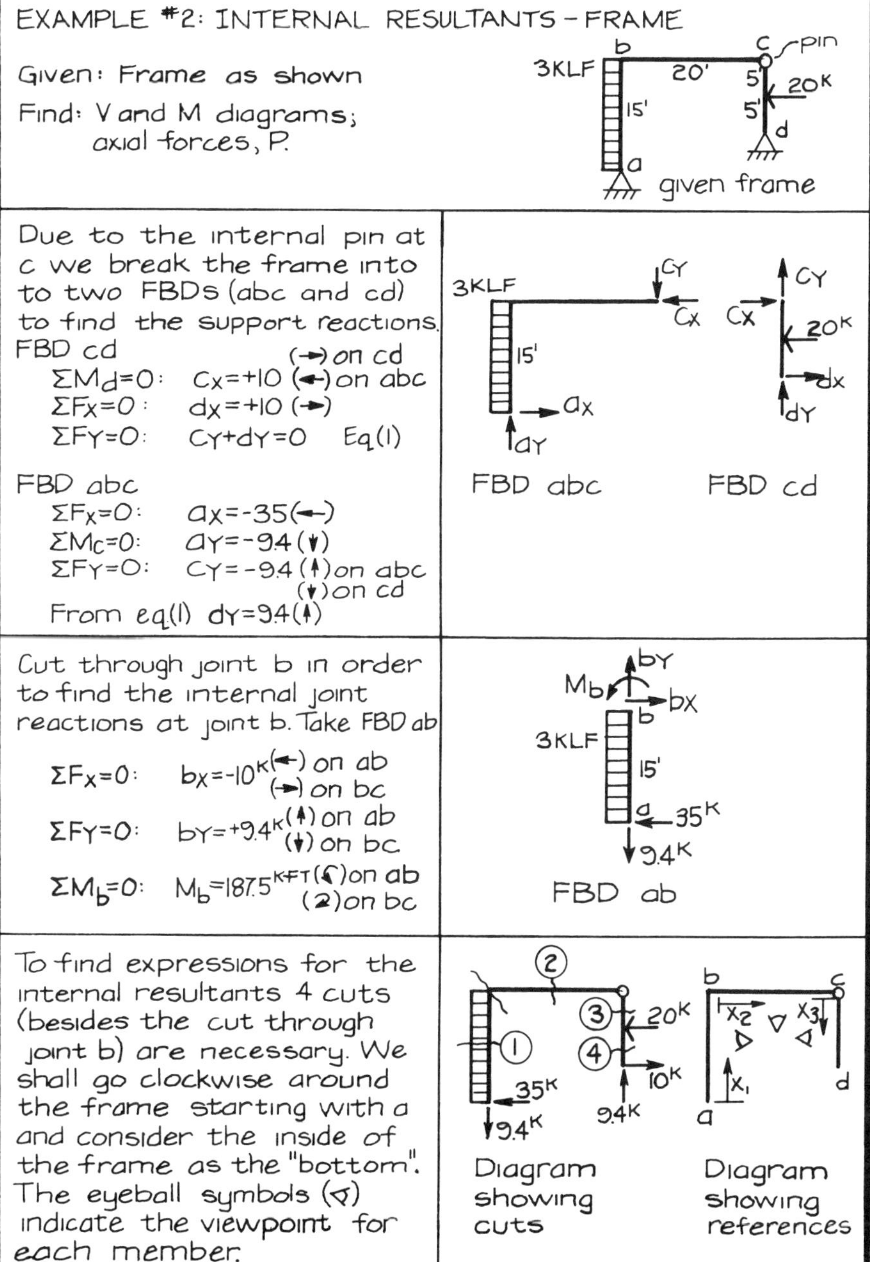

EXAMPLE #2: INTERNAL RESULTANTS – FRAME

Given: Frame as shown

Find: V and M diagrams; axial forces, P.

Due to the internal pin at c we break the frame into two FBDs (abc and cd) to find the support reactions.

FBD cd

$\Sigma M_d = 0$: $C_X = +10$ (→) on cd, (←) on abc

$\Sigma F_X = 0$: $d_X = +10$ (→)

$\Sigma F_Y = 0$: $C_Y + d_Y = 0$ Eq.(1)

FBD abc

$\Sigma F_X = 0$: $a_X = -35$ (←)

$\Sigma M_c = 0$: $a_Y = -9.4$ (↓)

$\Sigma F_Y = 0$: $C_Y = -9.4$ (↑) on abc, (↓) on cd

From eq.(1) $d_Y = 9.4$ (↑)

Cut through joint b in order to find the internal joint reactions at joint b. Take FBD ab

$\Sigma F_X = 0$: $b_X = -10^K$ (←) on ab, (→) on bc

$\Sigma F_Y = 0$: $b_Y = +9.4^K$ (↑) on ab, (↓) on bc

$\Sigma M_b = 0$: $M_b = 187.5^{K\text{-}FT}$ (↶) on ab, (↷) on bc

To find expressions for the internal resultants 4 cuts (besides the cut through joint b) are necessary. We shall go clockwise around the frame starting with a and consider the inside of the frame as the "bottom". The eyeball symbols (◁) indicate the viewpoint for each member.

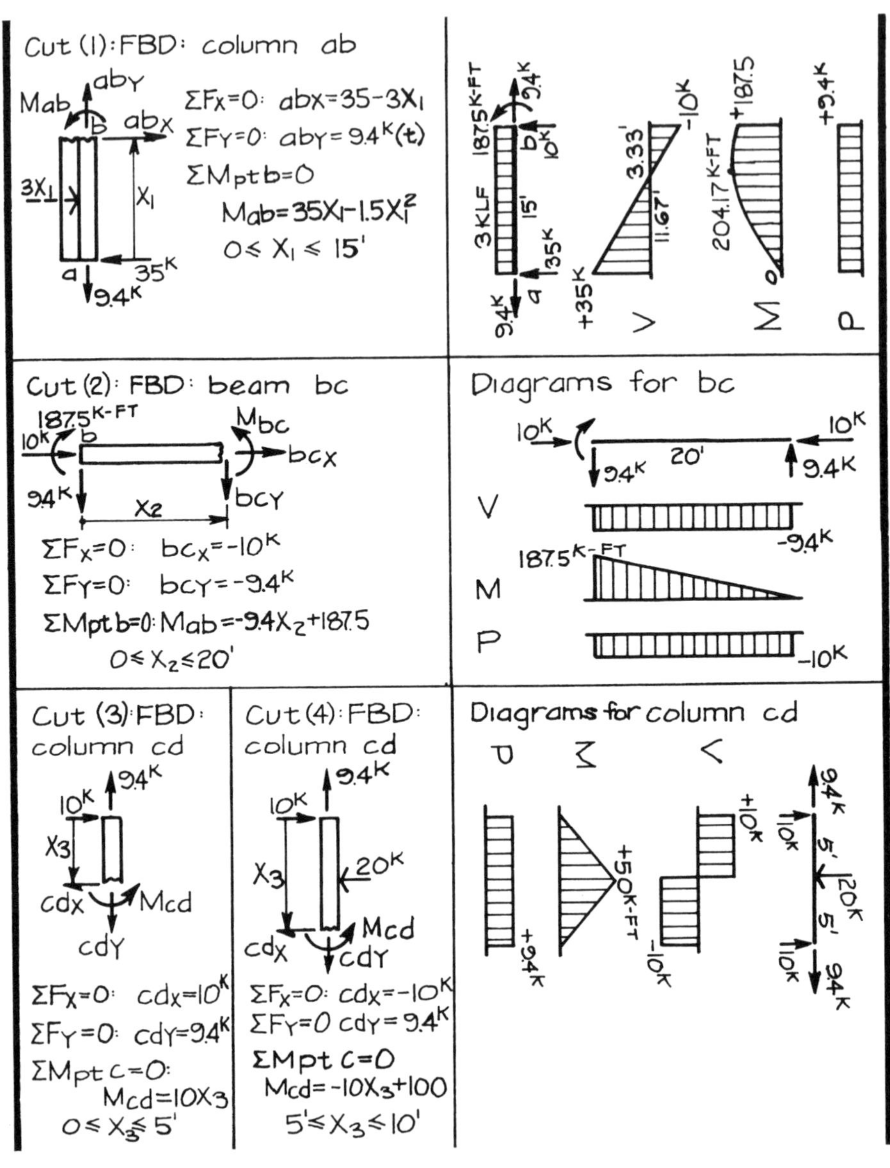
Cut (1): FBD: column ab
ΣFx=0: abx=35-3X1
ΣFy=0: aby=9.4K(t)
ΣMpt b=0
Mab=35X1-1.5X1²
0≤X1≤15'
Cut (2): FBD: beam bc
ΣFx=0: bcx=-10K
ΣFy=0: bcy=-9.4K
ΣMpt b=0: Mab=-9.4X2+187.5
0≤X2≤20'
Diagrams for bc
Cut (3): FBD: column cd
ΣFx=0: cdx=10K
ΣFy=0: cdy=9.4K
ΣMpt c=0:
Mcd=10X3
0≤X3≤5'
Cut (4): FBD: column cd
ΣFx=0: cdx=-10K
ΣFy=0 cdy=9.4K
ΣMpt c=0
Mcd=-10X3+100
5'≤X3≤10'
Diagrams for column cd

PROBLEMS

Problems 1–11

For the frames shown, find the shear force, bending moment and axial force diagrams.

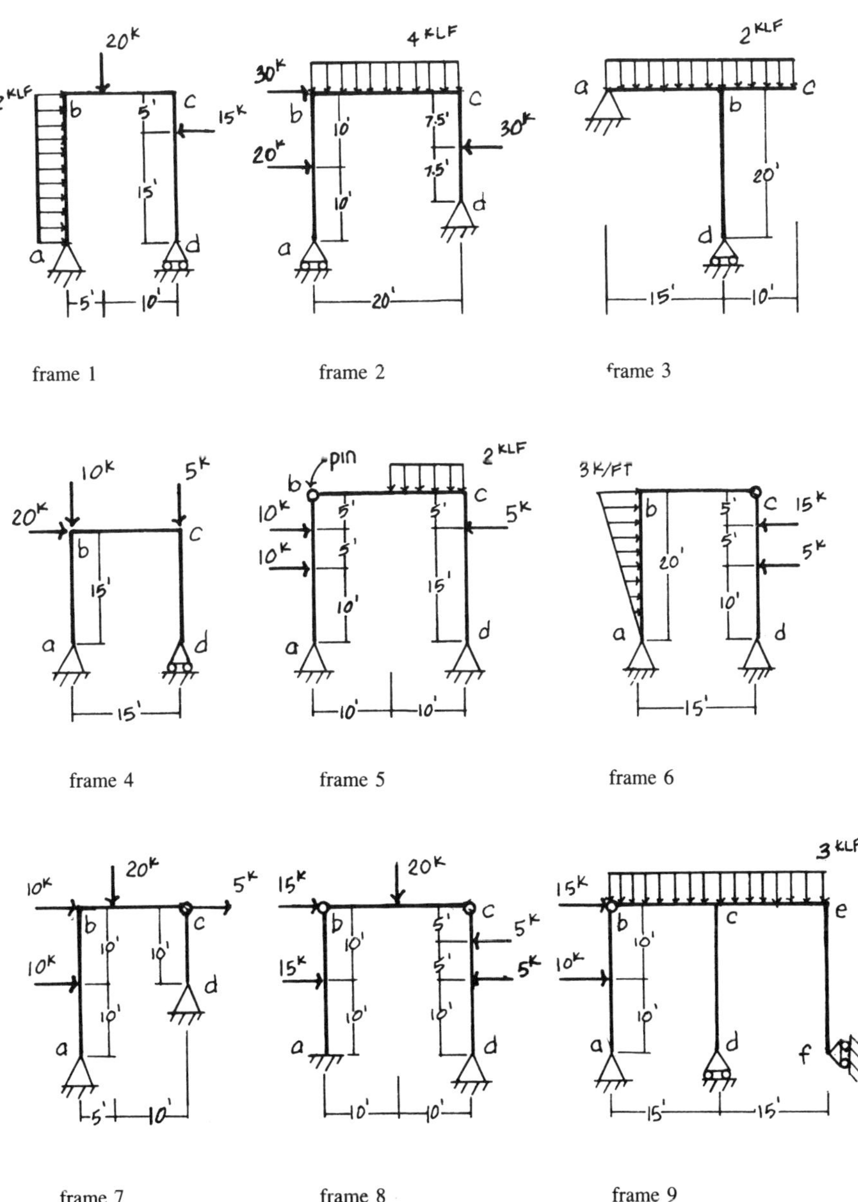

frame 1 frame 2 frame 3

frame 4 frame 5 frame 6

frame 7 frame 8 frame 9

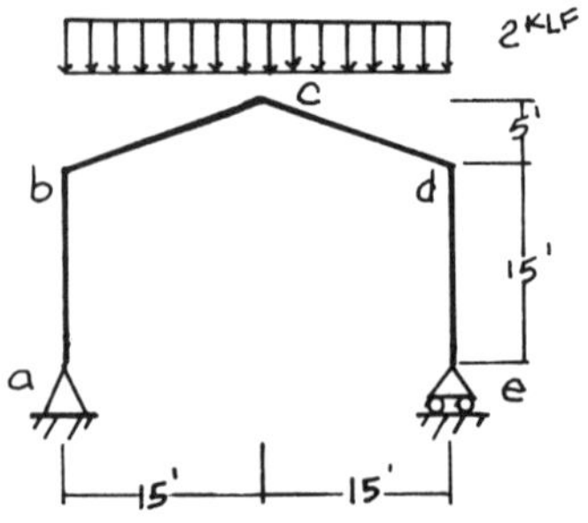

frame 10

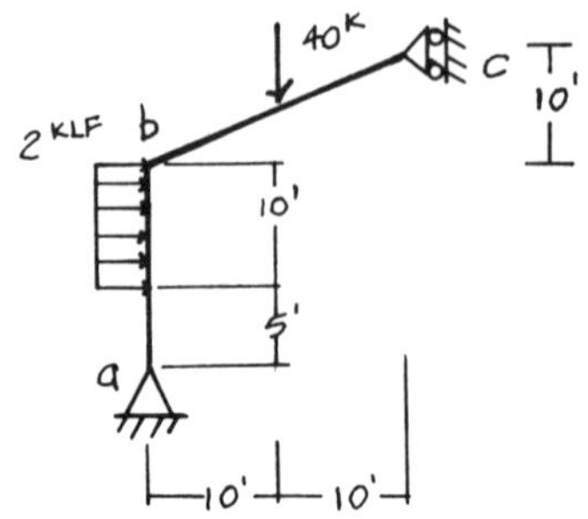

frame 11

5

DEFLECTIONS: MOMENT-AREA TECHNIQUE

INTRODUCTION

In Chapter 3 we learned about the relationships between bending moments and bending stresses and between bending moments and deflections. We drew shear force and bending moment diagrams which were graphical representations of the actions of forces on beams. In this chapter we will use these bending moment diagrams to determine deflections of beams and frames subjected to external loads. We should also note that the shear force in a beam or frame does contribute to the deflection, but, for shallow beams, this contribution is generally very small and in this text we shall neglect it and concentrate on deflections due to bending moments. In Chapter 3 we found deflections using direct integration. In this chapter we will develop a method which uses a form of graphical integration.

CURVATURE AND DEFLECTION

In order to examine the deflection of a beam let us begin, without loss of generality, by looking at the neutral axis of a simply-supported beam (Figure 5.1). Let us seek the deflection at a general point, c.

Keeping in mind the deflection diagram—Figure 5.2—we introduce a new term: the tangential deviation. The tangential deviation, $t_{q/p}$, is the vertical distance between the tangent line (to the deflection curve), drawn through point q, and point p on the deflection curve. (Note: The second subscript corresponds with the *point* on the deflection curve—the first subscript with the point at which the *tangent* line is

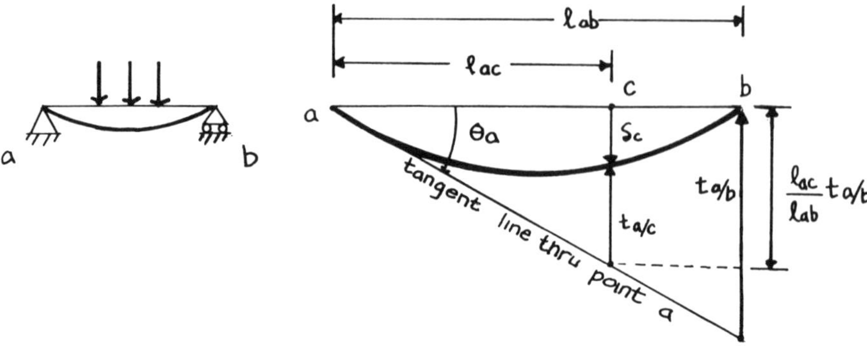

Figure 5.1 **Figure 5.2**

drawn.) The sign of $t_{q/p}$ is positive if point p (on the deflection curve) is above the tangent line through q. In Figure 5.2 both tangential deviations, $t_{a/b}$ and $t_{a/c}$, are positive as shown. (Note: It should be kept in mind that all deflections and slopes are extremely small in magnitude.) Referring again to Figure 5.2, we see that the deflection at point c (taken as positive downward, in the gravity direction) is

$$\delta_c = +\left[\left(\frac{\ell_{ac}}{\ell_{ab}}\right)t_{a/b} - t_{a/c}\right]. \tag{1}$$

Further, we note that the tangential deviation, $t_{a/b}$, can be put in terms of the tangent angle of the deflected curve, θ_a:

$$|t_{a/b}| = |\tan\,\theta_a|\ell_{ab} \simeq |\theta_a|\ell_{ab}.$$

Using the sign convention that an angle is positive if rotated counterclockwise from the horizontal reference axis we see

$$t_{a/b} = -\theta_a\ell_{ab}.$$

In this illustration it is important to realize that the reference points, a and b, are *support* points at which the deflections are known. We could just as easily have found δ_c by drawing the tangent line through point b (Figure 5.3). The sign conventions in use are summed up in Figure 5.4.

$$\delta_c = \left[\left(\frac{\ell_{cb}}{\ell_{ab}}\right)t_{b/a} - t_{b/c}\right]. \tag{2}$$

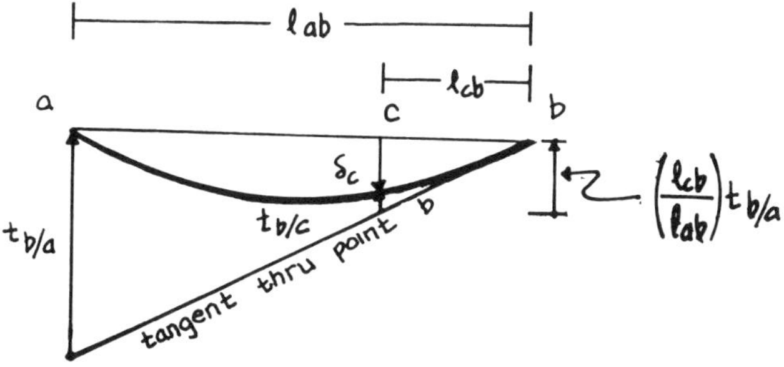

Figure 5.3

slope	tangential deviation	deflection
θ > 0	$t_{p/q} > 0$	$\delta_c > 0$
θ < 0	$t_{p/q} < 0$	$\delta_c < 0$

Figure 5.4

The *M*/*EI* Diagram

Thus far we have investigated the possibility of determining beam deflections (of the neutral axis) in terms of "tangential deviations." We will now develop a method to evaluate these tangential deviations; we will also determine means to evaluate the slope (tangent angle) at points on the deflected curve.

Referring to Chapters 2 and 3, we will state that the curvature, $1/\rho$, at a point on the deflected neutral axis is directly related to the bending moment at that point:

$$\frac{1}{\rho} = \frac{M}{EI} \qquad (E\text{: Young's modulus; } I\text{: moment of inertia}).$$

Recalling that the curvature is the rate of change (with respect to the arc distance) of the slope and that for small deflections the arc distance is approximated by the horizontal distance, we arrive at

$$\frac{1}{\rho}\left(\equiv \frac{d\theta}{ds} \simeq \frac{d\theta}{dx}\right) = \frac{M}{EI} \qquad \text{or} \qquad d\theta = \frac{M}{EI}\,dx.$$

Integrating from *left to right*, from two arbitrary points p and q, we get

$$\int_p^q d\theta = \int_p^q \frac{M}{EI}\,dx \qquad \text{or} \qquad \theta_q - \theta_p = \int_p^q \frac{M}{EI}\,dx. \tag{3}$$

We shall call Equation 3 the First Moment-Area Theorem. This theorem indicates that if we draw the portion of the "scaled" bending moment diagram (i.e., the M/EI Diagram) between points p and q we have a diagram representing the change of slope of the deflected beam between points p and q (Figure 5.5).

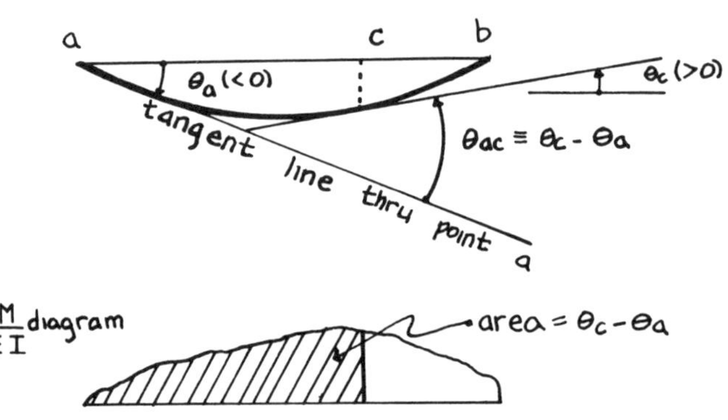

Figure 5.5

Now, if we construct two tangents to the deflected beam a small distance, dx, apart, "x-feet" from point a we see (Figure 5.6)

$$dt = \zeta\,d\theta = \zeta\left(\frac{M}{EI}\,dx\right), \quad (\zeta = \ell_{ac} - x).$$

Upon integrating from point a to c we get the tangential deviation $t_{a/c}$:

$$t_{a/c} = \int_a^c \zeta\left(\frac{M}{EI}\right) dx \equiv \ell_c\left[\text{area of } \frac{M}{EI}\right]_a^c, \tag{4}$$

in which ℓ_c = the distance from c to the centroid of the M/EI area (Figure 5.7).

Equation 4 is our Second Moment-Area Theorem; it indicates that the tangential deviation, $t_{a/c}$, is calculated by taking "moments" about point c of the area (between

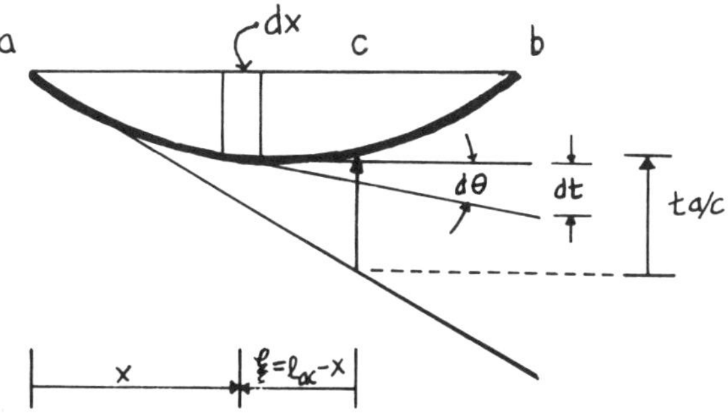

Figure 5.6

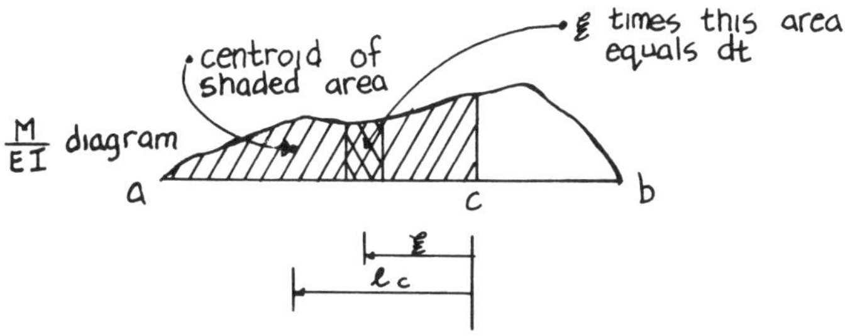

Figure 5.7

points *a* and *c*) of the *M/EI* diagram (Figure 5.7). Two important notes: 1) The "moments" are always taken from the *point* on the deflected curve (as opposed to the point where the tangent line is drawn). 2) The sign of $t_{a/c}$ is controlled by the sign of *M/EI*; i.e., in Equation 4, ℓ_c is always positive.

Let's take an example (Example 1) of a simply-supported beam with a concentrated load applied at some point other than the center. You may wish to check these results against those found by double integration in Chapter 3.

Example 2 shows a cantilever beam rigidly built into a wall at the left end and deflected by a single concentrated load at the free end.

Example 3 is a simply supported beam with a cantilever overhang having concentrated loads applied between the supports and at the end of the cantilever. The analysis is a combination of the two preceding examples, where the line has a slope other than zero at the left end. The cantilever overhang of Example 3 has a nonzero slope at *b*.

These three examples pretty well illustrate all possible deflection curves for statically determinate beams as they are influenced by the supports. It must be emphasized that in all cases supports are taken as reference points. The loads, of course, may change resulting in different bending moment diagrams, but the principles remain the same.

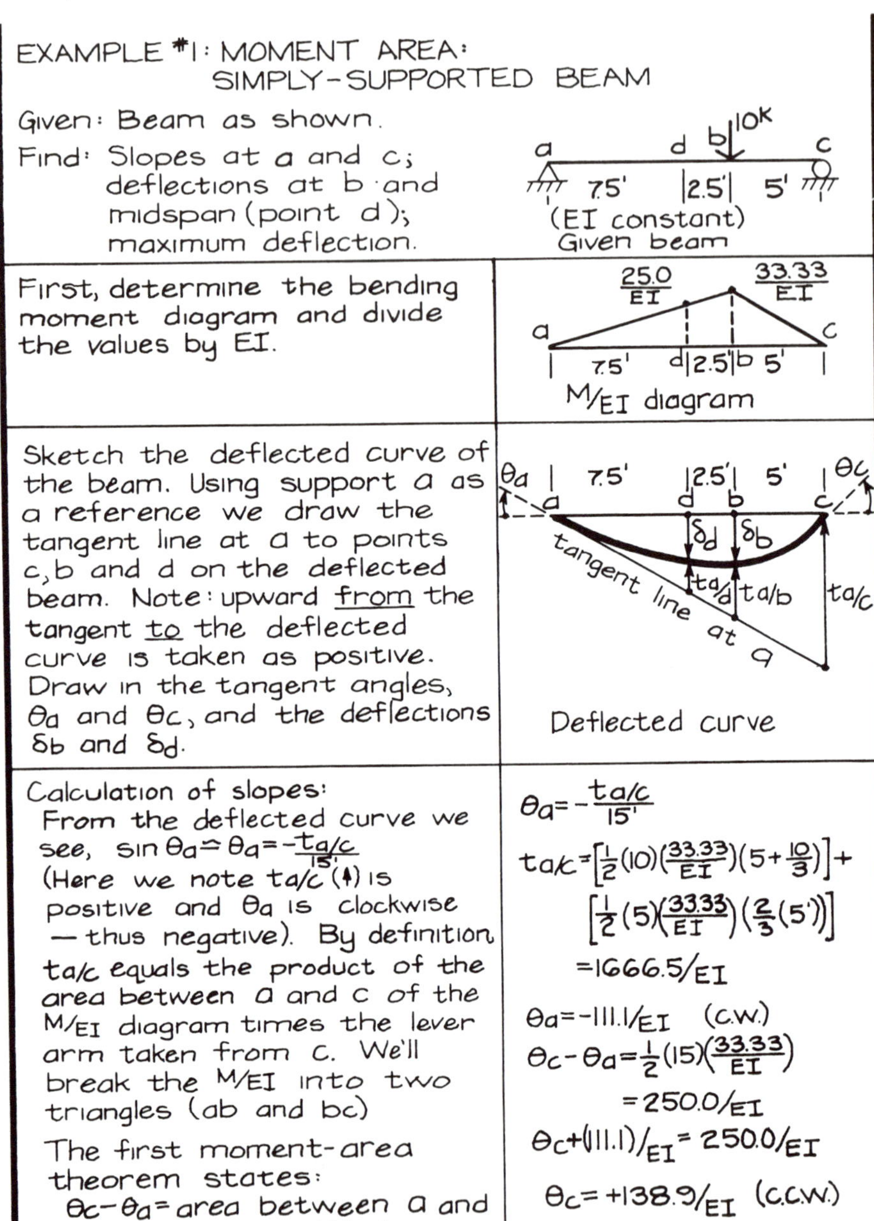
EXAMPLE #1: MOMENT AREA: SIMPLY-SUPPORTED BEAM
Given: Beam as shown.
Find: Slopes at a and c; deflections at b and midspan (point d); maximum deflection.
10K
a
d
b
c
7.5'
2.5'
5'
(EI constant)
Given beam
First, determine the bending moment diagram and divide the values by EI.
25.0/EI
33.33/EI
7.5'
2.5'
5'
M/EI diagram
Sketch the deflected curve of the beam. Using support a as a reference we draw the tangent line at a to points c, b and d on the deflected beam. Note: upward from the tangent to the deflected curve is taken as positive. Draw in the tangent angles, θa and θc, and the deflections δb and δd.
θa
θc
δd
δb
tangent line at a
ta/d
ta/b
ta/c
Deflected curve
Calculation of slopes:
From the deflected curve we see, sin θa ≐ θa = −ta/c / 15'
(Here we note ta/c (↑) is positive and θa is clockwise — thus negative). By definition ta/c equals the product of the area between a and c of the M/EI diagram times the lever arm taken from c. We'll break the M/EI into two triangles (ab and bc)
The first moment-area theorem states:
θc−θa = area between a and c of the M/EI diagram.
Knowing θa we calculate θc.
θa = −ta/c / 15'
ta/c = [½(10)(33.33/EI)(5 + 10/3)] + [½(5)(33.33/EI)(⅔(5'))]
= 1666.5/EI
θa = −111.1/EI (C.W.)
θc − θa = ½(15)(33.33/EI)
= 250.0/EI
θc + (111.1)/EI = 250.0/EI
θc = +138.9/EI (C.C.W.)

Calculation of deflections δ_b & δ_d: From the deflected curve we see, by similar triangles, that the vertical distance (↑) from the tangent line at a to the undeflected beam at b is $^{10}/_{15}\, t_{a/c}$. Subtract from this value the distance $t_{a/b}$ and δ_b (↓) is left. $t_{a/b}$ is calculated by taking the product of the area between a and b (of M/EI) times the lever taken from b. Similiarly, the deflection taken at midspan can be calculated.	$\delta_b = +\left[\frac{10}{15} t_{a/c} - t_{a/b}\right]$ $t_{a/c} = +1666.5/_{EI}$ $t_{a/b} = \left(\frac{1}{2}\right)(10)\left(\frac{33.33}{EI}\right)\left(\frac{10}{3}\right)$ $= +555.5/_{EI}$ $\delta_b = +555.5/_{EI}$ (↓) $\delta_d = +\left[\frac{7.5}{15} t_{a/c} - t_{a/d}\right]$ $t_{a/d} = \left[\frac{1}{2}(7.5)\left(\frac{25.0}{EI}\right)\right]\left[\frac{7.5}{3}\right]$ $= +234.38/_{EI}$ $\delta_d = +598.9/_{EI}$ (↓)
Calculation of the maximum deflection: The maximum deflection occurs at a point, call it *e*, at which the slope is zero. The use of the first theorem allows the finding of point *e*, then the deflection can be calculated. The first theorem tells us that the area (of M/EI) between point a and the unknown point *e* equals +111.1/EI. The sketch of this area (right) permits the calculation of the area in terms of the unknown distance to point *e*, noted as X. Once X is calculated we can calculate $\delta_e = \delta_{max}$ as we calculated δ_b and δ_d. Note the value of the moment at point is found to be 27.2.	$\theta_e^{\nearrow 0} - \theta_a = \text{Area}\frac{M}{EI}\Big]_a^e$ $-(-111.1/EI) = \text{Area}\frac{M}{EI}\Big]_a^e$ or $\text{Area}\frac{M}{EI}\Big]_a^e = +111.1/_{EI}$ a; X; e; $\frac{X}{10}\left(\frac{33.33}{EI}\right)$ $\text{Area}\frac{M}{EI}\Big]_a^e = \frac{1}{2}(X)\left(\frac{3.33X}{EI}\right)$ ∴ $\frac{1}{2}\left(\frac{3.33X^2}{EI}\right) = \frac{111.1}{EI}$ or X = 8.17' (from point a) $\delta_{e\,max} = +\left[\frac{8.17}{15} t_{a/c} - t_{a/e}\right]$ $t_{a/e} = \left[\frac{1}{2}(8.17)\left(\frac{27.2}{EI}\right)\right]\left[\frac{8.17}{3}\right]$ $= 302.6/EI$ $\delta_{e\,max} = +605.1/_{EI}$ (↓)

EXAMPLE #2: MOMENT-AREA: CANTILEVER BEAM

Given: Beam as shown

Find: Slope at point b; deflection at point b.

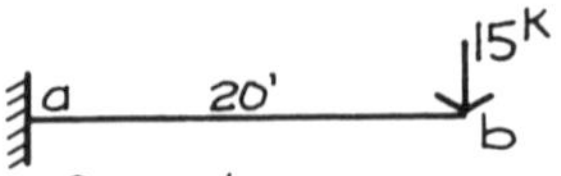

Given beam (EI constant)

First, draw the M/EI diagram

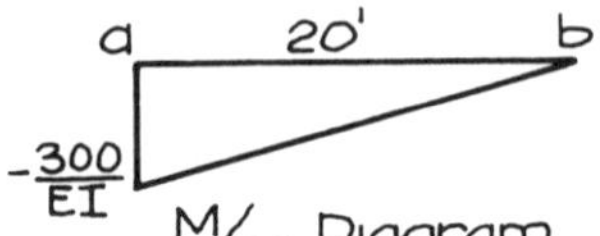

M/EI Diagram

Sketch the deflected curve. Due to the wall at a both the slope and deflection at a are zero. Thus the tangent line at point a is horizontal.

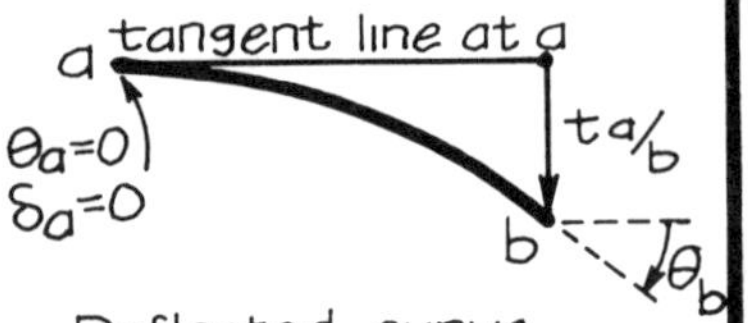

Deflected curve

Calculation of slope, θ_b:
The first moment-area theorem is used directly; that is, the difference between the slopes at a and b equals the area (of M/EI) between a and b.

$$\theta_b - \theta_a^{\,0} = \left[\frac{1}{2}(20)\left(\frac{-300}{EI}\right)\right]$$

$$\theta_b = -\frac{3000}{EI} \quad \text{(C.W.)}$$

Calculation of deflection, δ_b:
the second moment-area theorem is used directly. From the deflected curve we see that the magnitude of the deflection at b equals the magnitude of $t_{a/b}$. (Note: (Note: δ_b is ↓ and is taken as positive; $t_{a/b}$ is ↓ and is thus negative).
By definition $t_{a/b}$ equals the product of the area (of M/EI) between a and b times the lever taken from point b.

$$\delta_b = -t_{a/b}$$

$$t_{a/b} = \left[\frac{1}{2}(20)\left(\frac{-300}{EI}\right)\right]\left[\frac{2}{3}(20)\right]$$

$$= -40{,}000/EI$$

$$\delta_b = +40{,}000/EI \ (\downarrow)$$

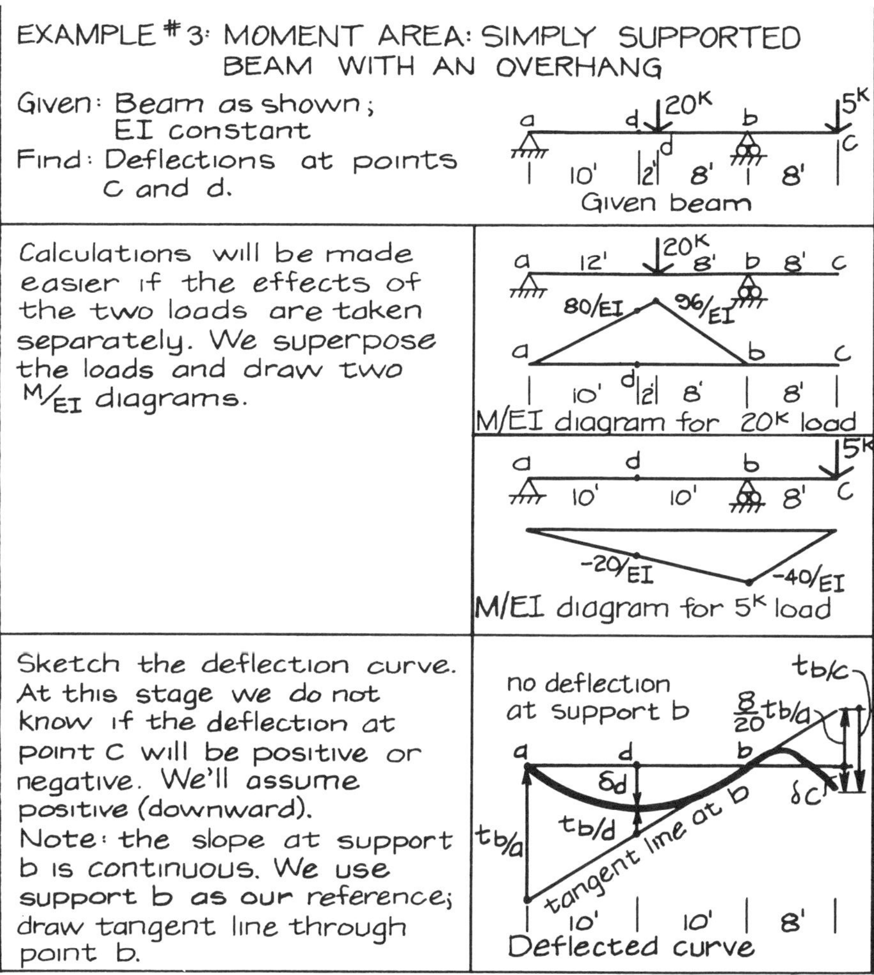

EXAMPLE #3: MOMENT AREA: SIMPLY SUPPORTED BEAM WITH AN OVERHANG

Given: Beam as shown; EI constant

Find: Deflections at points c and d.

Calculations will be made easier if the effects of the two loads are taken separately. We superpose the loads and draw two M/EI diagrams.

Sketch the deflection curve. At this stage we do not know if the deflection at point c will be positive or negative. We'll assume positive (downward).

Note: the slope at support b is continuous. We use support b as our reference; draw tangent line through point b.

Calculation of deflection δ_d:
This calculation is similar to those in Ex.#1 except here a right reference point is used. Theorem 2 is used for both $t_{b/a}$ and $t_{b/d}$. Both M/EI diagrams are used. The "levers" for $t_{b/a}$ are taken from a; the "levers" for $t_{b/d}$ are taken from d. Areas are broken into rectangles and triangles.

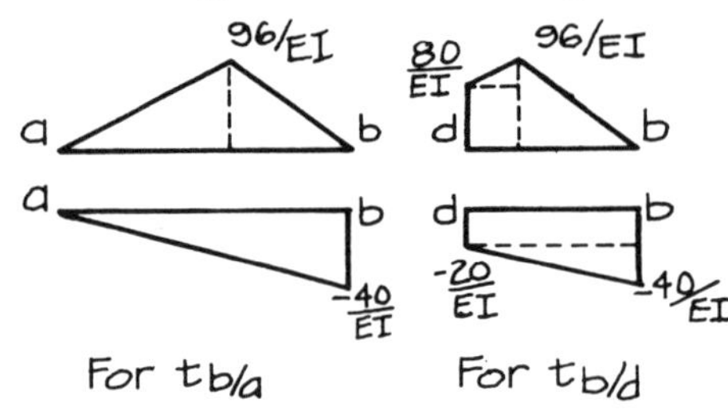

$$\delta_d = +\left[\frac{10}{20} t_{b/a} - t_{b/d}\right]$$

$$t_{b/a} = \left[\frac{1}{2}(12)\left(\frac{96}{EI}\right)\right]\left[\frac{2}{3}(12)\right] + \left[\frac{1}{2}(8)\left(\frac{96}{EI}\right)\right]\left[12+\frac{8}{3}\right] + \left[\frac{1}{2}(20)\left(\frac{-40}{EI}\right)\right]\left[\frac{2}{3}(20)\right]$$

$$t_{b/a} = +4906.67/EI \ (\uparrow)$$

$$t_{b/d} = \left[(2)\left(\frac{80}{EI}\right)\right][1] + \left[\frac{1}{2}(2)\left(\frac{16}{EI}\right)\right]\left[\frac{2}{3}(2)\right] + \left[\frac{1}{2}(8)\left(\frac{96}{EI}\right)\right]\left[2+\frac{8}{3}\right] + \left[10\left(\frac{-20}{EI}\right)\right][5] + \left[\frac{1}{2}(10)\left(\frac{-20}{EI}\right)\right]\left[\frac{2}{3}(10)\right]$$

$$= 306.64/EI$$

$$\delta_d = +2146.7/EI \ (\downarrow)$$

Calculation of deflection δ_c:
Still using point b as a reference the tangent line is extended to point c. By similar triangles the distance from the undeformed beam at c to the tangent line at b is $8/20\, t_{b/a}$. The distance from the tangent line at b to the deformed point c is $t_{b/c}$.
Note signs: δ_C is shown as positive (downward); $t_{b/a}$ is referenced from the tangent line to point a and is shown as positive (upward). $t_{b/c}$ is referenced from the tangent line to point C and is shown as negative (downward).

$$\delta_c = \left[\frac{8}{20} t_{b/a} + t_{b/c}\right]$$

$$t_{b/c} = \left[\frac{1}{2}(8)\left(\frac{-40}{EI}\right)\right]\left[\frac{2}{3}(8)\right] = -853.33/EI \ (\downarrow)$$

as previously found $t_{b/a} = +4906.67/EI \ (\uparrow) \therefore$

$$\delta_c = \left[\frac{8}{20}(4906.67/EI) + \left(\frac{-853.33}{EI}\right)\right] = +\frac{1109.33}{EI} \ (\downarrow)$$

STATICALLY DETERMINATE FRAMES

Moment-area may also be used to determine deflections of rigid frames, although as they increase in complexity, other methods are usually more convenient. Frames are analyzed in the same manner as beams. There are several points, however, which are not considered in the analysis of beams: 1) the joints are assumed to be rigid, i.e., the ends of two members which are oriented at 90° before bending remain at 90° after bending; 2) shortening of a member due to bending and axial loads are neglected; 3) changes in magnitude of a moment in a member, particularly a column, due to bending in the member causing an eccentricity in the axial load are neglected.

It is, of course, necessary to be able to construct the bending moment diagram for the frame and from this visualize the manner in which the frame deflects under load. The slope and position of one point on the structure must be known (or be obtainable) so as to have a reference point for the remainder of the structure.

Just as with beams, the change in slope between any two points on the frame may be represented by the M/EI diagram between those two points and deflections are determined from joint rotations and by finding the tangential deviation between tangents to the two points. Examples 4 and 5 determine slopes and deflections for points on two statically determinate frames.

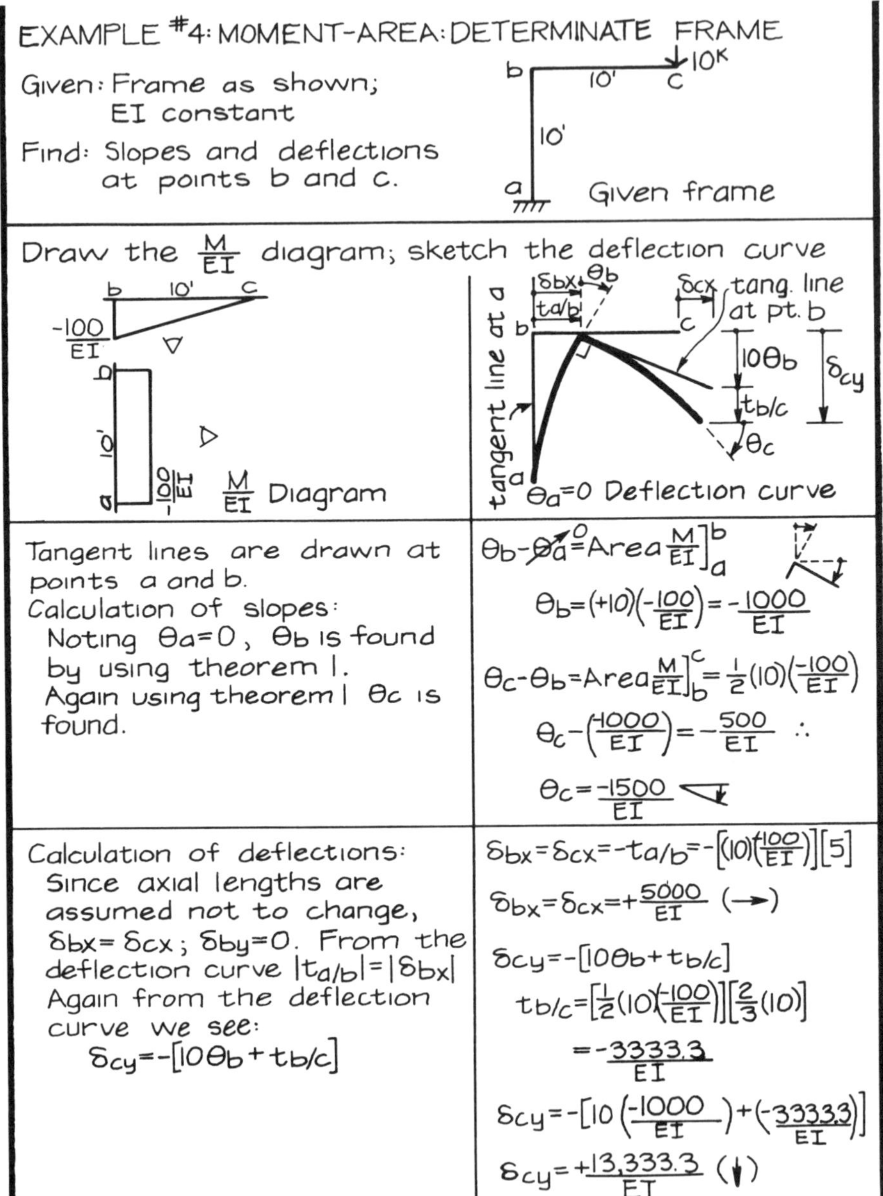

EXAMPLE #4: MOMENT-AREA: DETERMINATE FRAME

Given: Frame as shown; EI constant

Find: Slopes and deflections at points b and c.

Draw the $\frac{M}{EI}$ diagram; sketch the deflection curve

Tangent lines are drawn at points a and b.

Calculation of slopes: Noting $\theta_a = 0$, θ_b is found by using theorem 1. Again using theorem 1 θ_c is found.

$$\theta_b - \theta_a^{\,0} = \text{Area} \frac{M}{EI}\Big]_a^b$$

$$\theta_b = (+10)\left(-\frac{100}{EI}\right) = -\frac{1000}{EI}$$

$$\theta_c - \theta_b = \text{Area} \frac{M}{EI}\Big]_b^c = \frac{1}{2}(10)\left(\frac{-100}{EI}\right)$$

$$\theta_c - \left(\frac{-1000}{EI}\right) = -\frac{500}{EI} \quad \therefore$$

$$\theta_c = \frac{-1500}{EI}$$

Calculation of deflections: Since axial lengths are assumed not to change, $\delta_{bx} = \delta_{cx}$; $\delta_{by} = 0$. From the deflection curve $|t_{a/b}| = |\delta_{bx}|$. Again from the deflection curve we see:

$$\delta_{cy} = -[10\theta_b + t_{b/c}]$$

$$\delta_{bx} = \delta_{cx} = -t_{a/b} = -\left[(10)\left(\frac{-100}{EI}\right)\right][5]$$

$$\delta_{bx} = \delta_{cx} = +\frac{5000}{EI} \ (\rightarrow)$$

$$\delta_{cy} = -[10\theta_b + t_{b/c}]$$

$$t_{b/c} = \left[\frac{1}{2}(10)\left(\frac{-100}{EI}\right)\right]\left[\frac{2}{3}(10)\right]$$

$$= \frac{-3333.3}{EI}$$

$$\delta_{cy} = -\left[10\left(\frac{-1000}{EI}\right) + \left(\frac{-3333.3}{EI}\right)\right]$$

$$\delta_{cy} = \frac{+13{,}333.3}{EI} \ (\downarrow)$$

EXAMPLE #5: MOMENT-AREA: DETERMINATE FRAME

Given: Frame as shown
EI constant

Find: Slope and deflection at d.

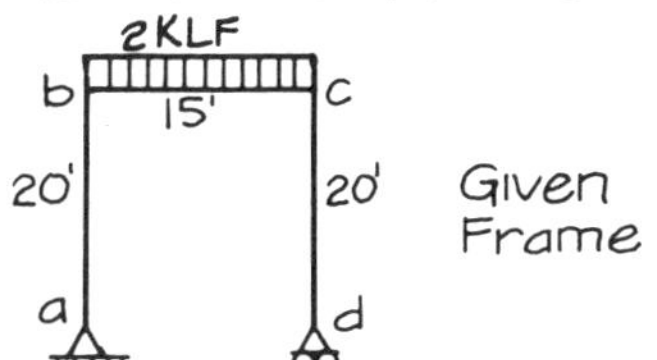

Draw $\frac{M}{EI}$ diagrams; sketch the deflection curve.

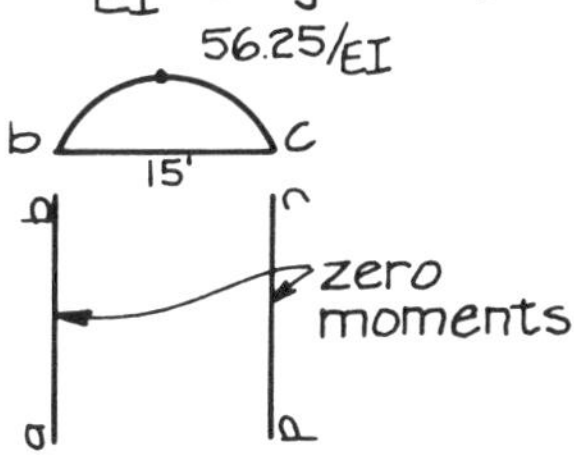

Note: columns ab and cd do not bend.

M/EI Diagram

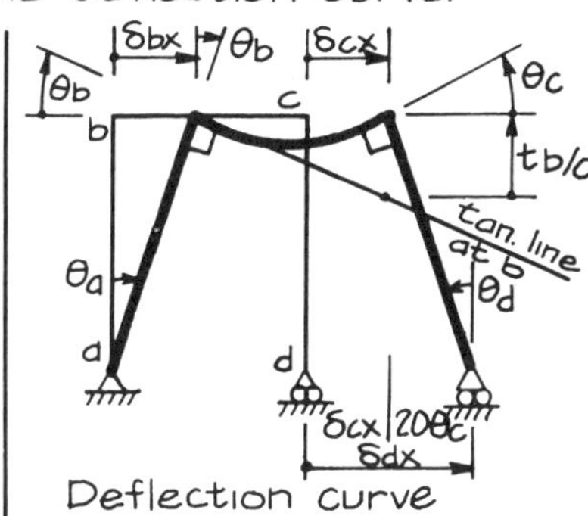

Deflection curve

Calculation of slopes:
Since there is no bending in columns, $\theta_d = \theta_c$ and $\theta_a = \theta_b$. We must work our way from point b to point d; draw a tangent line through point b on beam. We see then

$$\theta_b = -\frac{t_{b/c}}{15'}$$

θ_c is found by using theorem I.

Calculation of deflection δdx:
From the deflection curve we see:

$\delta_{dx} = \delta_{cx} + 20\theta_c$
where $\delta_{cx} = \delta_{bx} = -20\theta_a = -20\theta_b$

$\theta_d = \theta_c$ and $\theta_a = \theta_b$

$$\theta_b = \frac{-t_{b/c}}{15} = -\frac{\left[15\left(\frac{2}{3}\right)\left(\frac{56.25}{EI}\right)\right][7.5]}{15}$$

$$\theta_b = -\frac{281.25}{EI}$$

$$\theta_c - \theta_b = \text{Area}\,\frac{M}{EI}\Big]_b^c$$

$$= \left[15\left(\frac{2}{3}\right)\left(\frac{56.25}{EI}\right)\right]$$

$$\theta_c - \left(-\frac{281.25}{EI}\right) = +\frac{562.5}{EI}$$

$$\theta_c = \theta_d = +\frac{281.25}{EI}$$

Calculation of deflection δdx:
From the deflection curve we see

$\delta_{dx} = \delta_{cx} + 20\theta_c$
where $\delta_{cx} = \delta_{bx} = -20\theta_a = -20\theta_b$

$$\delta_{dx} = \delta_{cx} + 20\theta_c$$

$$\delta_{cx} = -20\theta_b = -20\left(\frac{-281.25}{EI}\right)$$

$$= +\frac{5625}{EI}\ (\rightarrow)$$

$$\delta_{dx} = \frac{5625}{EI} + 20\left(\frac{281.25}{EI}\right)$$

$$\delta_{dx} = \frac{11{,}250}{EI}\ (\rightarrow)$$

STATICALLY INDETERMINATE BEAMS

Methods used to determine deflections of beams are particularly useful in the analysis of statically indeterminate beams and simple rigid frames. These beams are supported in such a manner that the number of support reactions is greater than the number of equations of equilibrium available for the analysis. The difference between these two numbers is known as the degree of indeterminacy. General procedure for the analysis is to remove enough restraints so that the structure is statically determinate and stable and then to determine the magnitudes of the forces and/or the moments required to restore the structure to its original condition. This is generally called the force method or flexibility method of analysis. Moment-area may be used to determine the deflections after all reactions are found, although other methods are available.

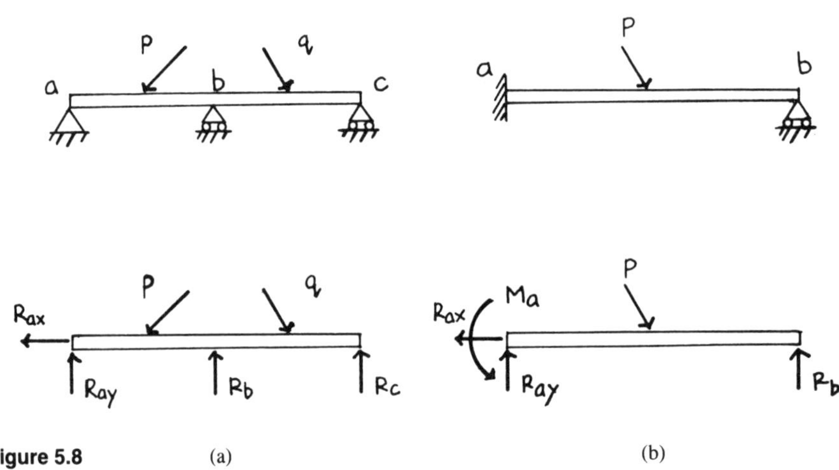

Figure 5.8 (a) (b)

Examples of statically indeterminate beams are illustrated in Figure 5.8. We shall investigate statically indeterminate frames in the next section. A free body diagram of each member in Figure 5.8 indicates that there are more unknowns to be determined (four) in each example than there are independent equations of equilibrium available (three) for use in the solution. Thus, an extra equation from a source other than equilibrium must be found somewhere or at least one of the support forces must be determined before it will be possible to determine the remainder.

Suppose we remove one of the vertical supports from each of these members and allow them to deflect at those points. Moment-area enables us to determine the deflections on the modified or "cutback" structures at the points at which the extra or "redundant" supports were located. In the first of these examples the support force at

b is arbitrarily selected as the redundant and when it is removed, the beam deflects vertically a distance δ_b (Figure 5.9a). The moment at a produced by having the beam built into the wall is arbitrarily selected as the redundant for the second example and its removal allows the left end to rotate through an angle θ_a, converting support a to a pin (Figure 5.9b). To restore the structures, the unknown reactions (R_b, M_a) must be reapplied. For the purpose of determining the magnitudes of these redundants, a force is applied at b and a moment at a on the otherwise unloaded structures as illustrated in Figure 5.9, and their deflections, δ_{b_2} and θ_{a_2} determined. We are now ready to set up an equation which will allow us to solve for the redundant reactions. In the case of the first example we can write

$$\delta_{b_1} - \hat{\delta}_{b_2} R_b = 0 \qquad \left(\text{Here } \hat{\delta}_{b_2} = \frac{\delta_{b_2}}{R_b}\right)$$

$$R_b = \frac{\delta_{b_1}}{\hat{\delta}_{b_2}},$$

and for the second

$$\theta_{a_1} - \hat{\theta}_{a_2} M_a = 0 \qquad \left(\text{Here } \hat{\theta}_{a_2} = \frac{\theta_{a_2}}{M_a}\right)$$

$$M_a = \frac{\theta_{a_1}}{\hat{\theta}_{a_2}}.$$

It is now possible, using the equations of equilibrium, to determine the remaining support forces, the actual shear force and bending moment diagrams, stresses, slopes and deflections for each of these beams. Examples 6 and 7 show the analysis of a statically indeterminate beam utilizing moment-area to determine the necessary deflections. Two different approaches are used to solve the same problem.

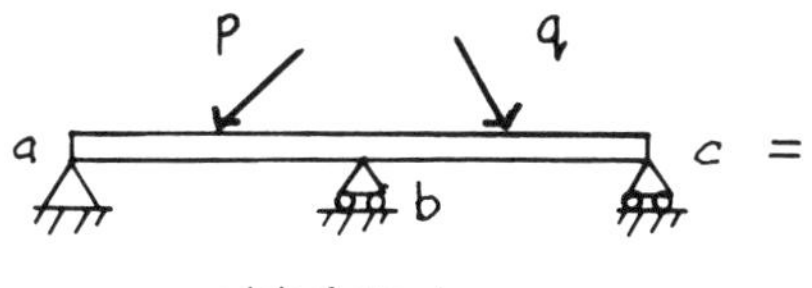

original structure

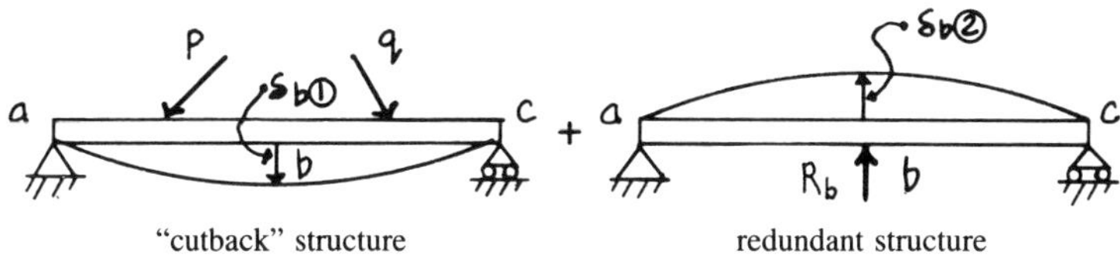

"cutback" structure

redundant structure

Figure 5.9a

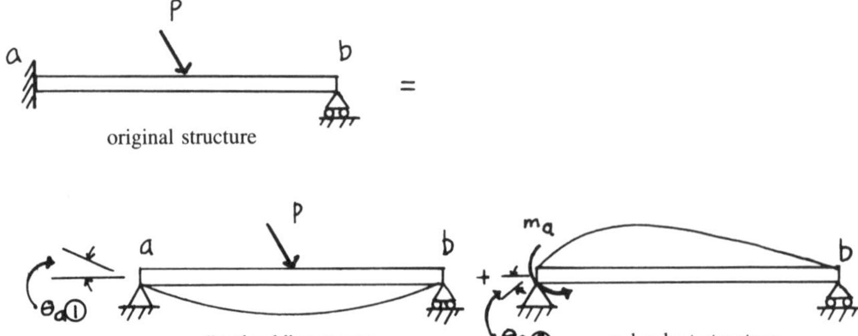

Figure 5.9b

It is important to realize that the cutback and redundant structures must have the exact same type of support conditions.

The preceding discussion considered only beams indeterminate to the first degree. As the degree of indeterminacy increases, the solution becomes somewhat more arithmetically difficult. In Figure 5.10 we have a beam which is statically indeterminate to the second degree with five unknown support forces and three equations of equilibrium available for the solution. If we select R_b and R_c as the redundants and remove them, the structure deflects as shown in Figure 5.10c and, using the moment-area technique, the deflections δ_{b_1} and δ_{c_1} may be determined. Now, one at a time, we place loads on the otherwise unloaded structure and determine the deflections due to these loads at points a and c. Note that the loads each cause small deflections at both points b and c and these are identified by the subscripts, i.e., δ_{b_3} means the deflection at b due to a load applied at c, etc. Now we can set up two equations, which must be solved simultaneously, to find the support forces R_b and R_c. Since, for the original structure (Figure 5.10a), the deflections at the supports b and c were zero, we can write

$$\delta_{b_1} - R_b\hat{\delta}_{b_2} - R_c\hat{\delta}_{b_3} = 0 \qquad \left(\text{Here } \hat{\delta}_{b_2} = \frac{\delta_{b_2}}{R_b};\ \hat{\delta}_{b_3} = \frac{\delta_{b_3}}{R_c}\right)$$

$$\delta_{c_1} - R_b\hat{\delta}_{c_2} - R_c\,\hat{\delta}_{c_3} = 0 \qquad \left(\text{Here } \hat{\delta}_{c_2} = \frac{\delta_{c_2}}{R_b};\ \hat{\delta}_{c_3} = \frac{\delta_{c_3}}{R_b}\right).$$

Note that the support force R_b contributes to the behavior at c as well as b just as the support force R_c contributes to the behavior at b as well as c. The solution of the above equations will yield R_b and R_c and the remainder of the reactions may then be determined using the available equations of equilibrium.

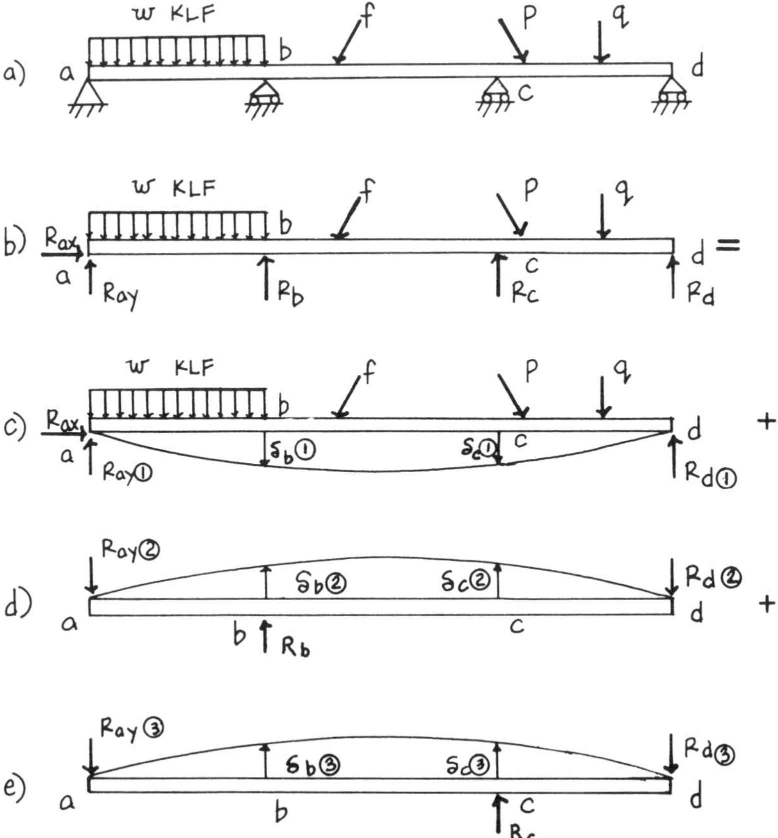

a)
w KLF
f
P
q
a
b
c
d
b)
w KLF
f
P
q
Rax
a
Ray
b
Rb
c
Rc
d
Rd
=
c)
w KLF
f
P
q
Rax
a
Ray①
b
δb①
δc①
c
d
Rd①
+
d)
Ray②
δb②
δc②
Rd②
a
b
Rb
c
d
+
e)
Ray③
δb③
δc③
Rd③
a
b
c
Rc
d

Figure 5.10

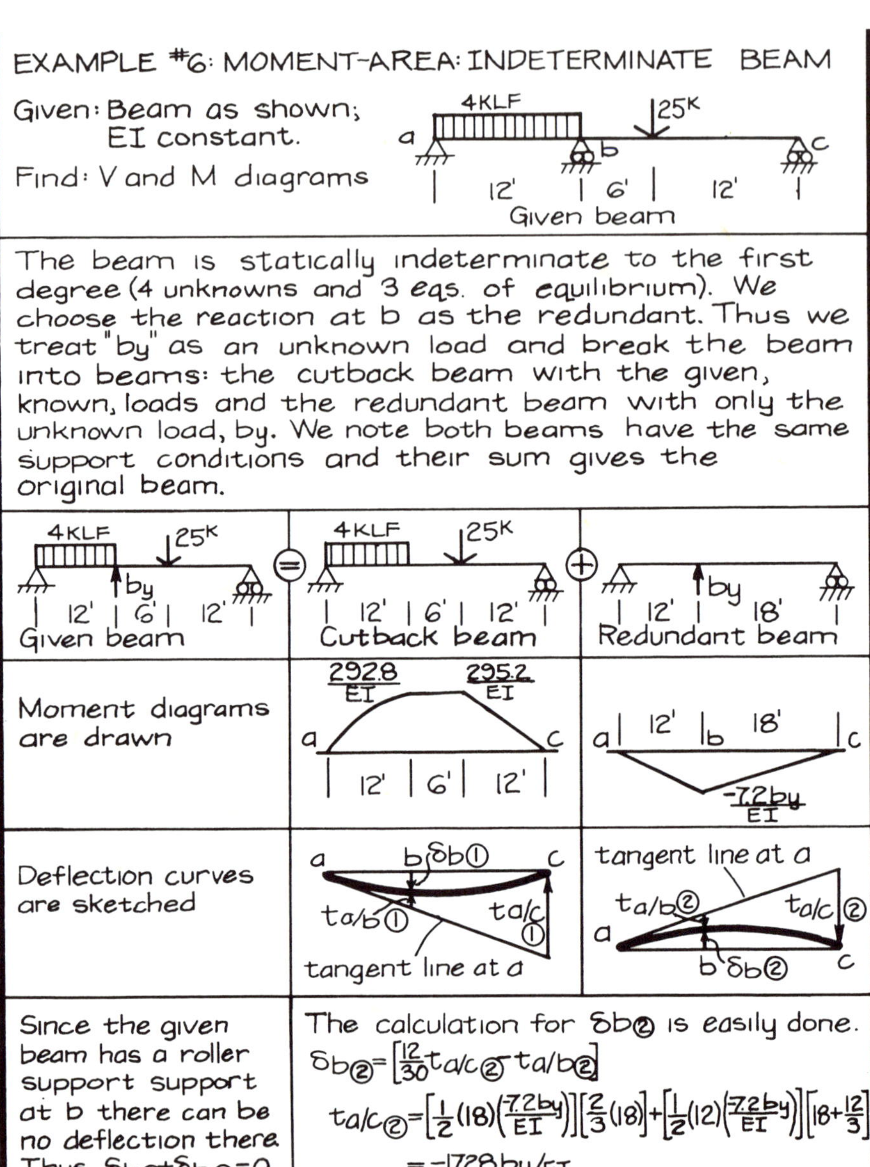

Since the given beam has a roller support support at b there can be no deflection there. Thus $\delta b_{①} + \delta b_{②} = 0$

From the deflection curves we see

$$\delta b_{①/②} = \left[\frac{12}{30} t_{a/c\,①/②} - t_{a/b\,①/②}\right]$$

The calculation for $\delta b_{②}$ is easily done.

$$\delta b_{②} = \left[\frac{12}{30} t_{a/c②} - t_{a/b②}\right]$$

$$t_{a/c②} = \left[\frac{1}{2}(18)\left(\frac{-7.2by}{EI}\right)\right]\left[\frac{2}{3}(18)\right] + \left[\frac{1}{2}(12)\left(\frac{-7.2by}{EI}\right)\right]\left[18 + \frac{12}{3}\right]$$

$$= -1728\,by/EI$$

$$t_{a/b②} = \left[\frac{1}{2}(12)\left(\frac{-7.2by}{EI}\right)\right]\left[\frac{12}{3}\right] = \frac{-172.8by}{EI}$$

$$\delta b_{②} = \left[\frac{12}{30}\left(\frac{-1728by}{EI} + \frac{172.8by}{EI}\right)\right] = -\frac{518.4by}{EI}\ (\uparrow)$$

The calculation of $\delta b_{①}$ from the above cutback beam M/EI diagram is not easily accomplished since the parabolic portion of the M/EI diagram has no true vertex (although the slope at b is close to zero). A useful technique is to separate the concentrated forces (loads and reactions) and the distributed load as shown below. Note that the two "separated beams" are not individually in equilibrium but their sum is.

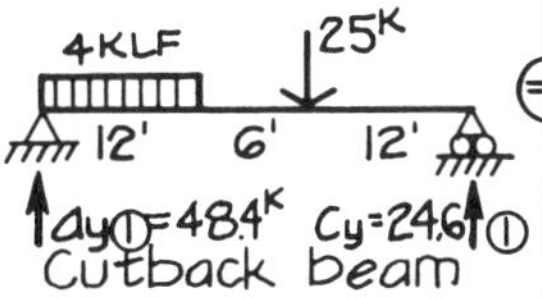

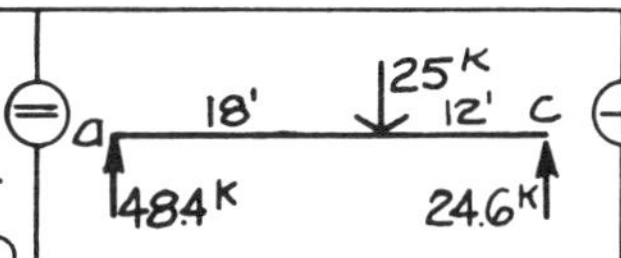

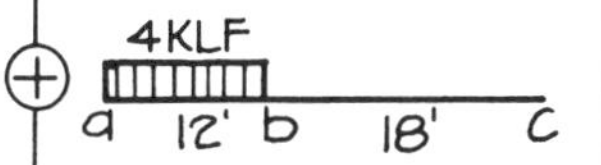

4KLF, 25K; 12', 6', 12'; $a_{y①} = 48.4^K$, $C_y = 24.6$ ①. Cutback beam = a, 18', 25K, 12', c; 48.4^K, 24.6^K + 4KLF; a, 12', b, 18', c

Draw the M/EI diagram for each load.

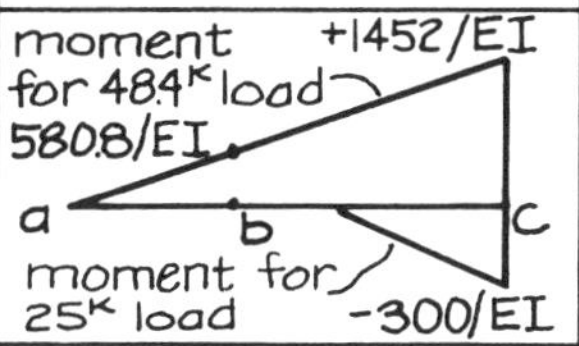

moment for 48.4^K load: +1452/EI; 580.8/EI; a, b, c; moment for 25^K load: −300/EI

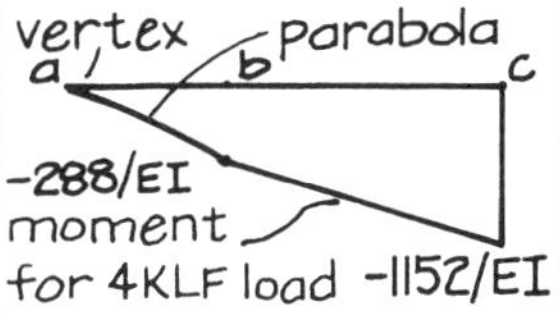

vertex, parabola; a, b, c; −288/EI; moment for 4KLF load: −1152/EI

From the above two M/EI diagrams the deflection $\delta b_{①}$ is found. Note the area and the centroid for a parabolic spandrel is as below.

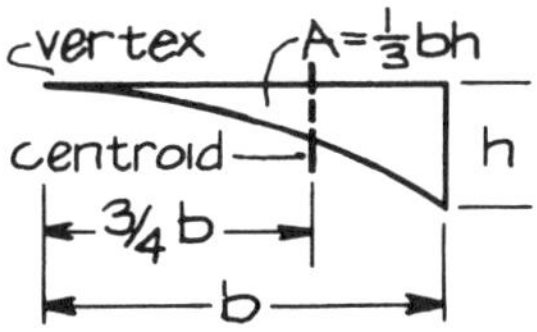

vertex; $A = \frac{1}{3}bh$; centroid; $\frac{3}{4}b$; b; h

$$\delta b_{①} = \left[\frac{12}{30} t_{a/c①} - t_{a/b①}\right]$$

$$t_{a/c①} = \left[\frac{1}{2}(12)\left(\frac{-300}{EI}\right)\right]\left[\frac{12}{3}\right] + \left[\frac{1}{2}(30)\left(\frac{1452}{EI}\right)\right]\left[\frac{30}{3}\right] + \left[\frac{1}{3}(12)\left(\frac{-288}{EI}\right)\right]\left[18 + \frac{12}{4}\right] + \left[(18)\left(\frac{-288}{EI}\right)\right][9] + \left[\frac{1}{2}(18)\left(\frac{-864}{EI}\right)\right]\left[\frac{18}{3}\right] = \frac{93096}{EI}$$

$$t_{a/b①} = \left[\frac{1}{2}(12)\left(\frac{580.8}{EI}\right)\right]\left[\frac{12}{3}\right] + \left[\frac{1}{3}(12)\left(\frac{-288}{EI}\right)\right]\left[\frac{12}{4}\right] = \frac{10{,}483.2}{EI} \quad \therefore$$

$$\delta b_{①} = +\frac{26{,}755.2}{EI} \ (\downarrow)$$

Thus we find the redundant reaction "by"

$$\delta b_{①} + \delta b_{②} = 0 \quad \text{or} \quad \frac{26755.2}{EI} - \frac{518.4 b_y}{EI} = 0 \Longrightarrow b_y = 51.61 \ (\uparrow)$$

With the redundant, by, we find the other support reaction by using statics on the given beam.

$\Sigma M_{pt\ a} = 0$: $6(48) + 18(25) - 30c_y - 12(51.61) = 0$

$c_y = +3.96^K (\uparrow)$

$\Sigma F_y = 0 \quad a_y = 17.45^K (\uparrow)$

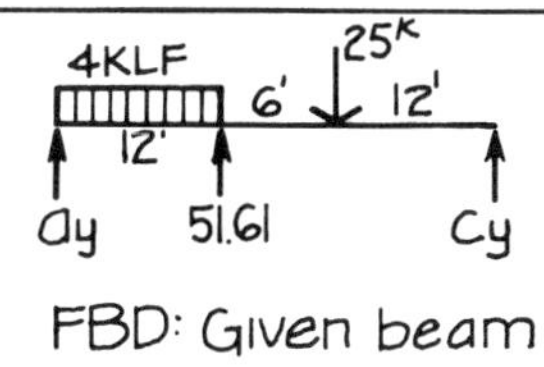

FBD: Given beam

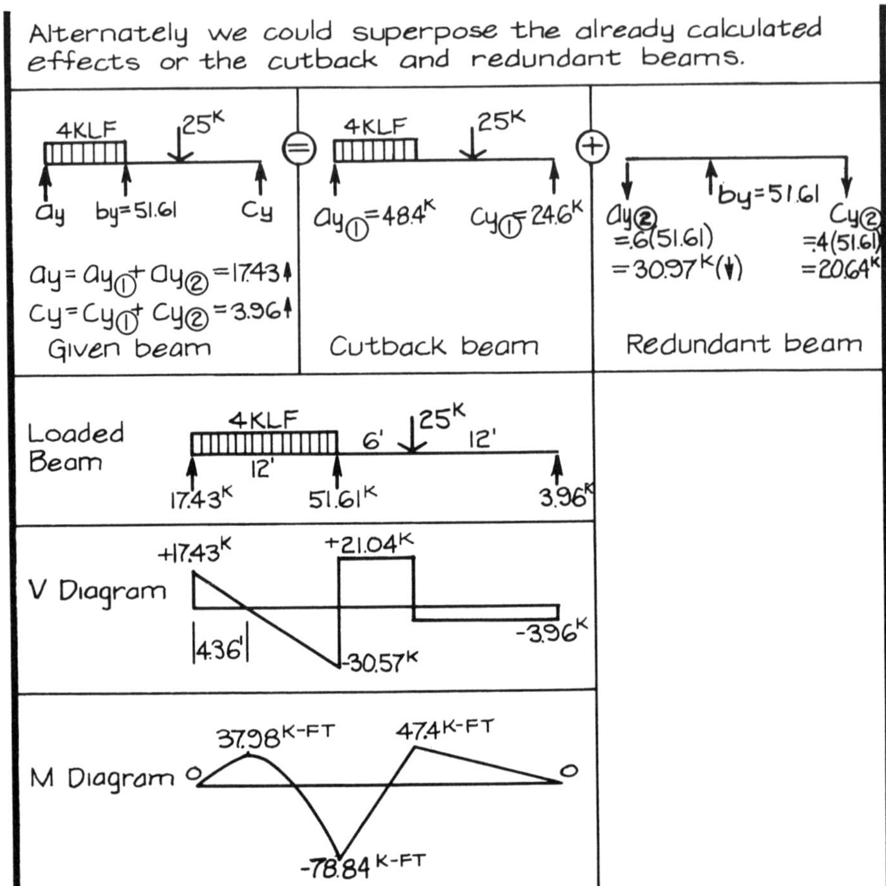

Alternately we could superpose the already calculated effects or the cutback and redundant beams.
4KLF
25K
ay
by=51.61
Cy
ay= ay① + ay② =17.43↑
Cy=Cy① + Cy② =3.96↑
Given beam
=
ay① =48.4K
Cy① =24.6K
Cutback beam
+
by=51.61
ay②
=.6(51.61)
=30.97K(↓)
Cy②
=.4(51.61)
=20.64K
Redundant beam
Loaded Beam
4KLF
12'
6'
25K
12'
17.43K
51.61K
3.96K
V Diagram
+17.43K
+21.04K
4.36'
-30.57K
-3.96K
M Diagram
0
37.98K-FT
47.4K-FT
0
-78.84 K-FT

EXAMPLE #7: MOMENT-AREA: INDETERMINATE BEAM

Given: Same beam as in example #6

Find: V & M diagrams using the moment at b as the redundant.

The procedure is essentially the same as in Ex. #6: Remove the indeterminacy and then find a force or moment which will restore the beam to its original condition. In this case, the negative moment at b is selected as the redundant and removed by placing an internal pin at b. Now, instead of being continuous, the ends of each segment rotate separately. A pair of internal restoring moments must be applied at b in order to correct the difference in slopes. The moment-area calculations are straight-foward.

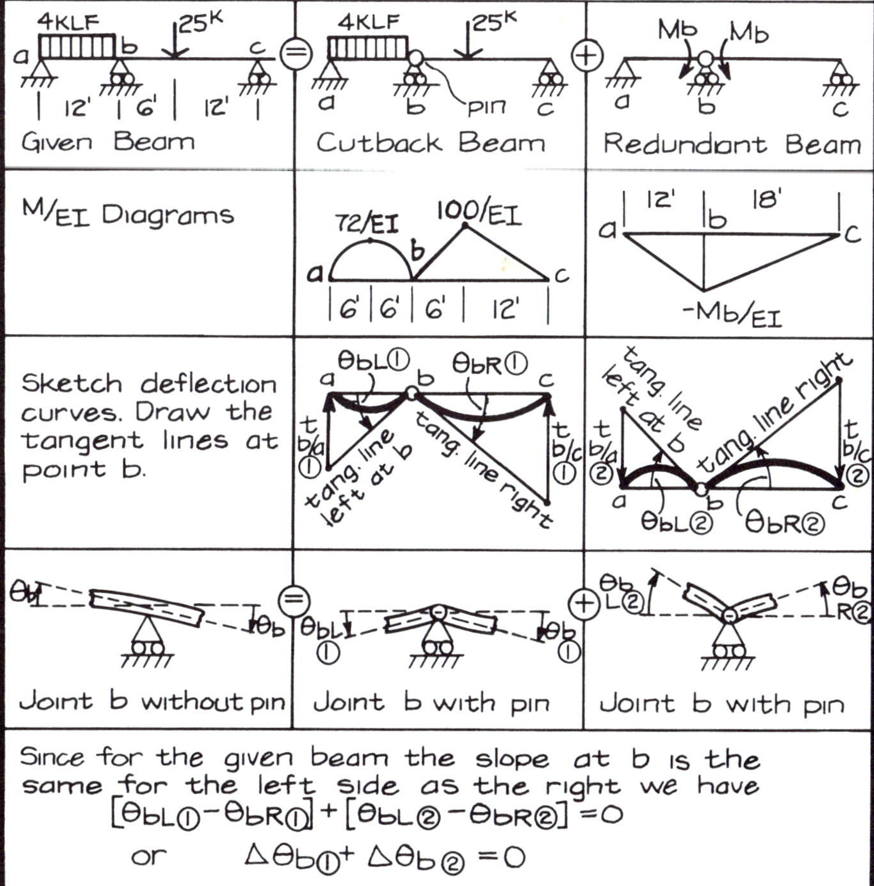

Since for the given beam the slope at b is the same for the left side as the right we have

$$[\theta_{bL①} - \theta_{bR①}] + [\theta_{bL②} - \theta_{bR②}] = 0$$

or

$$\Delta\theta_{b①} + \Delta\theta_{b②} = 0$$

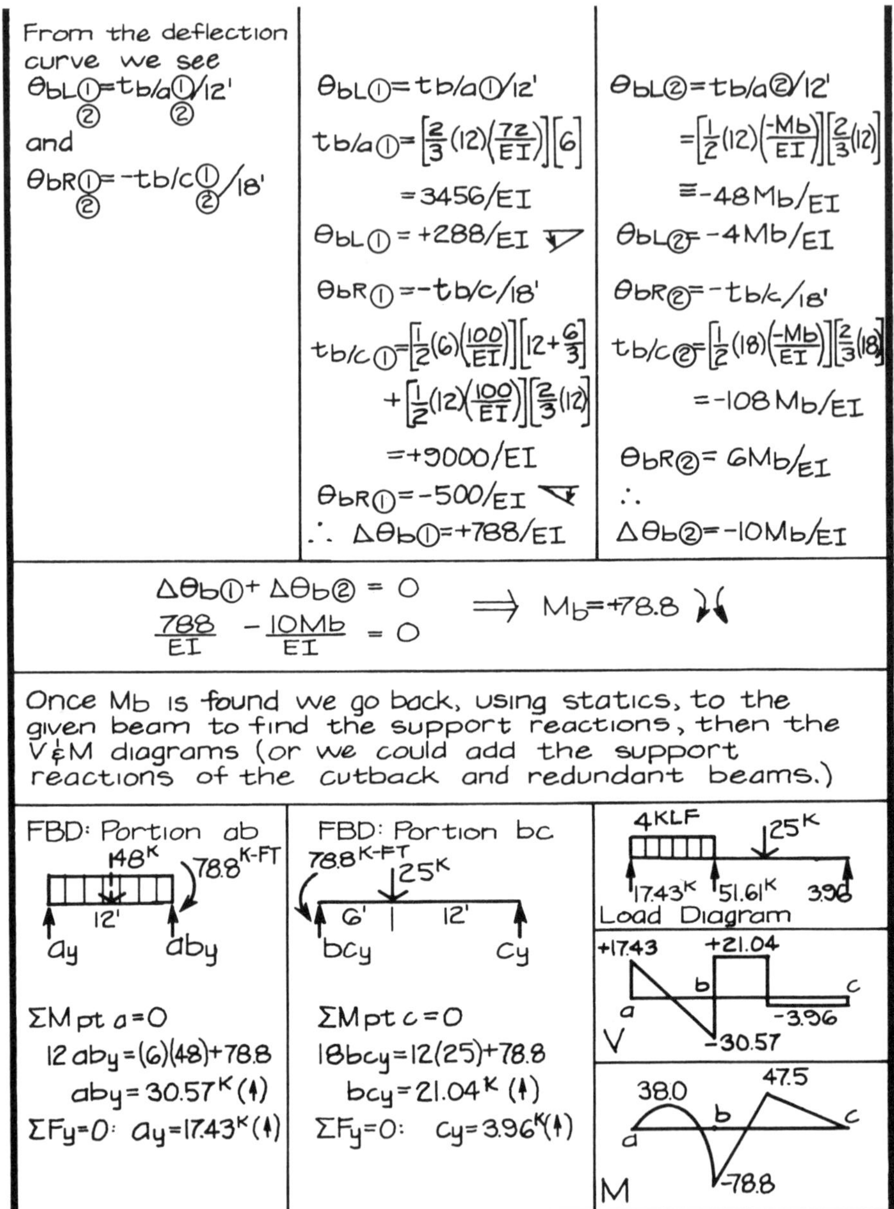

From the deflection curve we see
θbL①②=tb/a①②/12'
and
θbR①②=-tb/c①②/18'
θbL①=tb/a①/12'
tb/a①=[2/3(12)(72/EI)][6]
=3456/EI
θbL①=+288/EI
θbR①=-tb/c/18'
tb/c①=[1/2(6)(100/EI)][12+6/3]
+[1/2(12)(100/EI)][2/3(12)]
=+9000/EI
θbR①=-500/EI
∴ Δθb①=+788/EI
θbL②=tb/a②/12'
=[1/2(12)(-Mb/EI)][2/3(12)]
=-48Mb/EI
θbL②=-4Mb/EI
θbR②=-tb/c/18'
tb/c②=[1/2(18)(-Mb/EI)][2/3(18)]
=-108Mb/EI
θbR②=6Mb/EI
∴
Δθb②=-10Mb/EI
Δθb①+Δθb②=0
788/EI - 10Mb/EI = 0
⟹ Mb=+78.8
Once Mb is found we go back, using statics, to the given beam to find the support reactions, then the V&M diagrams (or we could add the support reactions of the cutback and redundant beams.)
FBD: Portion ab
48K
78.8K-FT
12'
ay
aby
ΣM pt a=0
12 aby=(6)(48)+78.8
aby=30.57K(↑)
ΣFy=0: ay=17.43K(↑)
FBD: Portion bc
78.8K-FT
25K
6'
12'
bcy
cy
ΣM pt c=0
18bcy=12(25)+78.8
bcy=21.04K (↑)
ΣFy=0: cy=3.96K(↑)
4KLF
25K
17.43K
51.61K
3.96
Load Diagram
+17.43
+21.04
b
c
a
-3.96
V
-30.57
47.5
38.0
b
c
a
M
-78.8

STATICALLY INDETERMINATE FRAMES

Moment-area may also be utilized in the analysis of statically indeterminate frames, particularly small simple orthogonal frames. Once the degree of indeterminacy goes beyond three or four, methods other than the force or flexibility method usually become more feasible. As before, the analysis begins by removing enough of the redundant supports so that the structure becomes statically determinate, but remains stable.

With the redundant reactions determined, the remaining three reactions at the supports may be computed using the equations of equilibrium. It is now possible to draw shear force and bending moment diagrams, determine stresses or find slopes and deflections for each member of the frame. Example 8 carries out in detail the solution for an indeterminate frame to the point of drawing the shear force and bending moment diagrams. All required deflections are determined by the moment-area method.

DEFLECTIONS OF NONPRISMATIC MEMBERS

This section is primarily concerned with the application of the moment-area technique to the determination of deflections of elastic nonprismatic beams and frames and with adapting this technique to the solution of some statically indeterminate structures. There are times when it is necessary to deal with nonprismatic members. Bending stresses, deflections, etc., are inversely related to the moments of inertia of the flexural member and in cases where there is a large variation in bending moment from one end of the member to the other, it is sometimes convenient to vary the moment of inertia in order to use the material in the member more efficiently. Varying the moment of inertia from point to point does not change the bending moment (as represented by the bending moment diagram) of a statically determinate structure, but it does change the M/EI diagram or the diagram which represents the manner in which the member bends. It will, however, change the moment diagram of a statically indeterminate structure because it will change the reactions.

Example 9 is an analysis of a statically indeterminate beam with a varying moment of inertia.

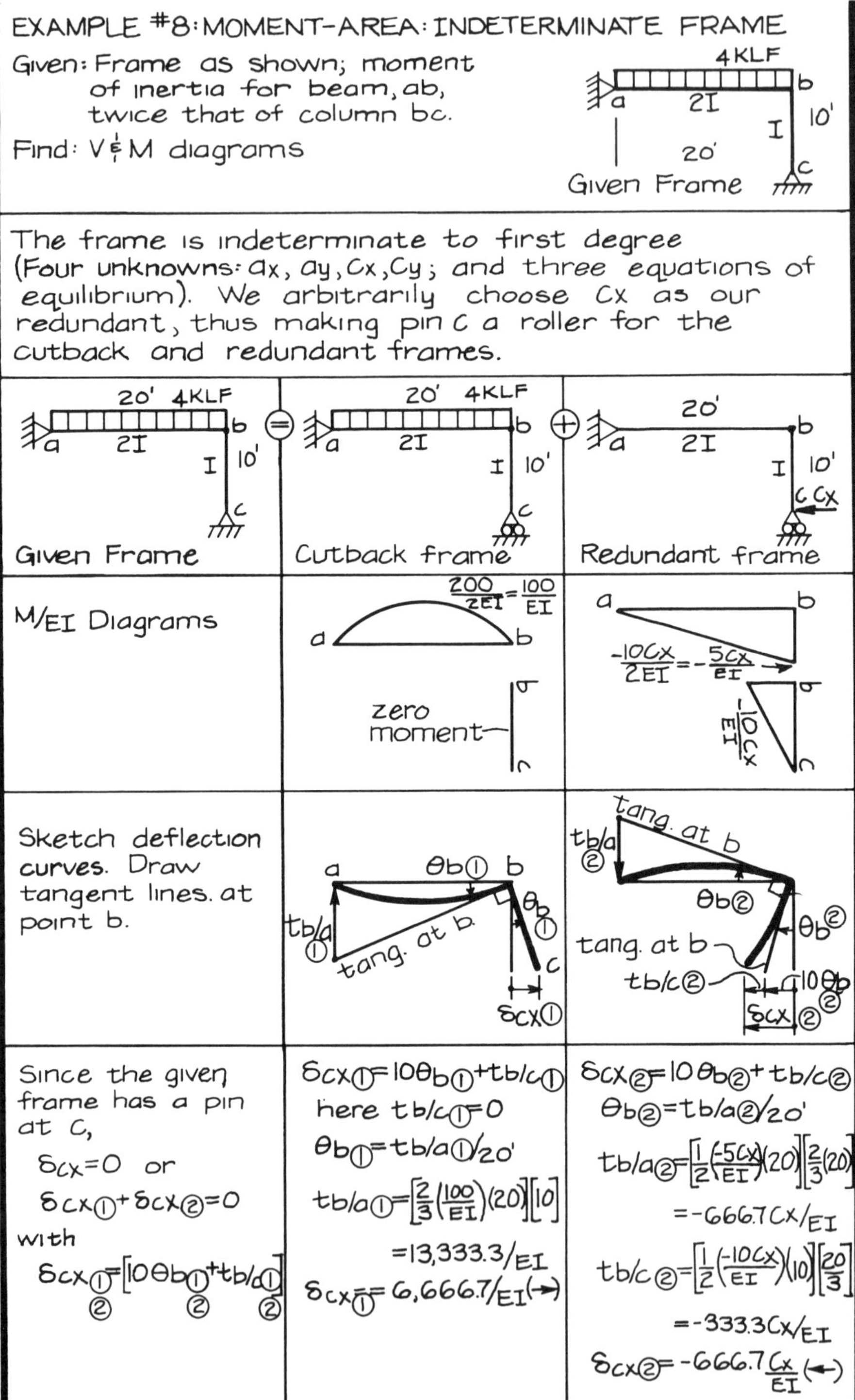

EXAMPLE #8: MOMENT-AREA: INDETERMINATE FRAME

Given: Frame as shown; moment of inertia for beam, ab, twice that of column bc.

Find: V & M diagrams

The frame is indeterminate to first degree (Four unknowns: a_x, a_y, C_x, C_y; and three equations of equilibrium). We arbitrarily choose C_x as our redundant, thus making pin C a roller for the cutback and redundant frames.

M/EI Diagrams

Sketch deflection curves. Draw tangent lines at point b.

Since the given frame has a pin at C,

$\delta_{cx} = 0$ or

$\delta_{cx①} + \delta_{cx②} = 0$

with

$\delta_{cx①,②} = [10\theta_{b①,②} + t_{b/c①,②}]$

$\delta_{cx①} = 10\theta_{b①} + t_{b/c①}$

here $t_{b/c①} = 0$

$\theta_{b①} = t_{b/a①}/20'$

$t_{b/a①} = \left[\frac{2}{3}\left(\frac{100}{EI}\right)(20)\right][10]$

$= 13{,}333.3/EI$

$\delta_{cx①} = 6{,}666.7/EI\ (\rightarrow)$

$\delta_{cx②} = 10\theta_{b②} + t_{b/c②}$

$\theta_{b②} = t_{b/a②}/20'$

$t_{b/a②} = \left[\frac{1}{2}\left(\frac{-5C_x}{EI}\right)(20)\right]\left[\frac{2}{3}(20)\right]$

$= -666.7C_x/EI$

$t_{b/c②} = \left[\frac{1}{2}\left(\frac{-10C_x}{EI}\right)(10)\right]\left[\frac{20}{3}\right]$

$= -333.3C_x/EI$

$\delta_{cx②} = -666.7\frac{C_x}{EI}\ (\leftarrow)$

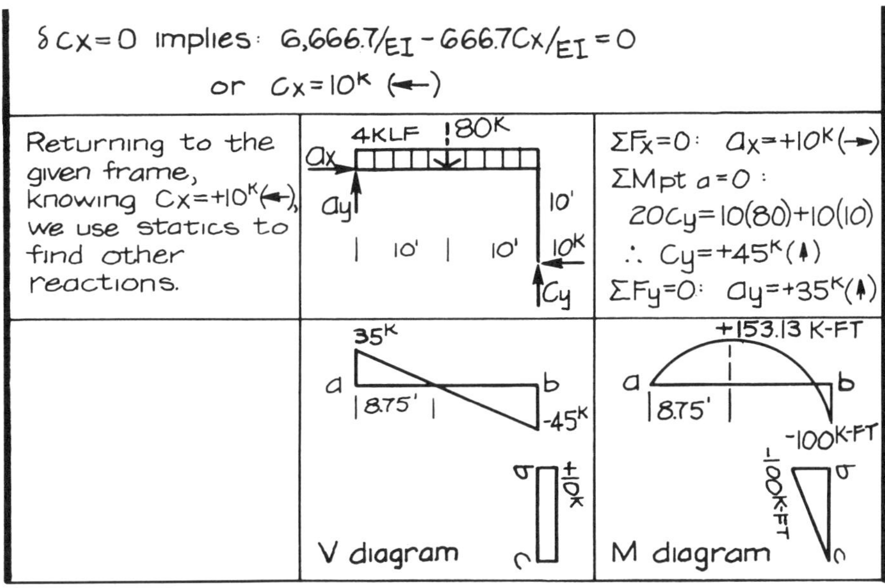
δ Cx = 0 implies: 6,666.7/EI − 666.7Cx/EI = 0
or Cx = 10K (←)
Returning to the given frame, knowing Cx = +10K (←), we use statics to find other reactions.
4KLF
80K
ax
ay
10'
10'
10'
10K
Cy
ΣFx=0: ax=+10K (→)
ΣMpt a=0:
20Cy=10(80)+10(10)
∴ Cy=+45K (↑)
ΣFy=0: ay=+35K (↑)
35K
a
b
8.75'
-45K
±10K
V diagram
+153.13 K-FT
8.75'
-100K-FT
-100K-FT
M diagram

EXAMPLE #9: MOMENT-AREA: INDETERMINATE BEAM WITH VARIABLE MOMENT OF INERTIA

Given: Beam as shown

Find: V & M Diagrams

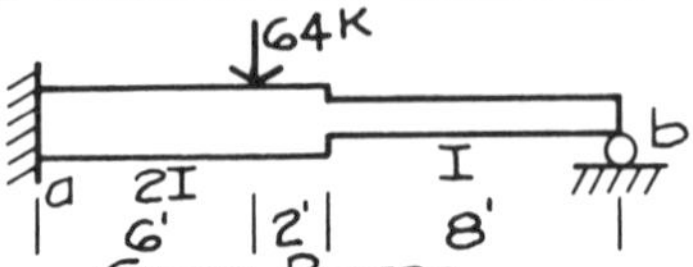

Given Beam

The beam is indeterminate to the first degree. We choose the vertical roller reaction, b_y, as our redundant.

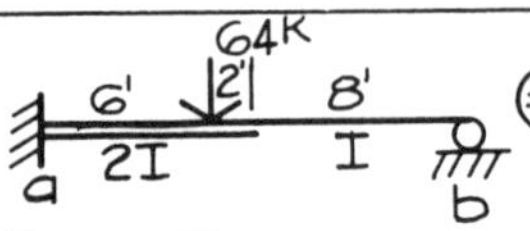

Given Beam

=

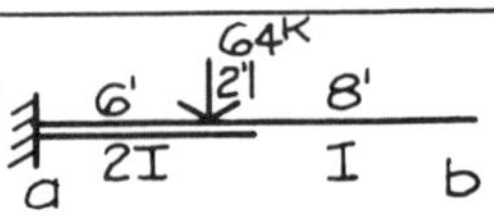

Cutback Beam

+

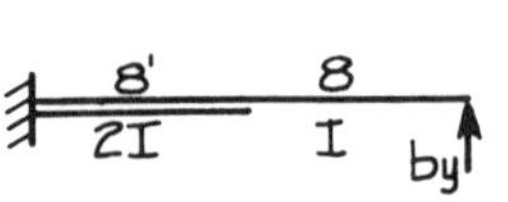

Redundant Beam

Draw the M/EI diagrams

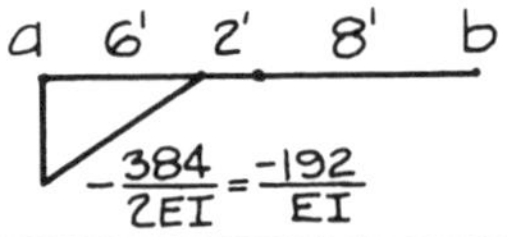

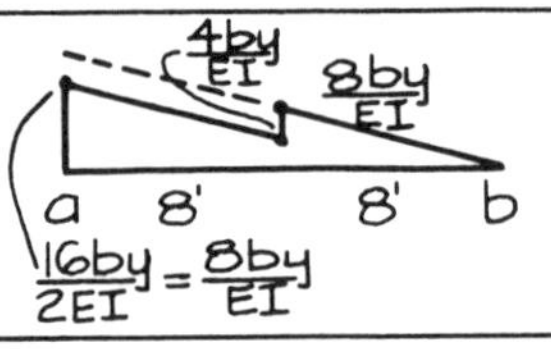

Sketch deflection curves. Draw a tangent line at a.

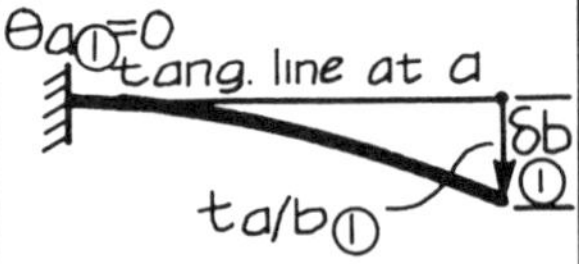

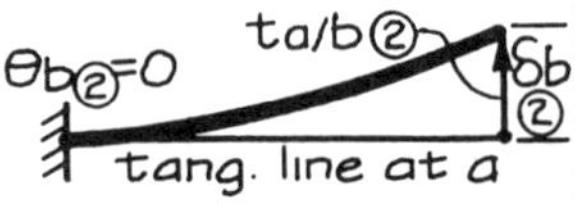

Since the given beam has a roller at b, the deflection there must zero, ie

$\delta b = 0$ or

$\delta b_{①} + \delta b_{②} = 0$

with

$\delta b_{①/②} = -t_{a/b ①/②}$

$\delta b_{①} = -t_{a/b①}$

$t_{a/b①} = \left[\frac{1}{2}(6)\left(\frac{-192}{EI}\right)\right] \times \left[10 + \frac{2}{3}(6)\right] = -\frac{8064}{EI} \therefore$

$\delta b_{①} = +\frac{8064}{EI} (\downarrow)$

$\delta b_{②} = -t_{a/b②}$

$t_{a/b②} = \left[\frac{1}{2}(8)\left(\frac{8by}{EI}\right)\right]\left[\frac{2}{3}(8)\right] + \left[8\left(\frac{4by}{EI}\right)\right]\left[8 + \frac{1}{2}(8)\right] + \left[\frac{1}{2}(8)\left(\frac{4by}{EI}\right)\right]\left[8 + \frac{2}{3}(8)\right]$

$t_{a/b②} = +\frac{768by}{EI} \therefore$

$\delta b_{②} = -\frac{768by}{EI} (\uparrow)$

$$\delta b_{①} + \delta b_{②} = 0 \text{ or } \frac{8064}{EI} - \frac{768by}{EI} = 0 \Rightarrow by = 10.5^K (\uparrow)$$

Once the redundant reaction is found, we return to the given beam. Using statics we find the other reactions.

$\Sigma F_y = 0$

$a_y = 53.5^K (\uparrow)$

$\Sigma M_{pt\ a} = 0$

$M_a + 16(10.5) - 6(64) = 0$

$M_a = 216^{K\text{-}FT}$ ↻

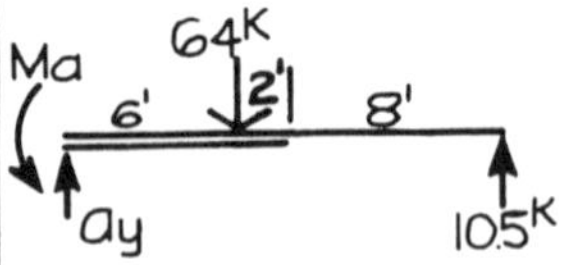

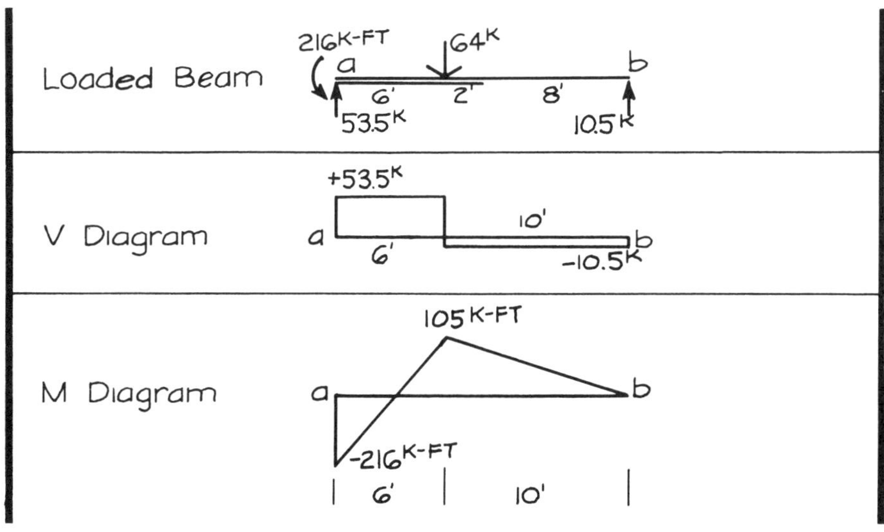
Loaded Beam
216K-FT
64K
a
b
6'
2'
8'
53.5K
10.5K
V Diagram
+53.5K
a
10'
b
6'
-10.5K
M Diagram
105K-FT
a
b
-216K-FT
6'
10'

OTHER METHODS

Throughout this chapter, the emphasis has been on the geometry of the deflected structure, both for finding deflections of statically determinate structures and in the analysis of indeterminate structures. Other methods, e.g., the "method of elastic weights" and the "conjugate beam method" are used to determine deflections (references 3 and 5). These are essentially moment-area but with an element of artificiality introduced by treating the angle change or *M/EI* diagram as a load on an imaginary beam. They do tend to simplify the analysis, however. Even so, the authors feel that it is important to develop an ability to visualize the geometry of a structure from the loads and the resulting *M/EI* diagrams and to calculate the deflections without resorting to tricks.

INDETERMINATE VS. DETERMINATE STRUCTURES

Upon looking closely at example and homework problems in this chapter, one may observe that, in general, for the same loading and same basic arrangement, statically indeterminate beams and frames are more rigid (they deflect less) and have lower maximum moments than the corresponding determinate structure. (See Figure 5.11, which is adapted from Example 6.) The latter point means the indeterminate structures can be designed with smaller members than analogous determinate structures, resulting in savings of materials of the structure itself. However, indeterminate structures require more rigid connections or more supports than determinate structures. These provisions can be costly.

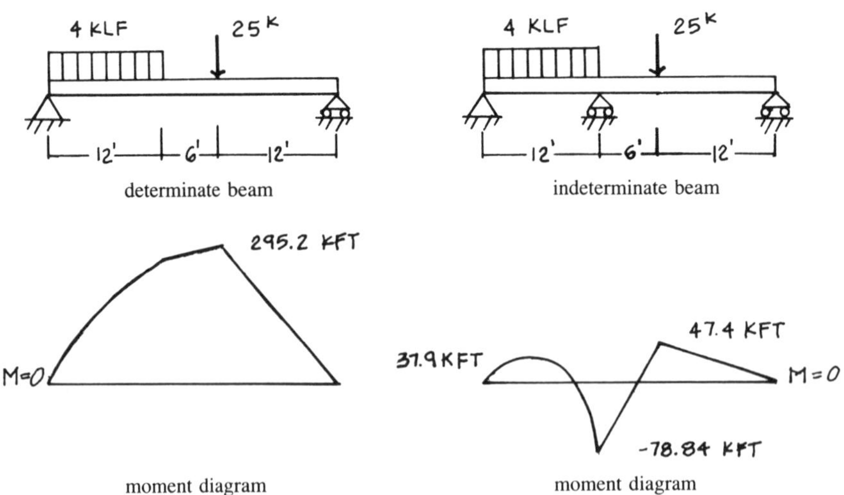

Figure 5.11

Also, indeterminate design inherently provides greater safety than determinate design. To wit, in Example 10, considering the determinate beam, if one of the supports totally fails, the beam itself fails. In the indeterminate design, the beam would remain in equilibrium.

If indeterminate structures generally result in more rigid, safer, and lighter structures than determinate structures, does this mean that in actual practice indeterminate structures are always employed? No.

First, we should emphasize that our pins, rollers, and walls are "ideal" supports. In reality all connections will resist rotation to some degree (as opposed to our notion of pins and rollers); further, no connection can completely resist rotation as we assume for wall connections. Thus, "real" connections are analyzed as pins or rollers if they don't restrain rotation "very much" or as walls if they restrain rotation "a lot." Here, professional judgment is called for.

Construction materials and techniques have much to do with choosing indeterminate or determinate design. In concrete construction, due to its monolithic character, rotation resisting connections are easily provided. Thus, most concrete design is indeterminate in nature. Rotation resisting joints in steel are often designed although they are usually costly due to additional labor and material. Contemporary steel construction is usually (but certainly not always) indeterminate. Timber construction on the other hand is usually determinate from the analytic point of view.

There are additional considerations to take into account with indeterminate design. Foundation and/or soil conditions can greatly affect the choice of indeterminate or determinate design. In particular, if differential settlement of the foundations is a real possibility, determinate design has the advantage. In Figure 5.12 we see that with settlement of one support in the determinate beam there is no bending (thus no stresses are induced). However, for the indeterminate beam settlement induces bending stresses. These stresses can be very large. Similar stresses can result due to temperature variations or due to "forced-fit" resulting from poor fabrication.

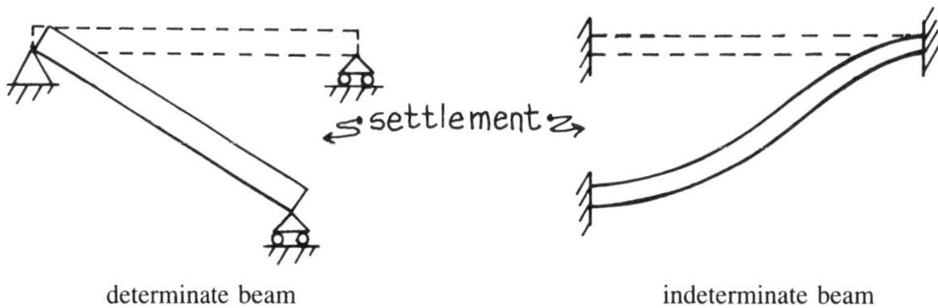

Figure 5.12

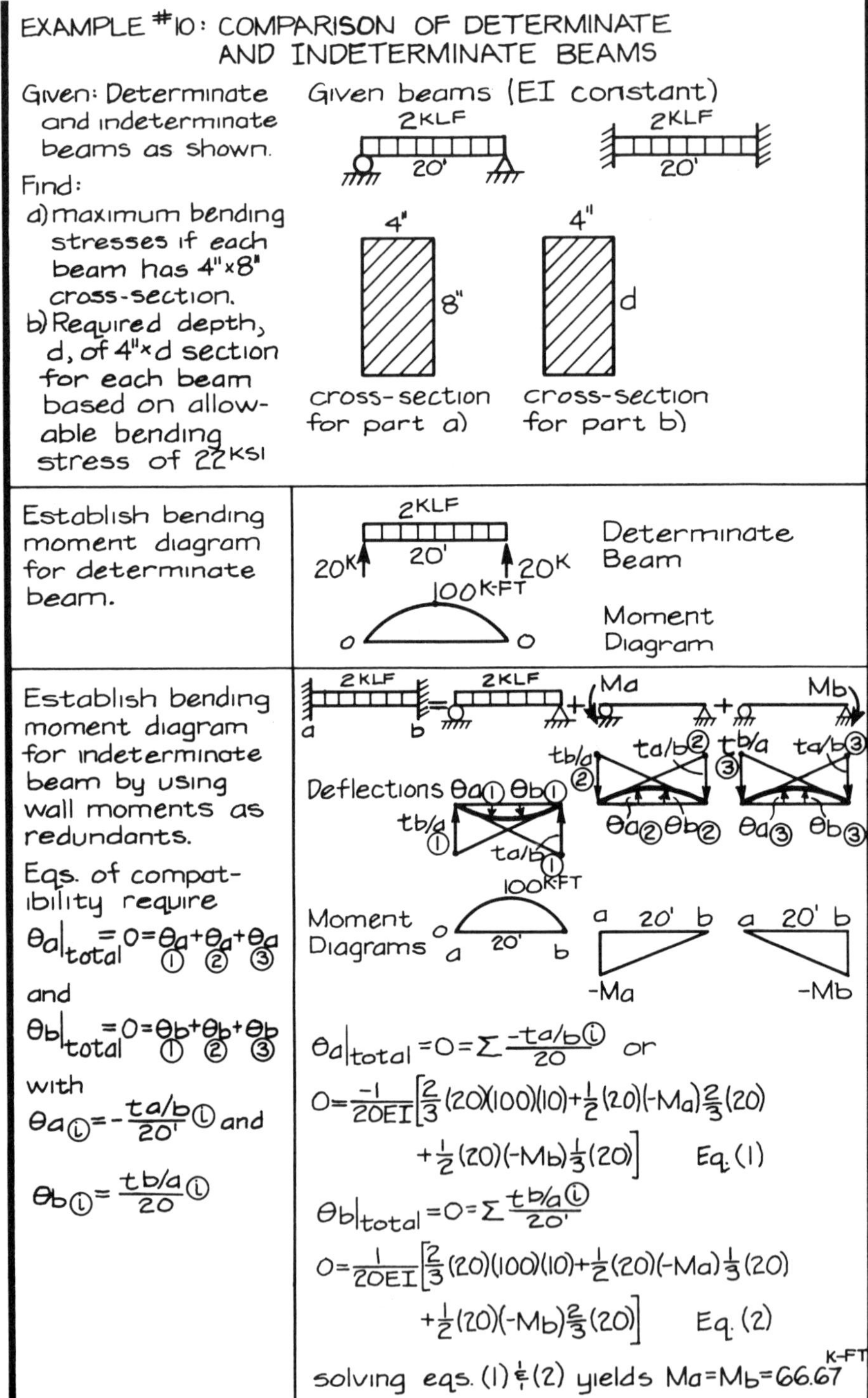

EXAMPLE #10: COMPARISON OF DETERMINATE AND INDETERMINATE BEAMS

Given: Determinate and indeterminate beams as shown.

Find:

a) maximum bending stresses if each beam has 4"×8" cross-section.

b) Required depth, d, of 4"×d section for each beam based on allowable bending stress of 22 KSI

Given beams (EI constant)

Establish bending moment diagram for determinate beam.

Establish bending moment diagram for indeterminate beam by using wall moments as redundants.

Eqs. of compatibility require

$\theta a|_{total} = 0 = \theta a_{①} + \theta a_{②} + \theta a_{③}$

and

$\theta b|_{total} = 0 = \theta b_{①} + \theta b_{②} + \theta b_{③}$

with

$\theta a_{①} = -\dfrac{ta/b_{①}}{20'}$ and

$\theta b_{①} = \dfrac{tb/a_{①}}{20}$

$\theta a|_{total} = 0 = \Sigma \dfrac{-ta/b_{①}}{20}$ or

$$0 = \frac{-1}{20EI}\left[\frac{2}{3}(20)(100)(10) + \frac{1}{2}(20)(-Ma)\frac{2}{3}(20) + \frac{1}{2}(20)(-Mb)\frac{1}{3}(20)\right] \quad \text{Eq. (1)}$$

$\theta b|_{total} = 0 = \Sigma \dfrac{tb/a_{①}}{20'}$

$$0 = \frac{1}{20EI}\left[\frac{2}{3}(20)(100)(10) + \frac{1}{2}(20)(-Ma)\frac{1}{3}(20) + \frac{1}{2}(20)(-Mb)\frac{2}{3}(20)\right] \quad \text{Eq. (2)}$$

solving eqs. (1) & (2) yields $Ma = Mb = 66.67^{K\text{-}FT}$

Recap of results	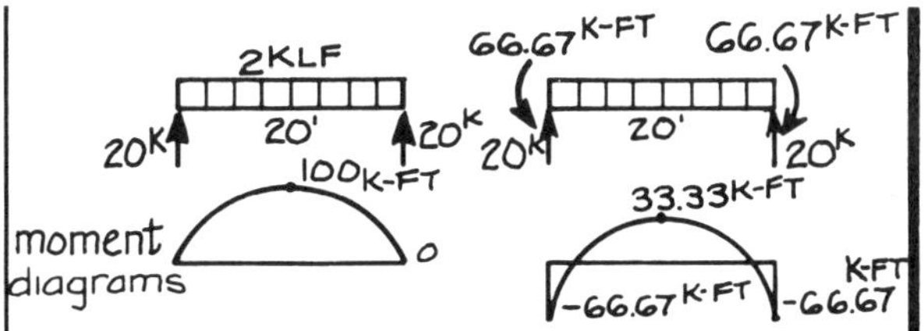 Det. Beam: 2KLF, 20', reactions 20^K, 20^K; moment diagram: $100^{K\text{-}FT}$ at midspan, 0 at ends. Indet. Beam: $66.67^{K\text{-}FT}$, $66.67^{K\text{-}FT}$ end moments, 20', reactions 20^K, 20^K; moment diagram: $33.33^{K\text{-}FT}$ at midspan, $-66.67^{K\text{-}FT}$ and $-66.67^{K\text{-}FT}$ at ends. moment diagrams Det. Beam Indet. Beam
Part (a): Find stresses $\sigma_{max} = \pm \frac{MY_{max}}{EI}$ Note: ratio of max. stress is same as ratio of max. moments, ie, .667.	$I = \frac{bh^3}{12} = \frac{(4)(8)^3}{12} = 170.67 \text{ in.}^4$ $Y_{max} = \frac{8}{2} = 4''$ $\sigma_{max \text{ det. beam}} = \frac{(100^{K\text{-}FT})(12''/_{FT})(4'')}{170.67 \text{ in}^4} = 28.12^{KSI}$ This stress occurs at midspan (top and bottom fibers). $\sigma_{max \text{ indet beam}} = \frac{(66.67^{K\text{-}FT})(12''/_{FT})(4'')}{170.67 \text{ in}^4} = 18.75^{KSI}$ This stress occurs at ends (top and bottom fibers).
Part (b): The section modulus is found as a function of beam depth Note: indet. beam cross-section 82% of det beam.	$S = \frac{I}{Y_{max}} = \frac{(4)(d)^3/12}{d/2} = \frac{2}{3}d^2$ thus $\sigma_{allow} = 22^{KSI} = \frac{M_{max}}{S_{required}} = \frac{M_{max}}{\frac{2}{3}d_{req}^2}$ $d_{req} = \sqrt{\frac{1.5 M_{max}}{\sigma_{allow}}} = \sqrt{\frac{1.5 M_{max}}{22^{KSI}}}$ For det. beam $d_{req.} = \sqrt{\frac{1.5(100)(12)}{22}} = 9.04''$ For indet. beam $d_{req} = \sqrt{\frac{1.5(66.67)(12)}{22}} = 7.39''$

PROBLEMS

Problems 1–12

For the beams shown find the slope and deflection at point *b*. (*EI* constant unless otherwise stated.)

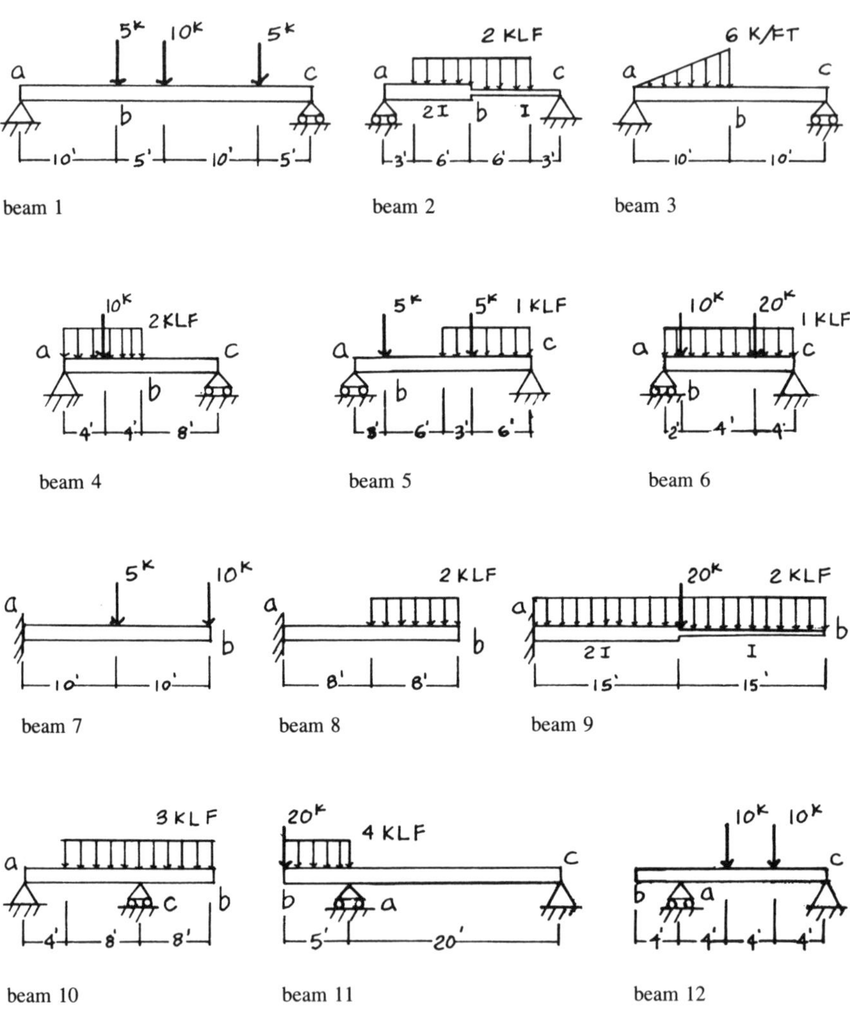

Problems 13–17

Find the maximum deflection for beams 1, 2, 3, 4, and 10.

Problems 18–29

Find the V and M diagrams.

Pr. 18: for beam 1 add a roller at point b.
Pr. 19: for beam 2 add a roller at point b.
Pr. 20: for beam 3 add a roller at point b.
Pr. 21: for beam 4 add a roller at point b.
Pr. 22: for beam 5 replace the pin at c by a wall.
Pr. 23: for beam 6 replace the pin at c by a wall.
Pr. 24: for beam 7 add a roller at point b.
Pr. 25: for beam 8 add a roller at point b.
Pr. 26: for beam 9 add a wall at point b.
Pr. 27: for beam 10 add a wall at point b.
Pr. 28: for beam 11 add a roller at point b.
Pr. 29: for beam 12 add a roller at point b.

Problems 30–35

For the frames shown find the slope and deflections at points b, c, and d (EI constant).

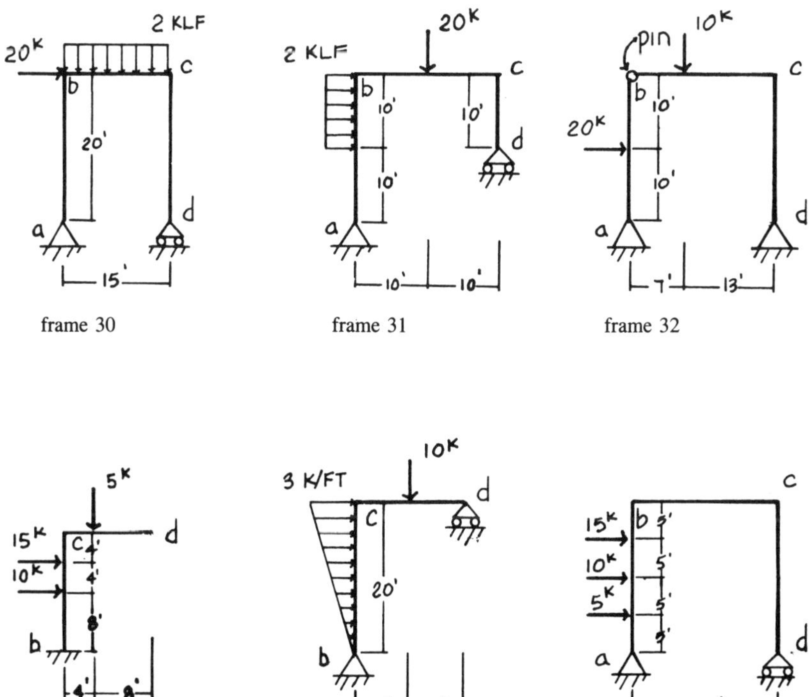

Problems 36–41

Find the V and M diagrams.

Pr. 36: for frame 30 replace roller at d by a pin.
Pr. 37: for frame 31 replace roller at d by a pin.
Pr. 38: for frame 32 add a vertical roller at point c.
Pr. 39: for frame 33 add a roller at point d.
Pr. 40: for frame 34 replace roller at d by a pin.
Pr. 41: for frame 35 add a vertical roller at point c.

Problems 42 and 43

Compare the maximum bending stresses for a 4″ wide by 10″ deep section.

Pr. 42: for the beams in Pr. 1 and Pr. 18.
Pr. 43: for the beams in Pr. 8 and Pr. 25.

Problems 44 and 45

Compare the required depths of rectangular beams with 4″ widths using as the design criterion an allowable bending stress of 22 ksi.

Pr. 44: the beams in Pr. 4 and Pr. 21.
Pr. 45: the beams in Pr. 7 and Pr. 24.

6

DEFLECTIONS: ENERGY METHODS

INTRODUCTION

Energy or work-energy methods permeate the field of mechanics. In fact, their base, conservation of energy, is itself a first principle or axiom of classical mechanics. In undergraduate physics and dynamics, work-energy methods are encountered and used in lieu of the familiar balances of linear momentum ("$F = ma$") and angular momentum ("$T = I\alpha$").

In structural analysis the first theorem of Castigliano and its continuation—virtual work, can be used to find deflections of any structure. These methods are naturally extended to reaction analysis for statically indeterminate structures.

We feel most strongly about energy methods for truss analysis—both determinate and indeterminate. These topics will be pursued extensively. Determinate beam and frame deflections will also be illustrated.

WORK AND STRAIN ENERGY

As previously mentioned, energy and work are related quantities. In fact, work is the transference of energy from one force system to another. Recall from physics that work was defined as the product of force and the corresponding (parallel) distance the force travels. If the force varies with distance then integration, of course, is necessary. Mathematically, work, W, is calculated as

$$W = \int_{r_1}^{r_2} \bar{F} \cdot d\bar{r}.$$

Let us look at the work done by internal forces, restricting our attention to isotropic, linearly elastic materials in equilibrium. Let's start by choosing a one-dimensional stress problem as shown in Figure 6.1. Here a small element is subjected to only normal stress, σ_x (or normal force, F_x).

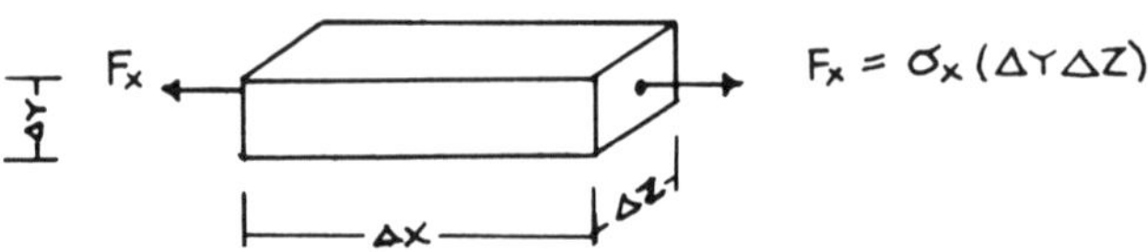

Figure 6.1

Since these normal stresses cause an axial elongation, e_x, work is done. The magnitude of the elongation depends upon the stress level, i.e., the elongation is a function of the stress; therefore, integration is necessary in order to calculate the work

$$W_{1-D} = \int F_x \, de_x = \int (\sigma_x \, \Delta y \, \Delta z) \, de_x.$$

Note that the change in length, e_x, is nothing more than the final axial strain, ε_x, times the original element length, Δx, i.e.,

$$e_x = \varepsilon_x \, \Delta x.$$

Thus the incremental elongation is

$$de_x = d\varepsilon_x \, \Delta x.$$

Then our work expression becomes

$$W_{1-D} = \int \sigma_x \, \Delta y \, \Delta z \, \Delta x \, d\varepsilon_x,$$

$$W_{1-D} = \int_0^{\varepsilon_x} \sigma_x \, d\varepsilon_x \, \Delta \, (\text{volume}).$$

The integration for the total work is from zero strain to the final strain, ε_x.

Let's define the strain energy, U, as the work per unit volume; thus, for this one dimensional case,

$$U_{1-D} = \int \sigma_x \, d\varepsilon_x.$$

We can easily extend this case to three dimensions, where three normal stresses give rise to three normal (lineal) strains and three shear stresses cause three shear strains (changes in angles). Thereby, the general form of strain energy is:

$$U_{\text{GEN}} = \int \sigma_x \, d\varepsilon_x + \int \sigma_y \, d\varepsilon_y + \int \sigma_z \, d\varepsilon_z + \int \tau_{xy} \, d\gamma_{xy} + \int \tau_{xz} \, d\gamma_{xz} + \int \tau_{yz} \, d\gamma_{yz}. \tag{1}$$

Hooke's law relates the normal strains to only normal stresses and shear strains to only shear stresses:

$$\varepsilon_x = \frac{1}{E}\,[\sigma_x - \mu(\sigma_y + \sigma_z)].$$

Thus for incremental normal strain,

$$d\varepsilon_x = \frac{1}{E}\,[d\sigma_x - (d\sigma_y + d\sigma_z)]. \tag{2a}$$

Similarly,

$$d\varepsilon_y = \frac{1}{E}\,[d\sigma_y - \mu(d\sigma_x + d\sigma_z)], \tag{2b}$$

$$d\varepsilon_z = \frac{1}{E}\,[d\sigma_z - \mu(d\sigma_x + d\sigma_y)]. \tag{2c}$$

Also, we find for incremental shear strains:

$$d\gamma_{xy} = \frac{1}{G}\,d\tau_{xy}, \tag{2d}$$

$$d\gamma_{xz} = \frac{1}{G}\,d\tau_{xz}, \tag{2e}$$

$$d\gamma_{yz} = \frac{1}{G}\,d\tau_{yz}. \tag{2f}$$

Here, of course, E, μ, and G are Young's modulus, Poisson's ratio, and the shear modulus, respectively.

Putting Equations 2(a–f) into Equation 1 and integrating yields the useful general form of strain energy for simple elastic materials.

$$U_{\text{GEN}} = \frac{1}{2E}\,(\sigma_x^2 + \sigma_y^2 + \sigma_z^2) + \frac{1}{2G}\,(\tau_{xy}^2 + \tau_{xz}^2 + \tau_{yz}^2) - \frac{2\mu}{E}\,(\sigma_x\sigma_y + \sigma_x\sigma_z + \sigma_y\sigma_z). \tag{3}$$

Work and Strain Energy—Structural Examples

With Equation 3 as a start, let's investigate the strain energy and work functions for structures of our immediate concern: trusses, beams, and frames.

Trusses

A truss consists of many members, each of which is ideally in an uniaxial state of stress: either tension or compression. Looking at any generic member and calling its long axis the x-axis, we get the following stress picture:

$$\sigma_x = \frac{P}{A},\ \sigma_y = \sigma_z = 0,$$

$$\tau_{xy} = \tau_{xz} = \tau_{yz} = 0.$$

Here, of course, P is the force in the member and A is the cross-sectional area. Putting this state of stress in Equation 3 yields:

$$U_{\text{truss member}} = \frac{P^2}{2EA^2}.$$

The work done by each truss member, of length ℓ, is found by integrating the strain energy over the volume of the member, then

$$W_{\text{truss member}} = \iiint \frac{P^2}{2EA^2}\, d\ell\, dA = \frac{P^2\ell}{2EA}. \tag{4}$$

Finally, the work done by the entire truss is simply the sum of the individual work done by each member; thus

$$W_{\text{truss}} = \frac{1}{2}\sum_{i=1}^{n} \frac{P_i^2\ell_i}{E_iA_i}, \tag{5}$$

where the sum is over all "n" truss members.

Beams and Frames

For beams, strength of materials tells us that bending and shear stresses exist. Choosing the x-axis (Figure 6.2) as the axis of the beam, we find

$$\sigma_x = \frac{My}{I_z},\ \sigma_y = \sigma_z = 0,$$

$$\tau_{xy} = \frac{VQ}{It},\ \tau_{xz} = \tau_{yz} = 0.$$

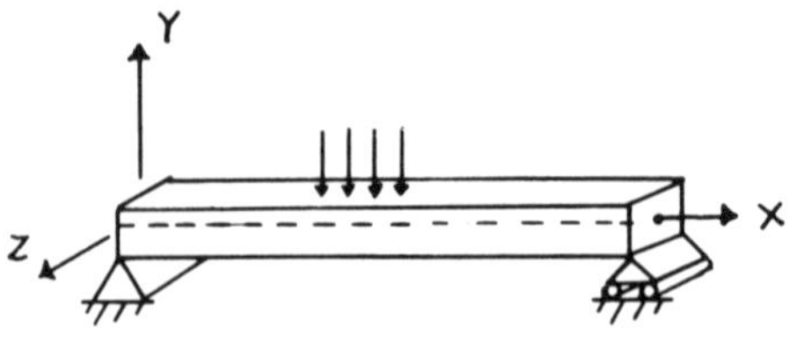

Figure 6.2

Here M and V are due directly to the loads and represent the local bending moment and the local shear force. The cross-sectional moment of inertia (about the z-axis) is I_z; t is the local thickness (in the z-direction), y is the vertical coordinate measured from the centroid of the cross-section and Q is the moment of area about the centroid of the cross-section.

Since shear stresses are usually much smaller in magnitude than bending stresses we shall neglect their effect in the strain-energy function. This, it should be noted, is consistent with what we did in deriving deflection relations through double integration and moment-area. Therefore,

$$U_{\text{beam}} \simeq \frac{1}{2E}\frac{M^2y^2}{I_z^2}.$$

The work is found by integrating the strain energy over the volume of the beam:

$$W_{\text{beam}} = \iiint U_{\text{beam}}\,dx\,dy\,dz = \frac{1}{2}\iiint \frac{M^2y^2}{EI_z^2}\,dx\,dy\,dz.$$

We note that since the local bending moment depends only upon the loading, it is independent of coordinates y and z. Thus,

$$W_{\text{beam}} = \frac{1}{2}\int_0^{\ell}\frac{M^2}{EI_z}\left(\iint_A y^2\,dy\,dz\right)dx.$$

We recognize the last integral as nothing more than the moment of inertia about the z-axis,

$$\iint y^2\,dy\,dz = \iint y^2\,dA \equiv I_z.$$

Finally, the work for the beam is found to be

$$W_{\text{beam}} = \frac{1}{2}\int_0^{\ell}\frac{M^2}{EI_z}\,dx. \tag{6}$$

The integration is over the total beam length. Of course, if the beam is prismatic, then I_z comes outside the integral; if the beam is homogeneous then E comes outside the integral.

In frame analysis both the axial stresses due to axial forces and the shear stresses are negligible compared to the bending (axial) stresses. Thus Equation 6 is also used for the frames. For completeness we reiterate:

$$W_{\text{frame}} = \frac{1}{2}\int_0^{\ell}\frac{M^2\,dx}{EI_z}. \tag{7}$$

CASTIGLIANO'S FIRST THEOREM

Now that we've formed strain energy and work expressions the obvious question arises: What do we do with them? Answer: Through Castigliano's first theorem we see that deflections can be related to the work done by internal stress. To approach this theorem, published in 1876 by the Italian railway engineer Albert Castigliano, we again look at a one-dimensional stress case. Consider a horizontal bar loaded with a concentric horizontal force, P_x (Figure 6.3).

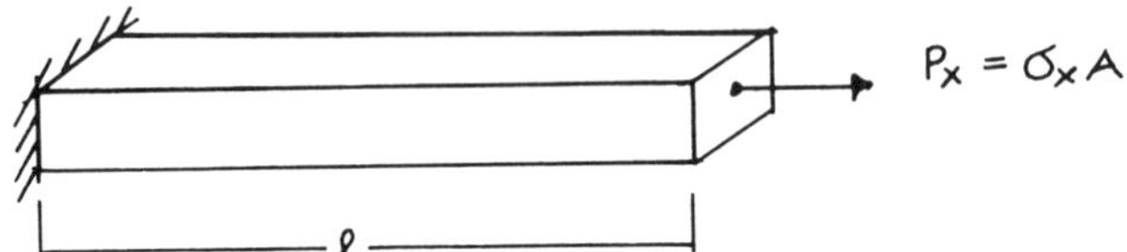

Figure 6.3

When a small load is applied a small elongation takes place. As the load increases, the elongation increases until the final elongation, e_x, corresponding to the total load P_x, is reached as shown (Figure 6.4) on the curve. If the material is linearly elastic, this curve is a straight line.

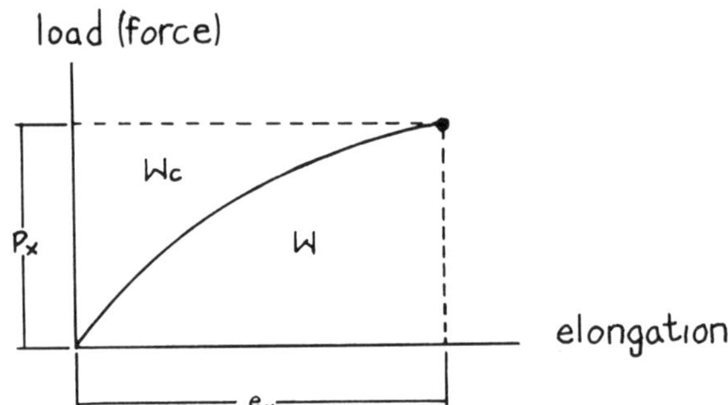

Figure 6.4

From our previous discussion it is clear that the area under the curve is the work done by the load, W. We'll call the area above the curve the complimentary work, W_c. For a linearly elastic material, W equals W_c.

With a small increase, ΔP_x, of the load a corresponding small increase, Δe_x, in the elongation results. Figure 6.5, for a linearly elastic material, greatly amplifies these "small" changes.

Since $W = W_c$ then

$$\Delta W = \Delta W_c.$$

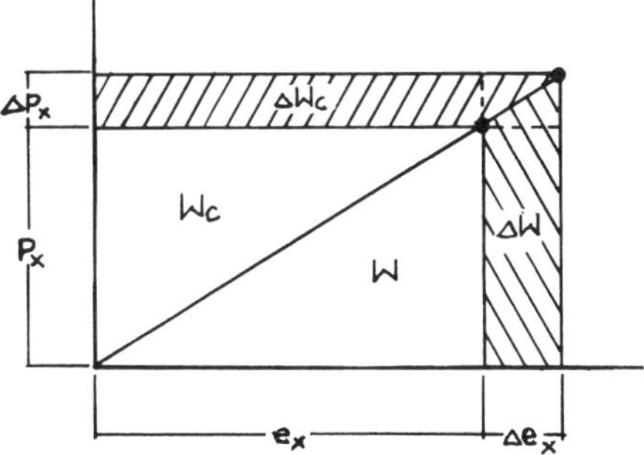

Figure 6.5

From the figure we see

$$\Delta W = P_x\,\Delta e_x + \frac{\Delta P_x\,\Delta e_x}{2} \simeq P_x\,\Delta e_x,$$

$$\Delta W_c = e_x\,\Delta P_x + \frac{\Delta P_x\,\Delta e_x}{2} \simeq (\Delta P_x)e_x.$$

Here the second-order incremental terms are neglected. Thus

$$\Delta W \equiv P_x\,\Delta e_x = \Delta W_c \equiv (\Delta P_x)e_x,$$

from which we select $\Delta W = \Delta P_x e_x$. Going to the limit, we arrive at

$$e_x = \frac{\partial W}{\partial P_x}.$$

This says that the axial elongation (deflection) can be evaluated by taking the partial derivative of the work function with respect to the applied axial load.

Generalizing this, we reach Castigliano's First Theorem: For a linearly elastic body in equilibrium, the rate of change of the work with respect to any applied load (force or moment) yields the deformation (deflection or rotation) at the point of application of that load in the direction of the load. The real impact of the theorem is felt when a total truss is investigated—although it should be pointed out that Castigliano's Theorem is for us a stepping stone to the much more easily applied technique called virtual work. First, an example using Castigliano's First Theorem:

Note in the following truss examples that the actual statics equations are not explicitly formed. We do not mean to imply here that statics is not important—on the contrary, for virtual work and for all structural analysis statics is fundamental. Its importance cannot be over-emphasized. The authors, in solving these virtual work examples, did draw careful free-body diagrams (using the method of joints); they did write out all equations of equilibrium. But in order to bring to the fore the new ideas of this chapter (namely, virtual work) the statics calculations were deleted. We strongly feel, however, that students should show all their work, including the statics portion.

EXAMPLE #1: CASTIGLIANO'S METHOD
TRUSS DEFLECTION

Given: Truss as shown

Find: The horizontal deflection at point d.
Area, A, and modulus of elasticity, E, are same for all members.

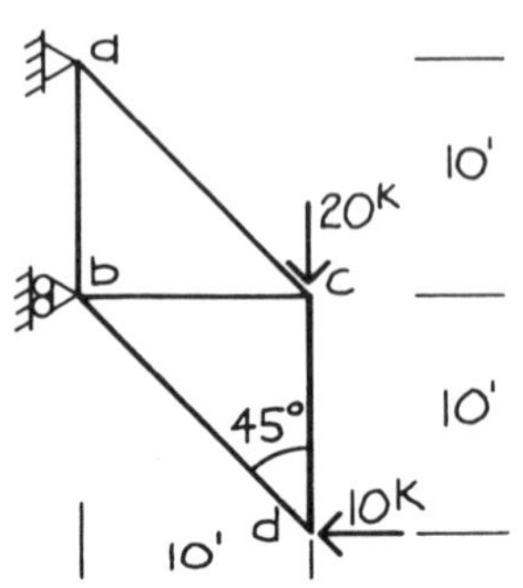

Since the derivative of the work function must be taken it is necessary to make the 10K horizontal load at point d into a parameter, P, (shown below). The internal force in each member is then found, by statics, in terms of P and the other load. The results of the statical analysis are shown in the chart below.

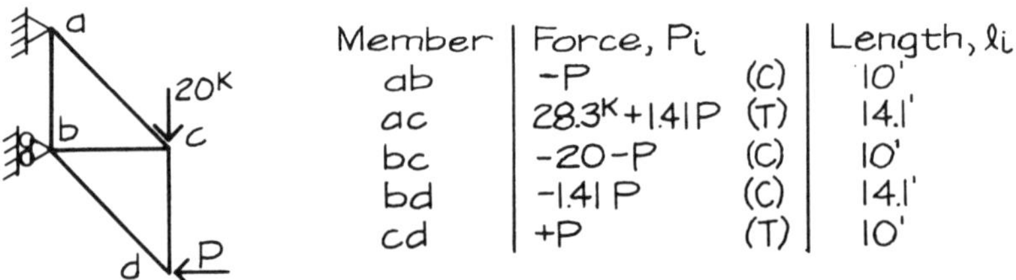

Member	Force, P_i		Length, ℓ_i
ab	$-P$	(C)	10'
ac	$28.3^K + 1.41P$	(T)	14.1'
bc	$-20-P$	(C)	10'
bd	$-1.41P$	(C)	14.1'
cd	$+P$	(T)	10'

The total work for the truss is then formed.

$$W_{\substack{total\\truss}} = \Sigma \frac{P_i^2 \ell_i}{2A_iE_i} = \frac{1}{2AE}\Big[(-P)^2(10) + (28.3+1.41P)^2(14.1) + (-P-20)^2(10) + (-1.41P)^2(14.1) + (P)^2(10)\Big]$$

$$= \frac{1}{2AE}\left[86.1P^2 + 1{,}525.3P + 15{,}292.5\right]$$

Now the derivative of W with respect to P is taken and the result is evaluated for the actual value of P.

$$\frac{\partial W}{\partial P} = \frac{1}{2AE}\left[172.2P + 1525.3\right]$$

$$\left.\frac{\partial W}{\partial P}\right|_{P=10} = \frac{1623.6}{AE}$$

Thus the horizontal deflection at point d is:

$$\delta_{dx} = +\frac{1623.6}{AE} \text{ (K-FT) } (\leftarrow)$$

The positive sign means the deflection is in the same direction as the parameter, P; i.e., δ_{dx} is to the left.

The major problems with Castigliano's method arise due to necessity of the parametric load; it makes the statics cumbersome. We're required to square terms and then take derivatives before we can evaluate. These are awkward steps. If we were seeking the vertical deflection at d we'd have to invent a vertical parametric load, Q, at d; go through the process of forming W in terms of Q; taking $\partial W/\partial Q$ and finally set $Q = 0$. In this final stage all the terms that involve Q will not contribute to δ_{d_y} since Q is, in fact, zero. There must be an easier way! There is: Virtual work.

VIRTUAL WORK: TRUSSES

Virtual work is just a perceptive variation of Castigliano's Theorem. If we look at the previous example and generalize we see all internal forces are of the form

$$P_i = \alpha_i P + M_i.$$

Here α_i is the portion of the parametric load seen in each member and M_i is the force due to loads other than the parameter P.

Thus the work is of the form

$$W = \sum_{i=1}^{n} \left[\frac{(\alpha_i^2 P^2 + 2\alpha_i M_i P + M_i^2)\ell i}{2A_i E_i} \right]$$

and

$$\frac{\partial W}{\partial P} = \sum_{i=1}^{n} \left[\frac{(2\alpha_i^2 P + 2\alpha_i M_i)\ell_i}{2A_i E_i} \right].$$

So the deflection, e, can be written

$$e \equiv \delta = \left.\frac{\partial W}{\partial P}\right|_P = \sum_{i=1}^{n} \left.\left[\frac{(\alpha_i)(\alpha_i P + M_i)\ell_i}{A_i E_i} \right]\right|_P.$$

Looking at our result carefully we see that the first bracket, α_i, represents the force in each member due to only a one unit parametric load. The second bracket (with P as a number) is exactly the force in each member due to the original loading situation. Thereby, we can find deflections by multiplying the corresponding forces in two trusses: the actual truss with all the real loads and the "virtual" truss consisting of a unit load at the point of concern in the direction of concern.

Succinctly, we can state this result as

$$\delta = \sum_{i=1}^{n} \left(\frac{P_i p_i \ell_i}{A_i E_i} \right), \tag{8}$$

where P_i is the force in each member of the actual truss and p_i is the force in each member in the virtual truss.

Let's look at the same truss in Example 1, this time using virtual work.

EXAMPLE #2: VIRTUAL WORK: TRUSS DEFLECTION

Given: Truss as shown
Area, A, and modulus of elasticity, E, are same for all members.

Find: The horizontal deflection at point d.

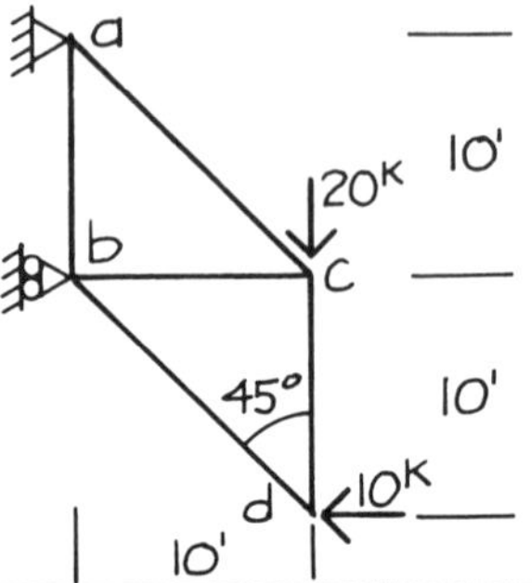

Calculate the force in each member for the given truss; call these P_i. The results are shown below with lengths.

Member	Force, P_i	Length, ℓ_i
ab	-10^K (C)	10'
ac	$+42.4^K$ (T)	14.1'
bc	-30^K (C)	10'
bd	-14.1^K (C)	14.1'
cd	$+10^K$ (T)	10'

Calculate the force in each member of the virtual truss, p_i, which consists of the same truss with only a unit horizontal load at point d. The results are shown below.

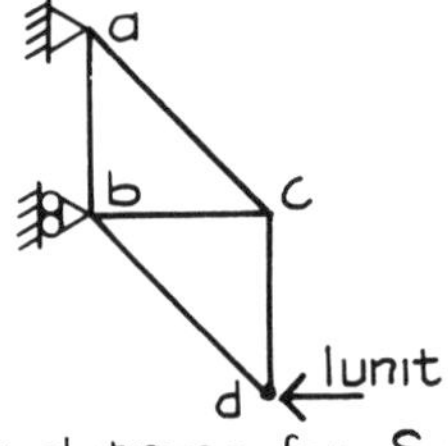

Virtual truss for δ_{dx}

Member	Force, p_i	
ab	-1.0	(C)
ac	1.41	(T)
bc	-1.0	(C)
bd	-1.41	(C)
cd	1.0	(T)

Make final calculation of the horizontal deflection at point d.

$$\delta_{dx} = \sum_{1}^{5} \frac{P_i p_i \ell_i}{A_i E_i} = \frac{1}{AE}\Big[(-10)(-1)(10) + (42.4)(1.41)(14.1) + (-30)(-1)(10)$$

$$+ (-14.1)(-1.41)(14.1) + (10)(1)(10)\Big]$$

$$\delta_{dx} = +\frac{1623.3}{AE}\ (\text{K-FT})\ (\leftarrow)$$

Again, the positive sign means the deflection is in the same direction as the virtual force; therefore δ_{dx} is to the left.

EXAMPLE #3: VIRTUAL WORK: TRUSS DEFLECTION

Given: Truss as shown;
A and E are same for all members.

Find: The vertical deflection at point d.

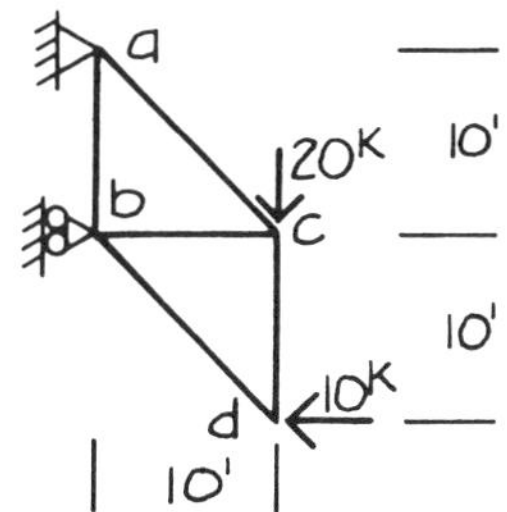

Even though no real <u>vertical</u> force is applied at point d virtual work will allow an easy calculation of the vertical deflection there.

Most of the calculations have already been made. We use the P_i values of the real truss as in example #2.

Now calculate the virtual member forces, q_i, considering a unit vertical load at point d. The results of the statics analysis are shown below.

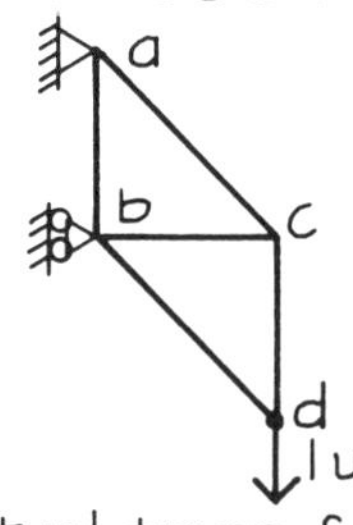

Virtual truss for δ_{dY}

Member	Force, q_i	
ab	0	
ac	1.41	(T)
bc	-1.0	(C)
bd	0	
cd	+1.0	(T)

Make final calculation of the vertical deflection at point d.

$$\delta_{dY} = \sum_{1}^{5} \frac{P_i q_i l_i}{A_i E_i} = \frac{1}{AE}\Big[(-10)(0)(10) + (42.4)(1.41)(14.1) + (-30)(-1)(10)$$

$$+ (-14.1)(0)(14.1) + (10)(1)(10)\Big]$$

$$\delta_{dY} = \frac{+1243.0}{AE} \text{ (K-FT) } (\downarrow)$$

VIRTUAL WORK: INDETERMINATE TRUSSES

The method of virtual work is easily extended to indeterminate trusses, most conveniently through the technique of superposition. In superposition we add the effects of only statically determinate trusses. Keep in mind that for statically determinate trusses we usually seek deflections; for indeterminate trusses we usually seek the forces in the members.

Note that there are two types of truss indeterminacy: external and internal.

External indeterminacy means there are more available support reactions than equilibrium requires. For a simple two dimensional truss if there are four or more available external support reactions the truss is externally indeterminate (Figure 6.6).

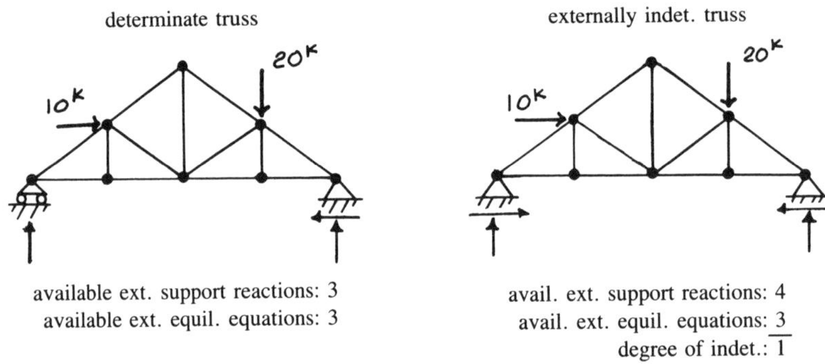

Figure 6.6

On the other hand, internal indeterminacy implies there are more truss members than equilibrium or stability requires. A simple two-dimensional truss would be internally indeterminate if the number of members (m) is greater than twice the number of joints (j) less three:

$$m > 2j - 3.$$

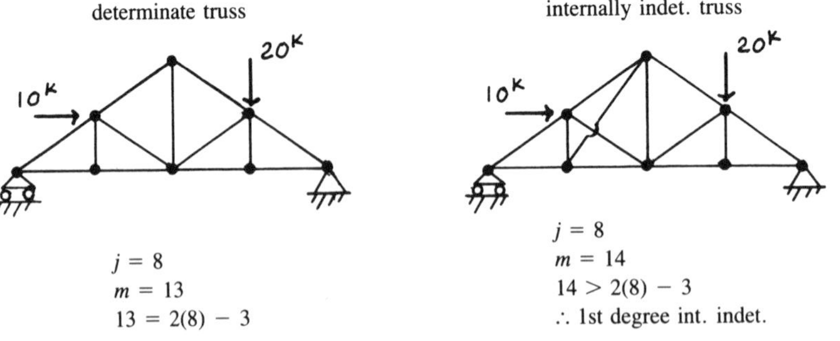

Figure 6.7

Now for the analysis. The basic idea is to choose a "redundant(s)." The number of redundants matches the degree of indeterminacy. A "redundant" is just what the name implies, a force that is not needed for purely equilibrium purposes—it is redundant as far as equilibrium is concerned. For external indeterminacy a redundant is a reaction component; for internal indeterminate problems a redundant is an internal truss member. In general, for any problem there will be several choices for redundants. In this respect the choice is arbitrary (as long as the resulting trusses are stable). The value of a redundant, of course, cannot be determined from statics alone. If we consider a redundant as a load (albeit unknown) we can separate the truss into several similar trusses: a truss with only the actual loads and trusses with loads consisting of only the redundants. Using virtual work, the deflection at the point of application of the redundant can be found for each truss. Using the original indeterminate problem, we know the true value of the deflection at the point of application of the redundant. Thus the deflections for each separated truss are related. This gives us an equation for the unknown redundant. After the equation is solved for this redundant, we use its value to determine internal forces and reactions. This process is best seen through examples.

EXAMPLE #4: VIRTUAL WORK: EXTERNALLY INDETERMINATE TRUSS:

Given: Truss as shown; A and E are same for all members.

Find: Reaction and member forces for the truss shown.

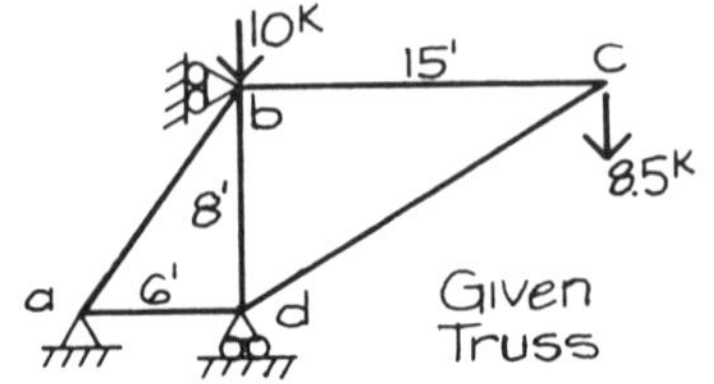

The free body diagram of the truss is shown here.

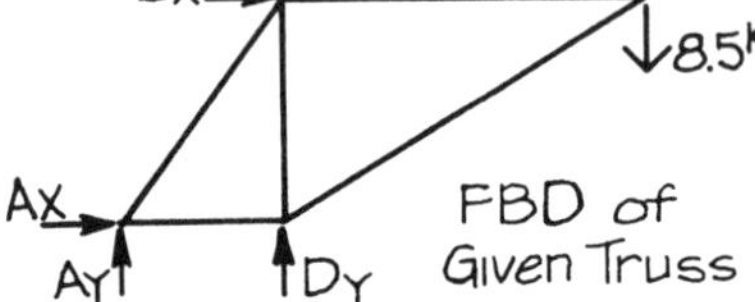

Here we have four external reactions but only three equations of equilibrium. Which support can we do without as far as equilibrium is concerned?
Answer: Any. Let's arbitrarily choose Bx (call it Q) as our redundant. Considering Bx as a load, we see a truss with three loads: two known (10^K and 8.5^K) and one unknown. We form trusses such that their sum corresponds to the given truss.

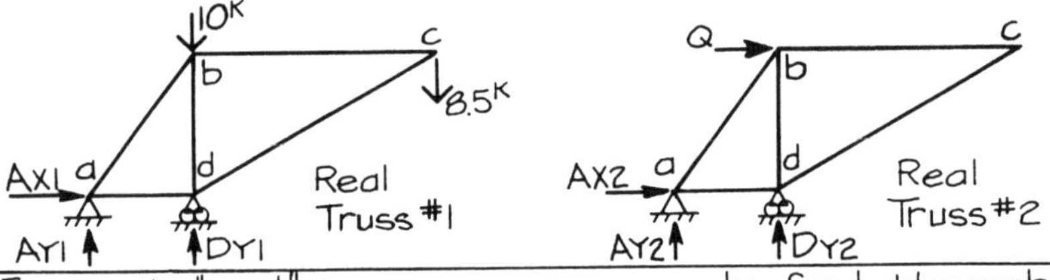

For each "real" truss we can easily find through statics the reactions and internal forces. Call these internal forces P_i for real truss #1 and Q_i for #2. The results of statics are shown below.

Member	Force, P_i		Force, Q_i		Length, ℓ_i
ab	$+26.7^K$	(T)	$+1.67Q$	(T)	10'
ad	-16.0^K	(C)	0		6'
bc	$+16.0^K$	(T)	0		15'
bd	-31.3^K	(C)	$-1.33Q$	(C)	8'
cd	-18.1^K	(C)	0		17'
Reaction	$i=1$		$i=2$		
Ax_i	0		$-Q$	(←)	
Ay_i	-21.3	(↓)	$-1.33Q$	(↓)	
Dy_i	$+39.8$	(↑)	$+1.33Q$	(↑)	

Note the given indeterminate truss prevents horizontal deflection at point b. Thus the deflections at point b for trusses #1 (real) and #2 (real) must sum to zero. That is:

$$\delta_{bx1} + \delta_{bx2} = 0$$

Note: The point of application of the redundant always provides the "compatible deflection relation."

We can find δ_{bx1} and δ_{bx2} by using the same virtual truss — a determinate truss with a unit horizontal load at point b. That is, virtual truss #1 corresponding to real truss #1 is identical to virtual truss #2 corresponding to real truss #2.

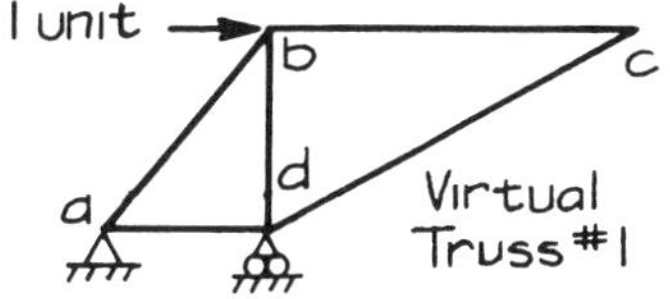

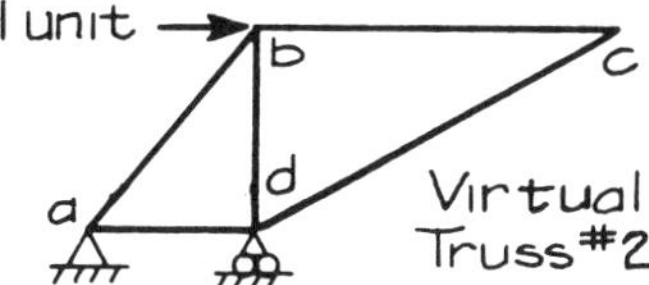

It is also clear, by inspection, that the forces in real truss #2 are exactly Q times the forces in the virtual trusses. If we call these virtual truss forces p_i, then:

$$Q_i = Qp_i,$$

or tabulating these values:

Member	Virtual Force, p_i	
ab	+1.67	(T)
ad	0	
bc	0	
bd	−1.33	(c)
cd	0	

Thus we can now calculate the corresponding horizontal deflection at point b.

$$\delta_{bx1} = \sum_1^5 \frac{P_i p_i l_i}{A_i E_i} = \frac{1}{AE}\left[(26.7)(1.67)(10) + (-31.3)(-1.33)(8)\right] = \frac{778.9}{AE}^{\text{K-FT}} (\rightarrow)$$

$$\delta_{bx2} = \sum_1^5 \frac{Q_i p_i l_i}{A_i E_i} = \frac{Q}{AE}\Sigma p_i^2 l_i$$

$$= \frac{Q}{AE}\left[(1.67)^2(10) + (-1.33)^2(8)\right] = \frac{42.0\,Q}{AE} \quad (\rightarrow)$$

Therefore: $778.9 + 42.0\,Q = 0$ or $Q = -18.5^K (\leftarrow)$

Thereby we can easily calculate the actual forces $(P_i + Q_i)$ and reactions $(A_x = A_{x1} + A_{x2}$, etc.).

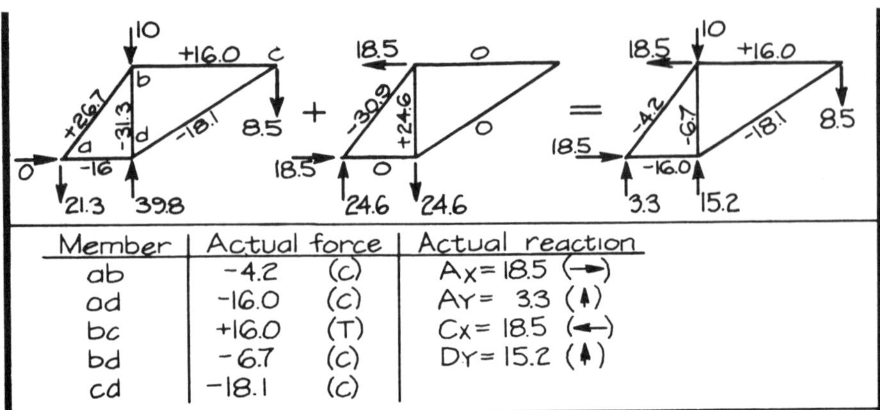

Member	Actual force		Actual reaction
ab	−4.2	(C)	A_X = 18.5 (→)
ad	−16.0	(C)	A_Y = 3.3 (↑)
bc	+16.0	(T)	C_X = 18.5 (←)
bd	−6.7	(C)	D_Y = 15.2 (↑)
cd	−18.1	(C)	

It might be interesting to note that for this case only two members are affected by the redundant support at point b. Force ab changes from a tensile force of 26.7^K to a compressive force of 4.2^K while bd is reduced from a 31.3^K compressive force to 6.7^K (compressive). Clearly the support forces are radically altered.

Point of caution: Both real trusses must be, of course, in equilibrium and have identical support conditions.

There are a few variations in the virtual work process when internal indeterminacy, as compared to external indeterminacy, is considered. The most important variation is that redundants are internal members. The degree of indeterminacy fixes the number of required redundants and the number of necessary compatibility of deflection equations. The basic idea of these compatibility equations is that when a redundant internal member is imagined to be cut, the *relative* displacement at the two ends of this internal cut is, in fact, zero. It must be kept in mind that these "cut" redundants are elastic members and do work when loads are applied. That is to say, the redundants must be included in the work function. Let's look at an example of an internally indeterminate truss.

EXAMPLE #5: VIRTUAL WORK: INTERNALLY STATICALLY INDETERMINATE TRUSS

Given: Truss as shown; A and E are same for all members

Find: Reaction and member forces for the truss shown.

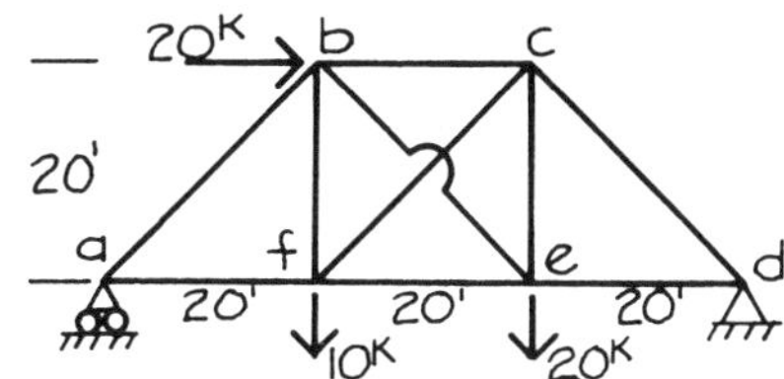

Degree of indeterminateness:

Number of members, $m = 10$

Number of joints, $j = 6$

$m \equiv 10 > 2j - 3 \equiv 9$ ∴

First degree indeterminate.

Either be or cf is a candidate for redundant. Arbitrarily choose cf. Call internal force Q.

Imagine member cf cut and superpose two trusses; one with known loads and one with unknown internal loads in cf, Q.

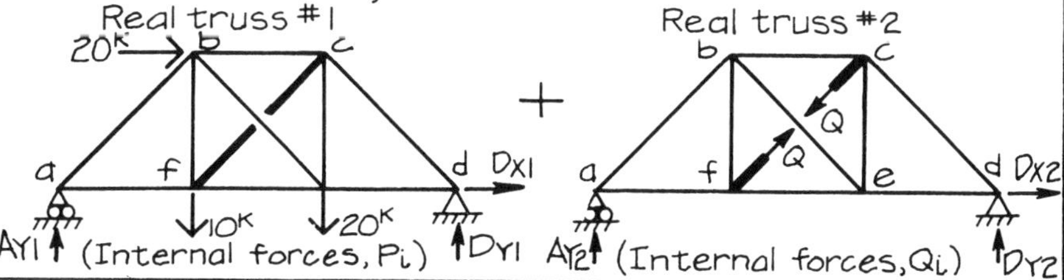

Since both real trusses are now determinate statics will give the internal forces — charted below. Note real truss #2 is loaded such that it is self-equilibriating, ie, no external support reactions are required.

Member	Force, P_i		Force, Q_i		Length, ℓ_i
ab	-9.43	(C)	0		$20\sqrt{2}'$
af	+6.67	(T)	0		20'
bc	-23.33	(C)	-.71Q	(C)	20'
be	-4.71	(C)	+Q	(T)	$20\sqrt{2}'$
bf	+10.00	(T)	-.71Q	(C)	20'
cd	-32.99	(C)	0		$20\sqrt{2}'$
ce	+23.33	(T)	-.71Q	(C)	20'
cf	0		+Q	(T)	$20\sqrt{2}'$
de	+3.33	(T)	0		20'
ef	+6.67	(T)	-.71Q	(C)	20'
Reaction	$i=1$		$i=2$		
AY_i	6.67	(↑)	0		
DX_i	-20.00	(←)	0		
DY_i	23.33	(↑)	0		

Now for the virtual trusses. Since we are seeking the relative displacement of the two ends of member cf we place unit loads at these two ends for both virtual trusses.

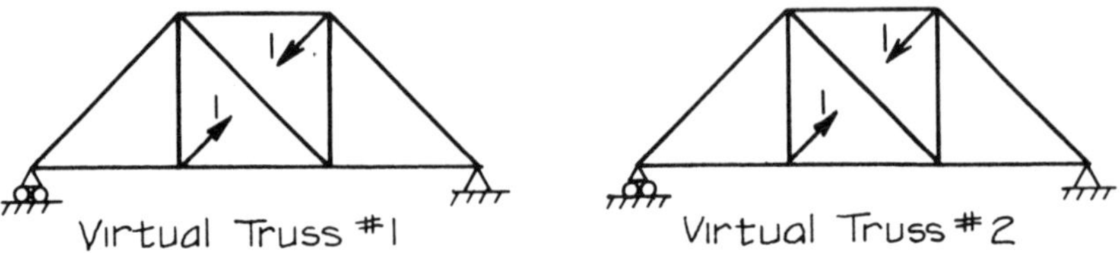

Virtual Truss #1 Virtual Truss #2

Again we see the virtual internal forces, p_i, are directly proportional to the internal forces in real truss #2, ie,

$$Q_i = Q p_i$$

Member	Force, p_i	
ab, af	0	
cd, ed	0	
be, cf	+1	(T)
bc, bf	-.71	(C)
ce, ef	-.71	(C)

Calculate the composite relative deflections.

$$\delta_{cf①} = \sum_{i=1}^{10} \frac{P_i p_i l_i}{A_i E_i} = \frac{20}{AE}\Big[0 + 0 + (-23.33)(-.71)(1) + (-4.71)(1)(\sqrt{2})$$
$$+ (10.00)(-.71)(1) + 0 + (23.33)(-.71)(1)$$
$$+ 0 + 0 + (6.67)(-.71)(1) = -\frac{369.93}{AE}$$

$$\delta_{cf②} = \sum_{1}^{10} \frac{Q_i p_i l_i}{A_i E_i} = \frac{Q}{AE} p_i^2 l_i$$
$$= \frac{20Q}{AE} = \left[2(1)^2(\sqrt{2}) + 4(-.71)^2(1)\right] = \frac{96.90\,Q}{AE}$$

Since $\Sigma \delta_{cf} = 0$ we get

$$-369.93 + 96.90\,Q = 0 \quad \text{or} \quad Q = +3.82^{K}\,(T)$$

With Q found we can now compute the Q_i values and thus find the actual internal forces $(P_i + Q_i)$ and the corresponding support reactions.

Member	Actual Force		Actual Reaction
ab	−9.43	(C)	AY = 6.67 (↑)
af	+6.67	(T)	DX = 20.00 (←)
bc	−26.03	(C)	DY = 23.33 (↑)
be	−.89	(C)	
bf	+7.30	(T)	
cd	−32.99	(C)	
ce	+20.63	(T)	
cf	+3.82	(T)	
de	+3.33	(T)	
ef	+3.97	(T)	

For trusses internally and externally indeterminate, a combination of the above processes must be used.

If deflections of indeterminate trusses are sought, virtual work becomes quite involved. For instance, say we are looking for the vertical deflection at point c of the truss in Example 4. First, the internal forces for the indeterminate actual truss are found—as we did in Example 4. This involves two real and one virtual truss. To determine δ_{cy} of this indeterminate truss, we put a unit load at point c of the indeterminate truss. But to find the internal forces for this *indeterminate* virtual truss we must use superposition and another (this time, determinate) virtual truss. This is a major "bookkeeping" problem; hence, computer methods are called for.

We must also keep in mind that analysis is rarely an end in itself. Analysis in engineering is the groundwork upon which design takes place. In particular, truss analysis is used to find the force in each member so that the member itself can be designed or sized. The cross-sectional area of the member is often chosen on the basis of allowable tensile stress or allowable buckling stress; these values of allowable stresses are generally dictated by codes. Therefore the cross-sectional area is found by

$$A = \frac{P}{\sigma_{\text{allowable}}}.$$

Here P is the force in the member found by analysis. For statically indeterminate trusses we encounter an interesting dilemma. As we've seen, statically indeterminate problems require equations of compatible deflections. But deflection values are dependent upon cross-sectional areas.

As previously indicated, in design, the areas are chosen after the internal forces are determined. That is, to find the internal forces the areas must be known; to find the areas, however, the internal forces must be known. Result: The chicken and egg problem. What can be done? Choose initial areas based on, say, the forces in a determinate truss similar to the indeterminate one. Using these areas virtual work will give the internal forces in the statically indeterminate truss. On these forces, revised areas are chosen for the members. Using these revised areas, go through another virtual work cycle, and so on until the virtual work process gives consistent values for the redundant.

VIRTUAL WORK: BEAMS AND FRAMES

To reiterate Castigliano's First Theorem: For a linearly elastic body in equilibrium the rate of change of the work with respect to any applied load (force or moment) yields the deformation (deflection or rotation) at the point of application of that load in the direction of the load.

For both beams and frames we found (Equations 6 and 7) that the work function is

$$W = \frac{1}{2}\int_0^{\ell} \frac{M^2\,dx}{EI}.$$

Therefore, to find, say, the deflection at point b, take the derivative of W with respect to the load P_b (zero or nonzero):

$$\delta_b = \frac{1}{2}\frac{\partial}{\partial P_b}\int_0^{\ell} \frac{M^2\,dx}{EI} = \int_0^{\ell} \frac{M}{EI}\frac{\partial M}{\partial P_b}\,dx.$$

The term $\partial M/\partial P_b$ represents the moment due to a unit load at point b—call this the virtual moment m. So we see for beams and frames a parallel expression to truss deflection,

$$\delta_b = \int_0^{\ell} \frac{Mm}{EI}\,dx, \tag{9}$$

where M is the moment for the given beam and m is the moment for the virtual beam, consisting of only unit force at point b. If the rotation at point b is required then m would be for a unit applied moment at point b.

We must be careful about limits of integration. In general, both M and m usually are made up of several expressions. Each expression is valid only over a portion of the beam. In integrating, care must be taken to insure that the expression for M is valid over the same interval as m. An example will make this point clear.

EXAMPLE #6: VIRTUAL WORK: BEAM DEFLECTION

Given: Beam as shown
E and I are constant

Find: The vertical deflection at point b for the cantilever beam.

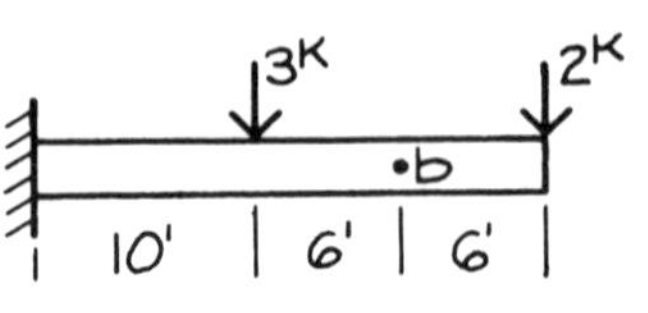

For the given beam we see two moment expressions must be written.

For the interval from the wall to the 3^K load we have:

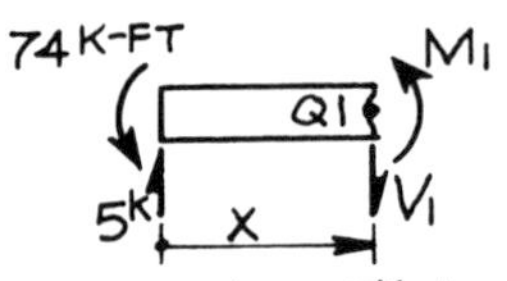

$\Sigma M_{Q1} = 0$ we get

$M_1 = 5X - 74$

$0 \leq X \leq 10'$

From the 3^K load to the beam end we have:

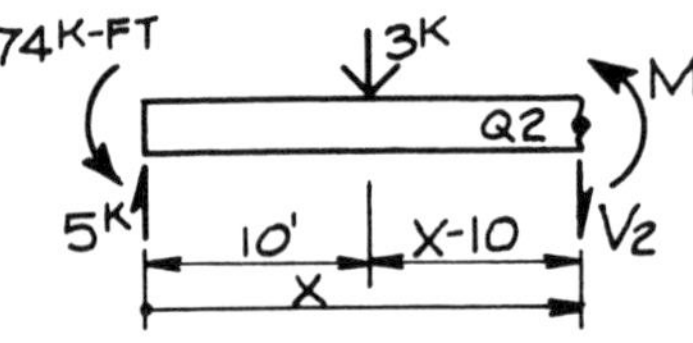

$\Sigma M_{Q2} = 0$ we get

$M_2 = 2X - 44$

$10' \leq X \leq 22'$

To find the deflection at point b we form a virtual beam as shown.

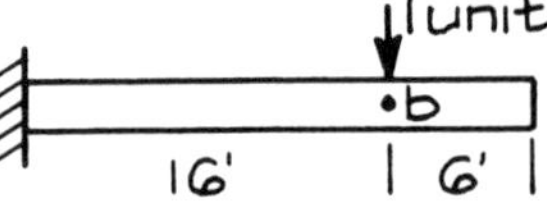

Virtual beam for δ_{bY}

Here two moment expressions are needed.

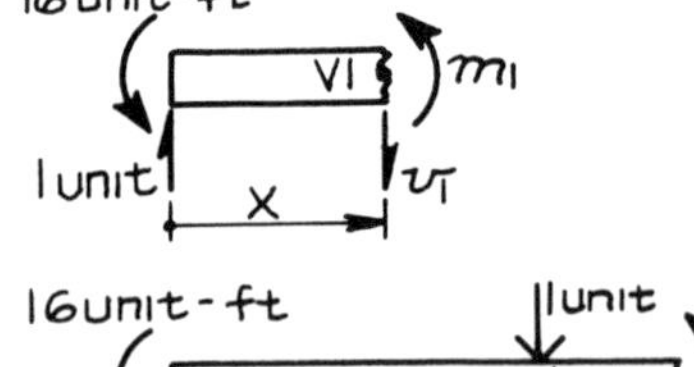

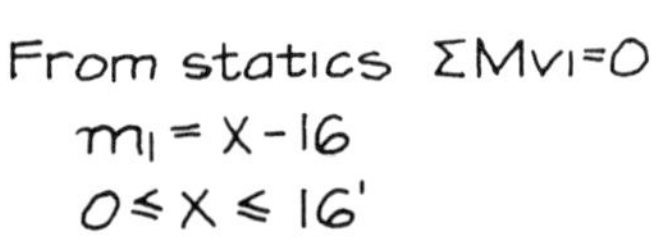

From statics $\Sigma M_{V1} = 0$

$m_1 = X - 16$

$0 \leq X \leq 16'$

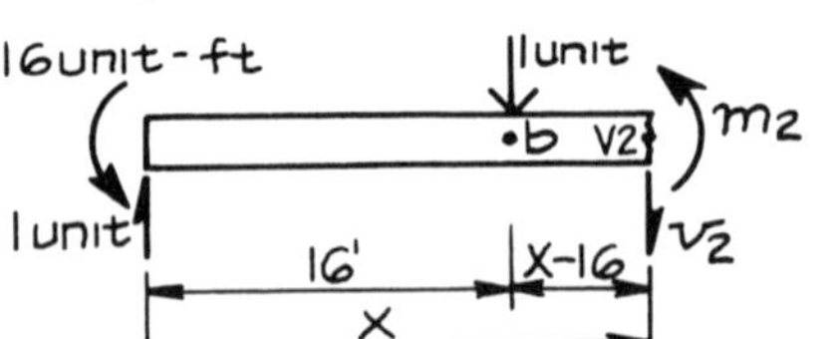

$\Sigma M_{V2} = 0$

$m_2 = 0$

$16' \leq X \leq 22'$

Now multiply the corresponding expressions to form δ_{bY}

$$\delta_{bY} = \frac{1}{EI}\int_0^{22'} Mm\,dx = \frac{1}{EI}\left[\int_0^{10} M_1 m_1\,dx + \int_{10'}^{16'} M_2 m_1\,dx + \int_{16'}^{22'} M_2 m_2\,dx\right]$$

$$\delta_{bY} = \frac{1}{EI}\left[\int_0^{10'} (5X-74)(X-16)\,dx + \int_{10'}^{16'} (2X-44)(X-16)\,dx + 0\right]$$

Carrying through the integration we get

$$\delta_{bY} = +\frac{6166.6}{EI}\,(\text{K-FT}^3)\ (\downarrow)$$

The positive sign means the deflection is consistent with the vertical virtual load, ie, downward.

Shown below are the intervals of validity of the moment expression; thus, we can easily see that three integrals are called for.

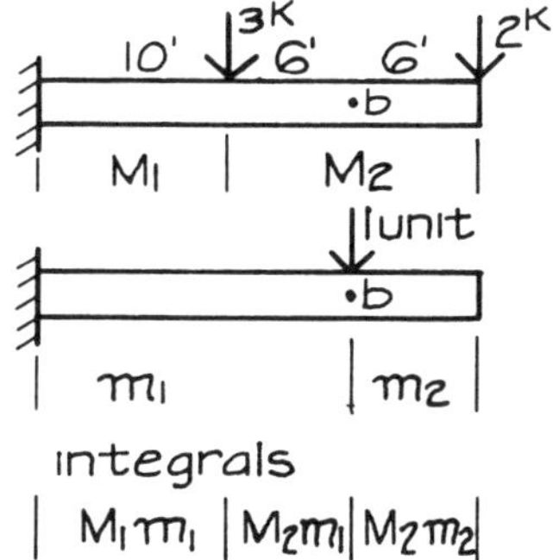

As one can see, the more concentrated loads on the given beam, the more integrals are required. Therefore, there is no advantage of virtual work for beams over moment-area. Our recommendation is to use moment-area for beam deflections. Thus, we shall not belabor virtual work for beams; we shall not present the use of virtual work for indeterminate beams even though this is a natural extension.

For frames we'll see that virtual work is just as easy to use as moment-area. Moment-area is generally awkward since usually three reference points are needed. Virtual work calculation can be extensive particularly for many concentrated loads.

The basis of virtual work for frames is the same as for beams, Equation (9):

$$\delta_b = \int_0^{\ell} \frac{Mm}{EI}\, dx.$$

Here integration is over the entire frame.

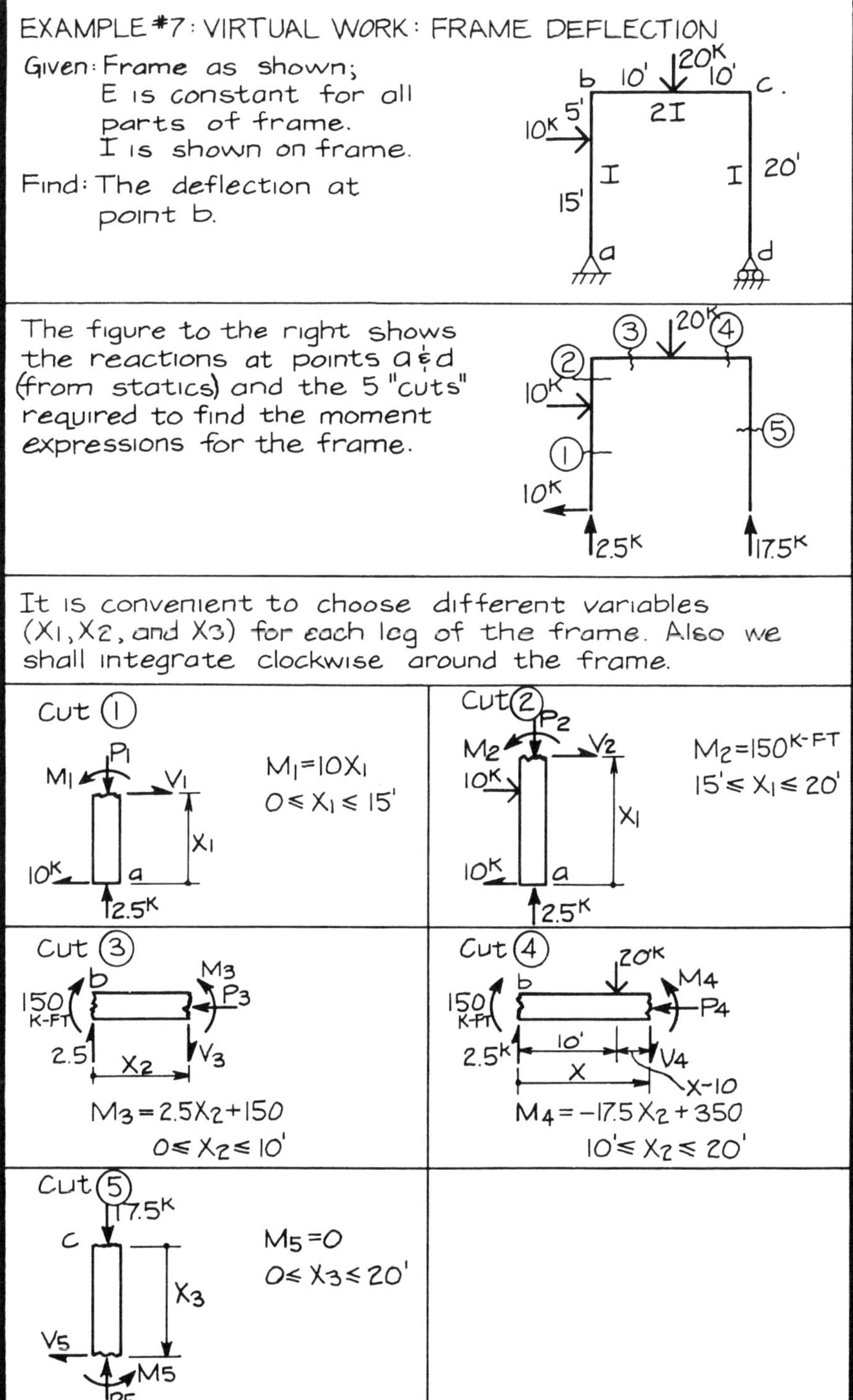
EXAMPLE #7: VIRTUAL WORK: FRAME DEFLECTION
Given: Frame as shown;
E is constant for all parts of frame.
I is shown on frame.
Find: The deflection at point b.
b 10' 20K 10' c
5' 10K 2I
I I 20'
15'
a d
The figure to the right shows the reactions at points a & d (from statics) and the 5 "cuts" required to find the moment expressions for the frame.
10K 2.5K 17.5K
It is convenient to choose different variables (X1, X2, and X3) for each leg of the frame. Also we shall integrate clockwise around the frame.
Cut 1
M1 = 10X1
0 ≤ X1 ≤ 15'
Cut 2
M2 = 150K-FT
15' ≤ X1 ≤ 20'
Cut 3
M3 = 2.5X2 + 150
0 ≤ X2 ≤ 10'
Cut 4
M4 = −17.5X2 + 350
10' ≤ X2 ≤ 20'
Cut 5
M5 = 0
0 ≤ X3 ≤ 20'

To find the deflection at point b we place a unit horizontal force at point b. (We should keep in mind that for frames axial changes in length are neglected; therefore, point b is assumed to deflect only horizontally).

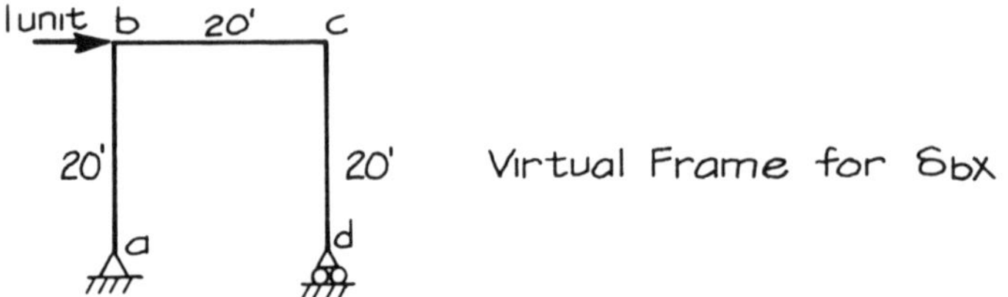

Virtual Frame for δ_{bx}

To find moment expressions for the entire frame we need only three cuts, one in each leg of the frame.

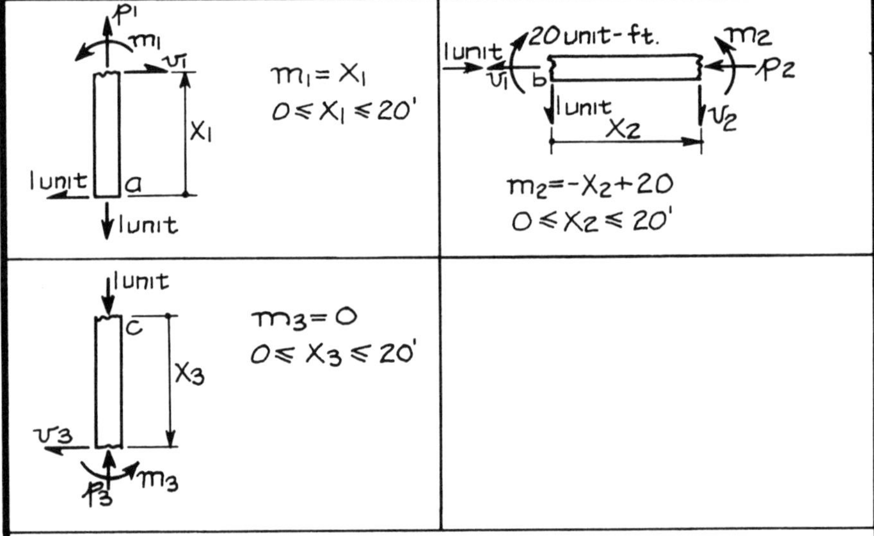

$m_1 = X_1$

$0 \leq X_1 \leq 20'$

$m_2 = -X_2 + 20$

$0 \leq X_2 \leq 20'$

$m_3 = 0$

$0 \leq X_3 \leq 20'$

Now form deflection expression

$$\delta_{bx} = \int \frac{Mm}{EI} dx = \left[\int_0^{15} \frac{M_1 m_1}{EI} dx_1 + \int_{15}^{20} \frac{M_2 m_1}{EI} dx_1 + \int_0^{10} \frac{M_3 m_2}{2EI} dx_2 + \int_{10}^{20} \frac{M_4 m_2}{2EI} dx_2 + \int_0^{20} \frac{M_5 m_3}{EI} dx_3 \right]$$

$$\delta_{bx} = \frac{1}{EI} \left[\int_0^{15} 10X_1^2 dx_1 + \int_{15} 150X_1 dx_1 + \int_0^{10} \frac{(2.5X_2 + 150)(-X_2 + 20)}{2} dx_2 + \int_{10}^{20} \frac{(-17.5X_2 + 350)(-X_2 + 20)}{2} dx_2 + 0 \right] = \frac{39375}{EI} \left(\text{K-FT}^3\right) (\longrightarrow)$$

EXAMPLE #8: VIRTUAL WORK: FRAME ROTATION

Given: Same frame as in example #7

Find: The rotation at point b

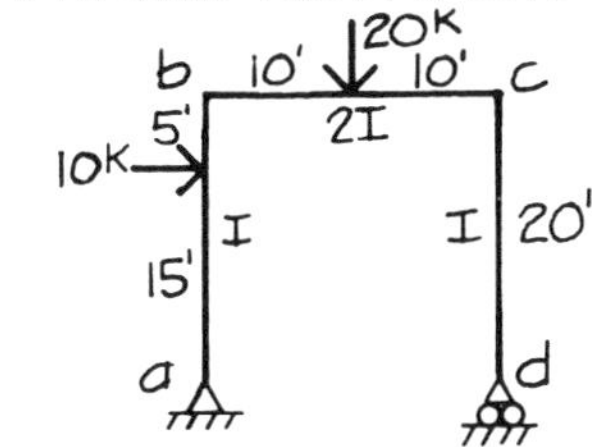

For rotation at point b we use the expression:

$$\theta_b = \int \frac{M\hat{m}}{EI}\,dx$$

Where M takes on exactly the same values as in example #7.

Here $\hat{m}$ is found for the virtual frame with a unit applied moment at point b.

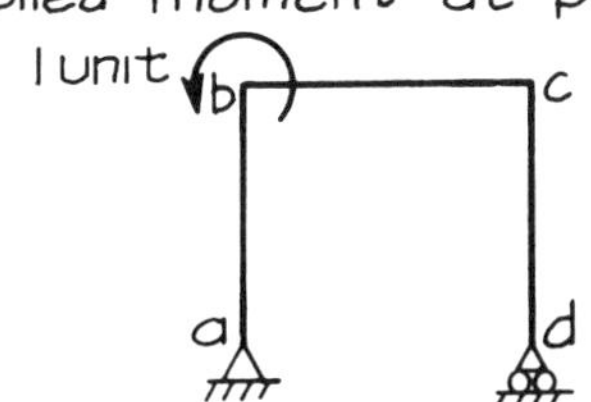

Virtual Frame for θ_b

Again three cuts are needed.

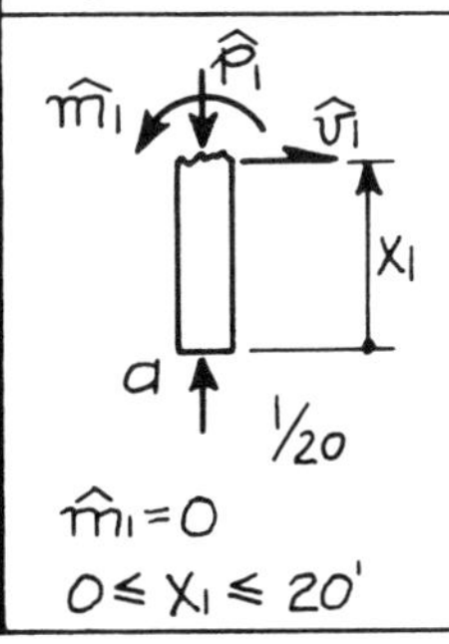

$\hat{m}_1 = 0$

$0 \leq X_1 \leq 20'$

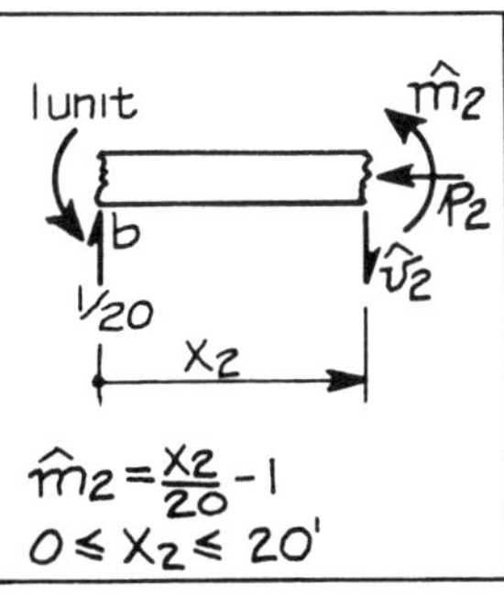

$\hat{m}_2 = \frac{X_2}{20} - 1$

$0 \leq X_2 \leq 20'$

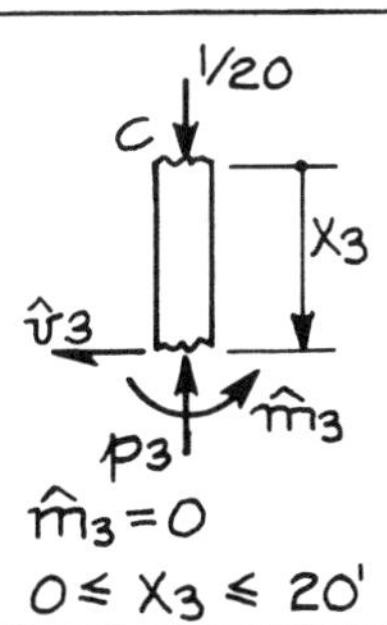

$\hat{m}_3 = 0$

$0 \leq X_3 \leq 20'$

Now the calculation of θ_b.

$$\theta_b = \int \frac{M\hat{m}\,dx}{EI} = \int_0^{15} \frac{M_1\hat{m}_1\,dx_1}{EI} + \int_{15}^{20} \frac{M_2\hat{m}_1\,dx_1}{EI} + \int_0^{10} \frac{M_3\hat{m}_2\,dx_2}{2EI}$$

$$+ \int_{10}^{20} \frac{M_4\hat{m}_2\,dx_2}{2EI} + \int_0^{20} \frac{M_5\hat{m}_3\,dx_3}{EI}$$

$$= \frac{1}{EI}\left[0 + 0 + \int_0^{10} \frac{(2.5x_2+150)(x_2/20-1)\,dx_2}{2}\right.$$

$$\left. + \int_{10}^{20} \frac{(17.5x_2+350)(x_2/20-1)}{2} + 0\right]$$

$$= \frac{-749.8}{EI} \text{ (K-FT}^2\text{)} \quad (\curvearrowright)$$

Here the negative sign means the rotation is opposite to the virtual moment.

Below is shown, greatly exaggerated pictorally, the results of examples #7 and #8.

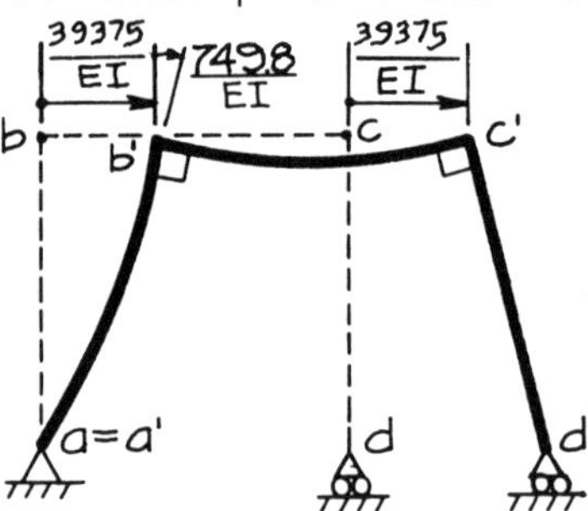

A few items to note about the deformed frame in Examples 7 and 8:

1) Since the frame is assumed to be rigidly joined at *b* and *c* these angles remain 90°.
2) Since axial elongations are neglected, point *c* moves the same amount as point *b*; also, no vertical deflections are found at *b* and *c*.
3) Member *cd* remains straight since no bending moments exist in the member due to the given loads, i.e., $M_5 = 0$.
4) Looking from the inside, members *ab* and *bc* are bent "positively," i.e., M_1, M_2, M_3 and M_4 are all positive bending moments.
5) For completeness we should find the movement of point *d*, i.e., δd_x. We leave this as an exercise.

Again, this process is easily extended to indeterminate frames but we prefer another method for such frames. The major complications of virtual work for indeterminate frames arise when the frame is indeterminate to a high degree while the method we shall investigate later—slope-deflection—avoids these complications. Slope-deflection is also easily applied to multi-indeterminate beams.

OTHER METHODS

For truss deflections there are available the graphical method of Williot and the Williot-Mohr Extension (Reference 1).

For indeterminant structures the energy method (Castigliano's Second Theorem) called least work is a variation of virtual work (Reference 5).

Virtual work can be extended to plastic analysis (Reference 3).

PROBLEMS

Problems 1–5

Find the vertical and horizontal deflection of point *a*. *E* same for all members. Cross-sectional areas: "A" for all tensile members and "2A" for all compressive members.

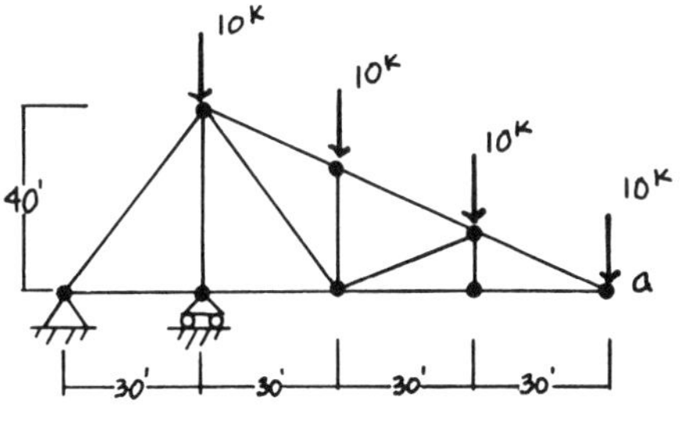

truss 1

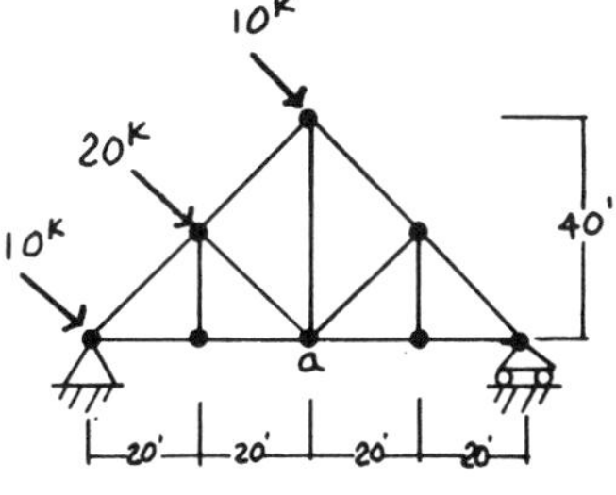

truss 2

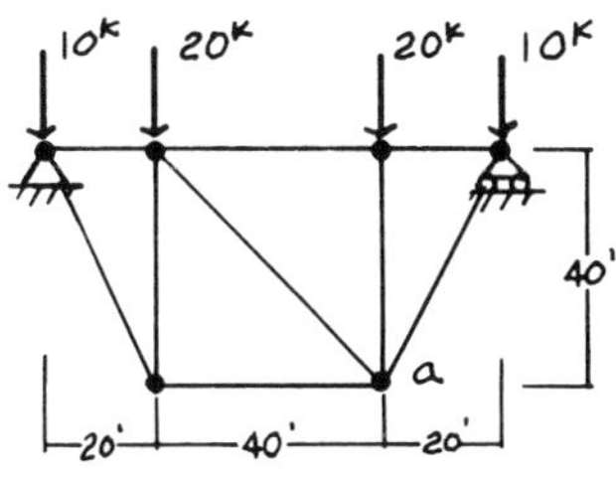

truss 3

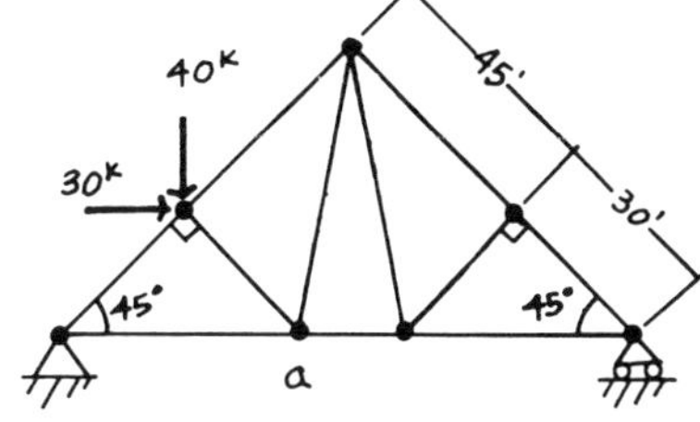

truss 4

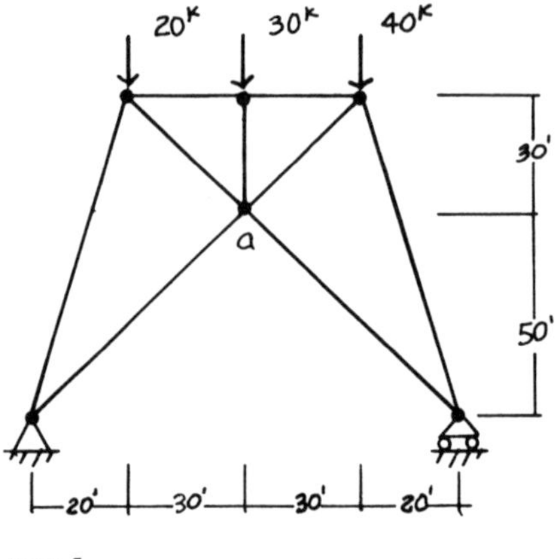

truss 5

Problems 6–10

Find the vertical deflections of joints *d* and *e*. $E = 29{,}000$ ksi for all members. Cross-sectional areas are to be found using the following design conditions: use standard steel pipe sections (Fig. 2.17).

For tension members:

Allowable stress is 21.6 ksi

For compression members (two conditions must be met):

1) Allowable load $= .52\ \pi^2\ EI/L^2$
2) Allowable stress $= 18.7$ ksi

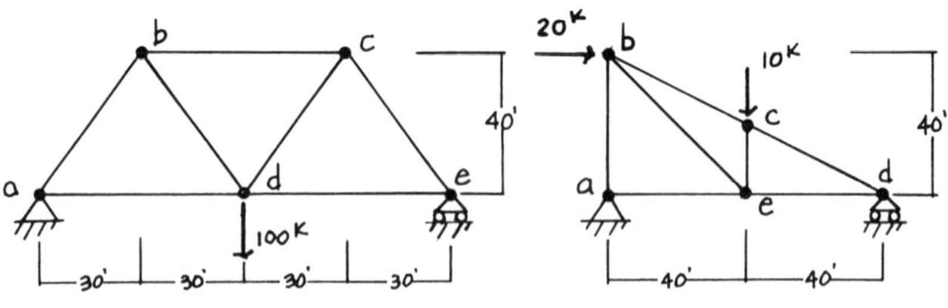

truss 6

truss 7

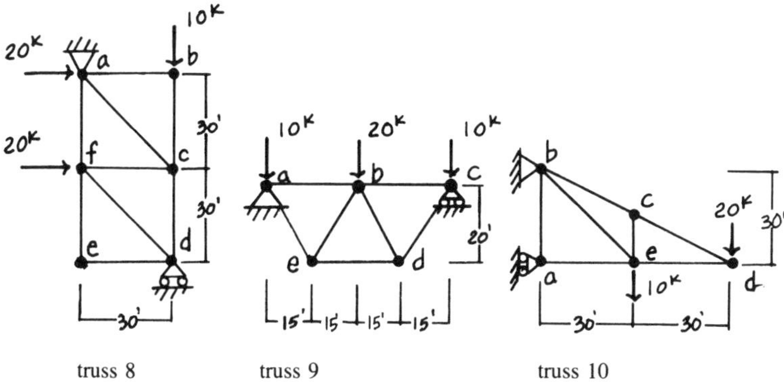

truss 8 truss 9 truss 10

Problems 11–20

Find the internal forces in the truss. *E* same for all members. Slashes indicate areas, i.e., –+– = A; –++– = 2A, etc.

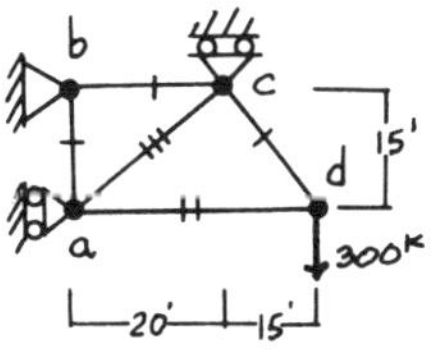

truss 11

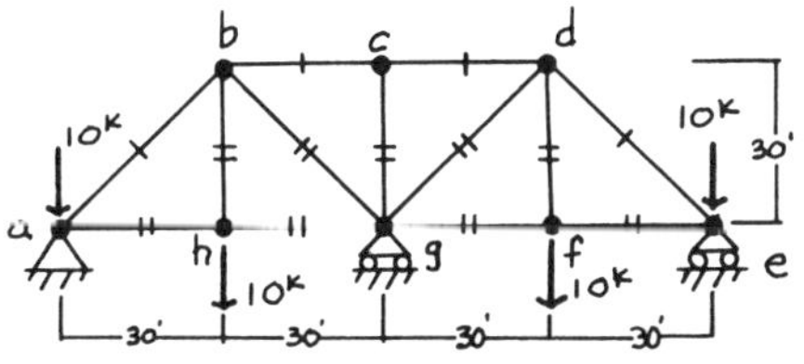

truss 12

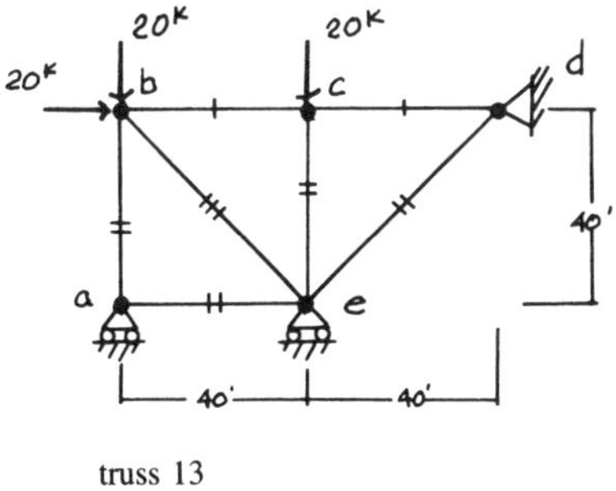

truss 13

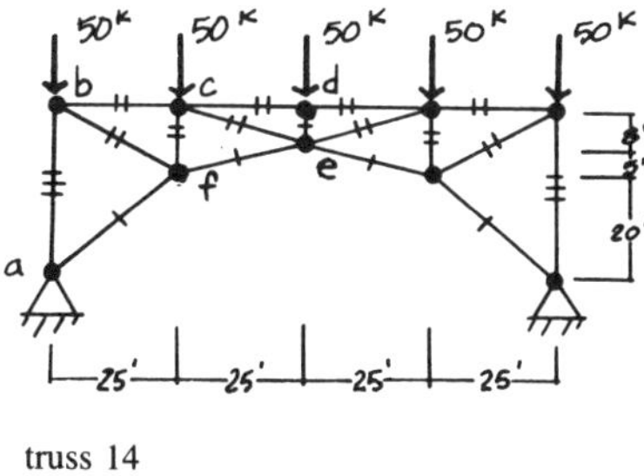

truss 14

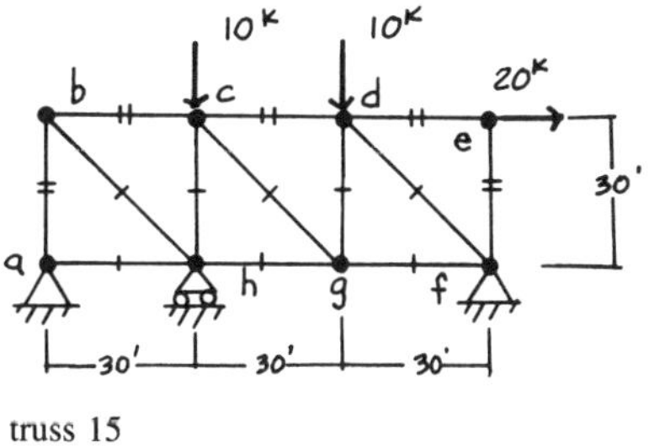

truss 15

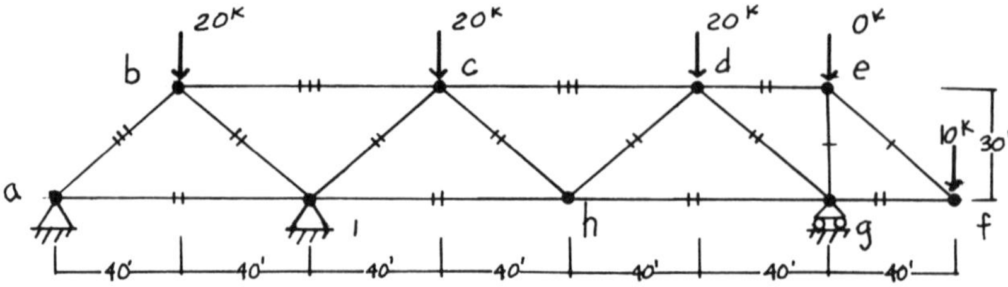

truss 16

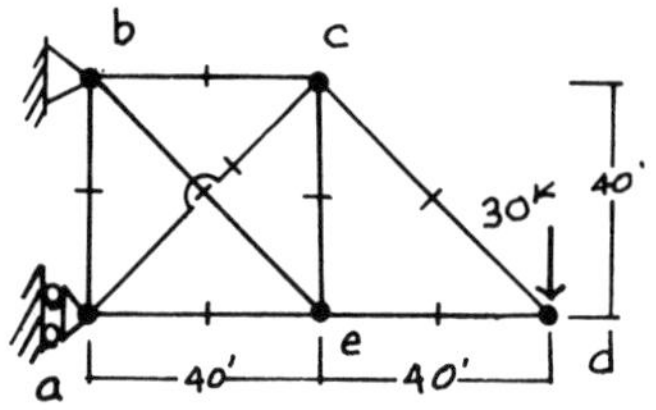

truss 17

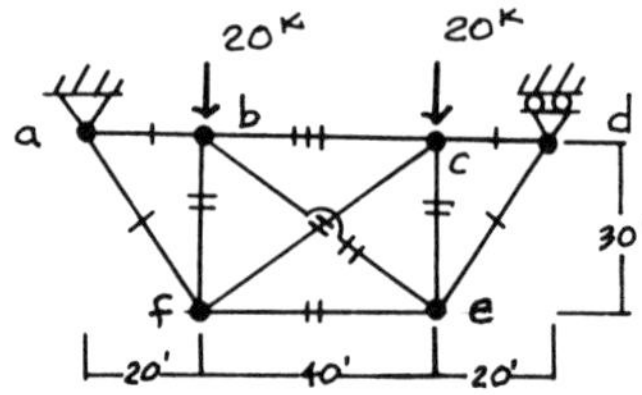

truss 18

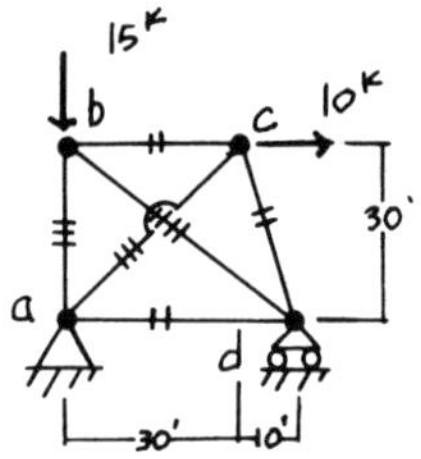

truss 19

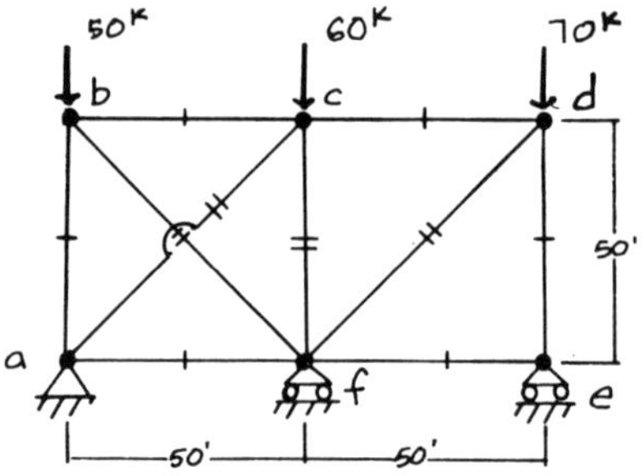

truss 20

Problems 21–26

Find the deflection at point *b* for each beam. *E* is constant; *I* values as shown.

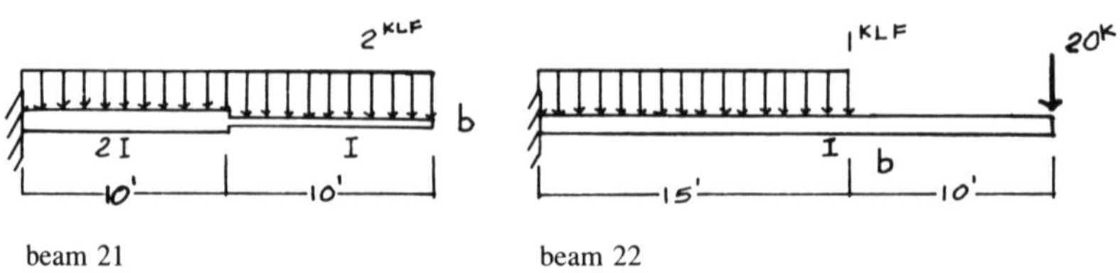

beam 21

beam 22

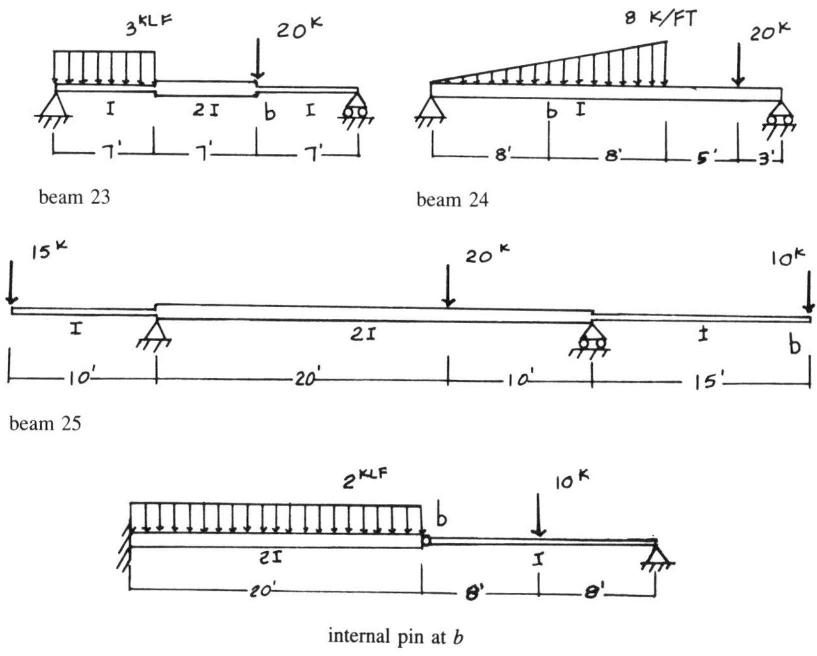

beam 23

beam 24

beam 25

internal pin at *b*

beam 26

Problems 27, 28, and 29

Find the rotation at point *b* for beams 21, 22, and 26.

Problems 30, 31, and 32

Find the rotation at the left support for beams 23, 24, and 25.

Problems 33, 34, and 35(a–c)

For beams 22, 24, and 25 find the maximum bending stresses (tensile and compressive) and the maximum shear stress for the cross-sections shown.

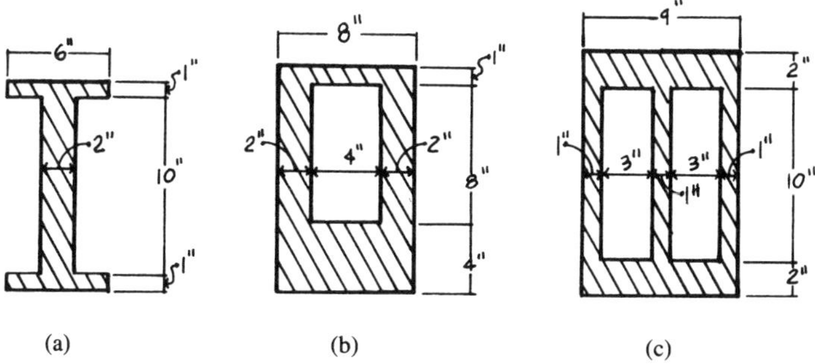

(a)

(b)

(c)

Problems 36 and 37

Find: (a) deflection at point *d*; (b) rotation at point *b*. *E* is constant.

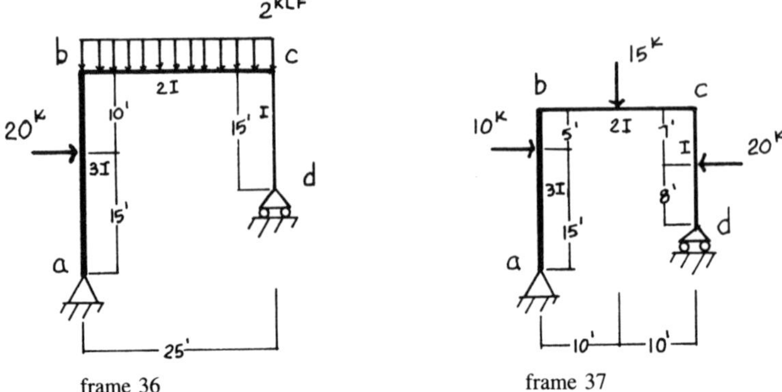

frame 36 frame 37

Problems 38 and 39

Find: (a) deflections at point *b*; (b) rotation at point *b*. *E* is constant.

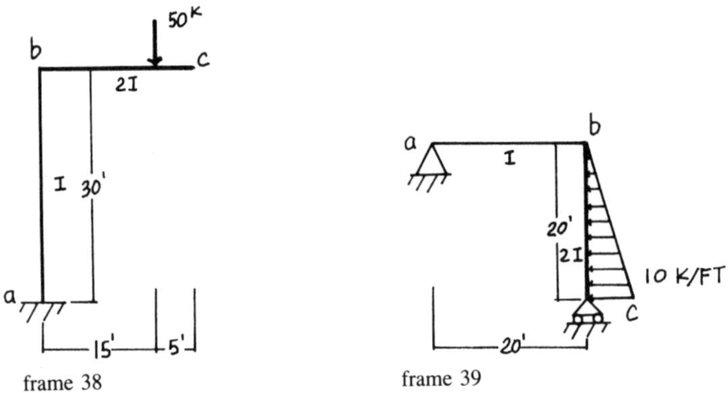

frame 38 frame 39

Problems 40 and 41

Find: (a) deflections at point *c*; (b) rotations at point *c*. *E* is constant.

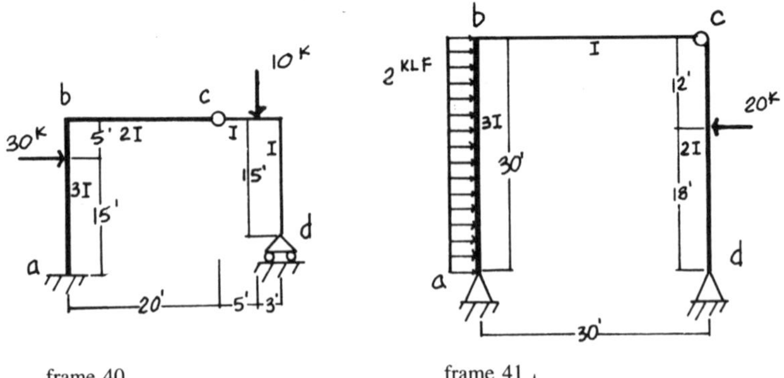

frame 40 frame 41

7

SLOPE-DEFLECTION

INTRODUCTION

The method called slope-deflection is a straightforward approach to the solution of statically indeterminate beams and frames. Basically, reactions are expressed in terms of displacements (joint rotations and chord rotations); these displacements are determined by the resolution of simultaneous equations formed from equilibrium conditions.

Slope-deflection is the first of several methods collectively called stiffness methods. The constituent members are initially treated as stiffer, i.e., more highly constrained, than they really are. The effects of these extra constraints are considered as the members are allowed to return to the real state.

The assumptions of slope-deflection are the same as those of moment-area and virtual work: elastic response and the consideration of only bending moments toward beam and frame deformation. This latter assumption means that the effects of shear on deflection are neglected and that axial deformations are taken to be zero. Further, we shall consider only prismatic members (EI constant) although the method can be extended to nonprismatic members (Reference 3).

Sign Conventions

The bending moment sign convention becomes cumbersome for complex frames; what the "bottom" of the member is can be ambiguous. Consequently we shall adopt a new convention, unambiguous for such cases, called the "fixed-end moment" sign convention. Simply stated, positive fixed-end moments act in a clockwise direction on the member (and counterclockwise on the joint). The accompanying sketches compare the two conventions. For angular deformation and chord rotation we will take clockwise rotations as positive.

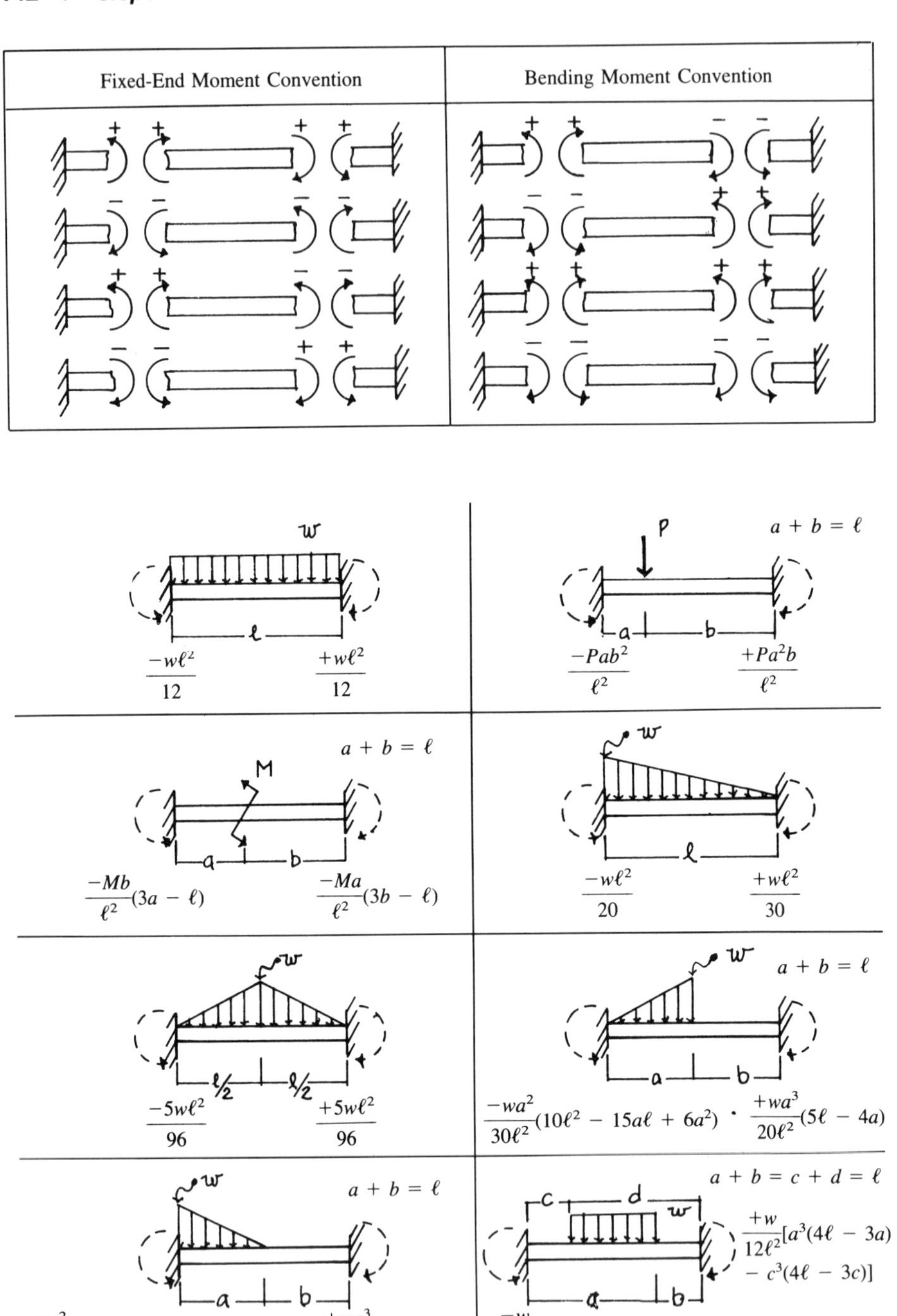

Chart 1 Fixed-end Moments (Prismatic Beams)

METHOD RATIONALE

Consider member *ab* of the frame shown in Figure 7.1a. Due to the applied loads on the frame, the joints of members *ab* will experience: 1) angular rotations at both joints (θ_a and θ_b, Figure 7.1c); 2) vertical (perpendicular to member *ab*) displacements at joints *a* and *b*, giving rise to a chord rotation, $\rho = (\delta_a - \delta_b)/\ell$. (Note: No relative horizontal displacement of *ab* takes place since members are considered not to change length.)

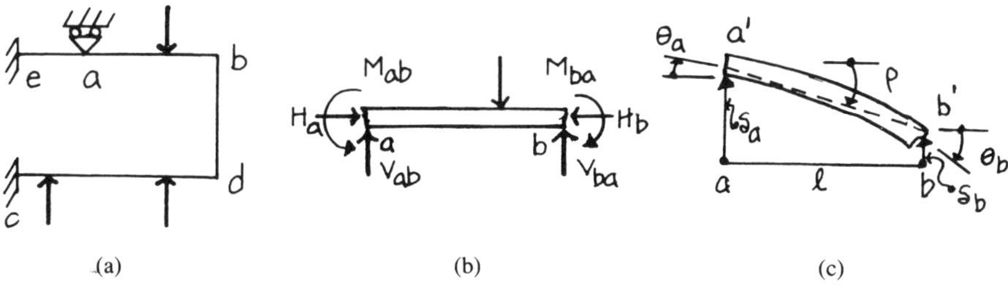

(a) (b) (c)

Figure 7.1

Conceptually we can consider these effects separately. First, consider the moments due to the applied load with both ends fixed; these moments are called "fixed-end moments" (Figure 7.2b). Then the effects of allowing each joint to rotate are considered (Figure 7.2c, d); finally, the chord rotation is allowed to occur (Figure 7.2e).

Next we shall derive the moment expressions for the respective stages.

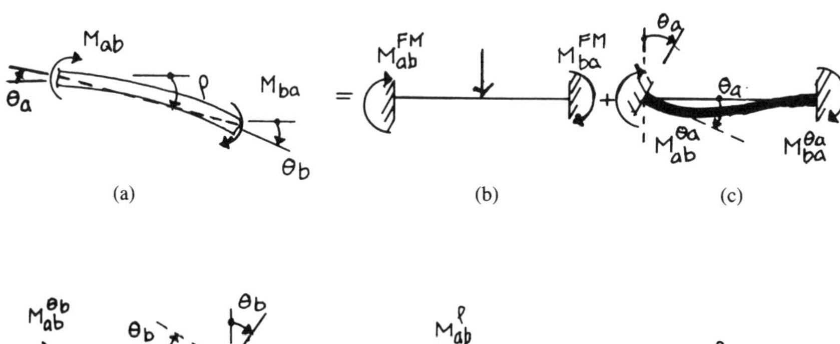

(a) (b) (c)

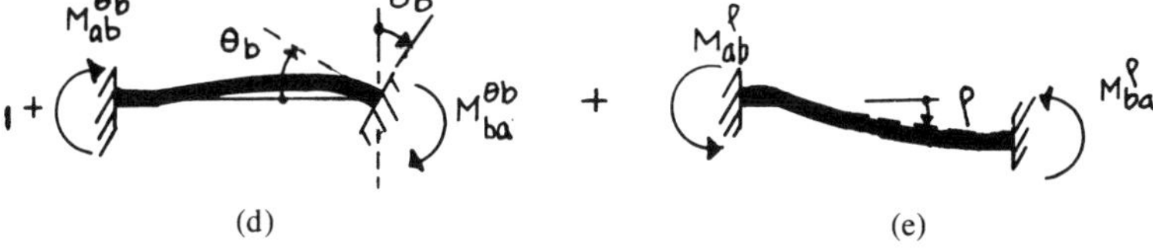

(d) (e)

Figure 7.2

Fixed-End Moments

Evaluating fixed-end moments, in reality, involves solving for the moment reactions of a 2° indeterminate beam. If the beam is, say, three spans and 4° indeterminate, we have to solve three 2° indeterminate beams to find the fixed-end moments (just one part of the slope-deflection process). Where is the saving in time and effort? By charts similar to Chart 1. Since all fixed-end moment problems involve, by definition, one span with fixed-ends, the common loading situations are easily tabulated.

To better understand Chart 1, find the fixed-end moments for a sample case. Consider the beam shown in Figure 7.3 with the known load, P. Use superposition with the moments at a and b as redundants and proceed by using moment-area and the bending moment sign convention to find these redundants. Imposing the compatibility relations $\Sigma\theta_a = 0$ and $\Sigma\theta_b = 0$ we get

$$\sum\theta_a = -\sum\frac{t_{a/b}}{\ell} = -\frac{P\ell^2}{16EI} - \frac{(M_{ab}^{FM})\ell}{3EI} + \frac{(M_{ba}^{FM})\ell}{6EI} = 0. \qquad \text{(a)}$$

$$\sum\theta_b = \sum\frac{t_{b/a}}{\ell} = +\frac{P\ell^2}{16EI} + \frac{(M_{ab}^{FM})\ell}{6EI} - \frac{(M_{ba}^{FM})\ell}{3EI} = 0. \qquad \text{(b)}$$

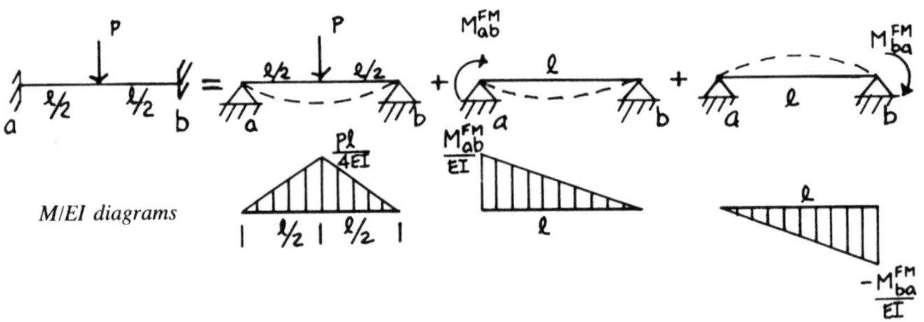

Figure 7.3

Solving (a) and (b) simultaneously, we find

$$M_{ab}^{FM} = -\frac{P\ell}{8} \text{ or C. C. W.}$$

$$M_{ba}^{FM} = +\frac{P\ell}{8} \text{ or C. W.}$$

Figure 7.4

Joint Rotations

Consider a beam, as shown in Figure. 7.5, where the left wall is forced to rotate through an angle, θ_a. We wish to find, in terms of the resulting angular rotation θ_a, the applied moment $M_{ab}^{\theta_a}$ causing the rotation and the reaction $M_{ba}^{\theta_a}$. In so doing we encounter two concepts of general use in all stiffness methods, the definition of "stiffness" and the "carry-over factor."

Stiffness

In general, stiffness is defined as the ratio of the load at a point to the corresponding deformation at that point. For our case, the beam stiffness, K_{ab}, is defined as $M_{ab}^{\theta_a}/\theta_a$.

Figure 7.5

Carry-over Factor

Carry-over factor for our beam is simply that fraction of the applied moment, $M_{ab}^{\theta_a}$, that is distributed to the far end, wall b. Thus CF $|_{ab} = M_{ba}^{\theta_a}/M_{ab}^{\theta_a}$ or

$$M_{ab}^{\theta_a}\ (\text{CF}\ |_{ab}) = M_{ba}^{\theta_a}.$$

Back to our problem of finding $M_{ab}^{\theta_a}$ and $M_{ba}^{\theta_a}$ in terms of θ_a: Since "wall" a rotates it can be modeled as a pin. Thus we choose $M_{ab}^{\theta_a}$ and V_a as redundants (Figure 7.6). Imposing $\Sigma\ \delta_a = 0$ and $\Sigma\ \theta_a = -\theta_a$ (the given rotation), we use moment-area (in which clockwise angles were taken as negative) and the bending moment sign convention

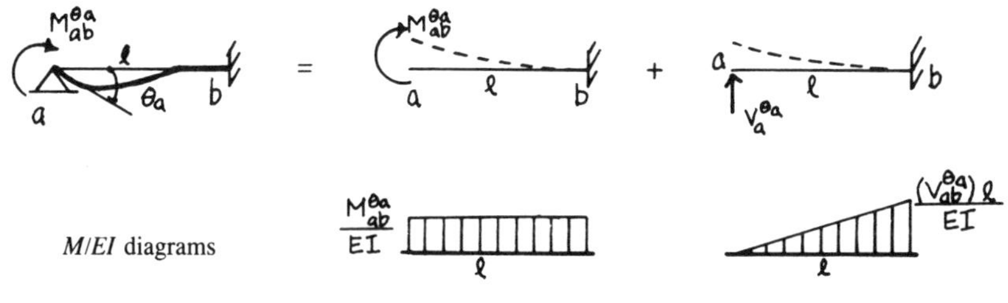

M/*EI* diagrams

Figure 7.6

$$\sum \delta_a = 0 = \sum t_{b/a}; \qquad \frac{(M_{ab}^{\theta_a})\ell^2}{2EI} + \frac{(V_a^{\theta_a})\ell^3}{3EI} = 0. \tag{c}$$

$$\sum \theta_a = \theta_a = -\sum \left(\text{area of } \frac{M}{EI}\right)_a^b; \qquad -\frac{(M_{ab}^{\theta_a})\ell}{EI} - \frac{(V_a^{\theta_a})\ell^2}{2EI} = -\theta_a. \tag{d}$$

Solving (c) and (d), we find $M_{ab}^{\theta_a} = +(4\theta_a EI/\ell)$, C. W. as assumed, and $V_a^{\theta_a} = -(6EI\theta_a/\ell^2)$; thus, from Figure 7.6, $M_{ba}^{\theta_a} = M_{ab}^{\theta_a} + (V_a^{\theta_a})\ell$. Then $M_{ba}^{\theta_a} = -(2EI\theta_a/\ell)$ (C. W.). Figure 7.7 sums up our results.

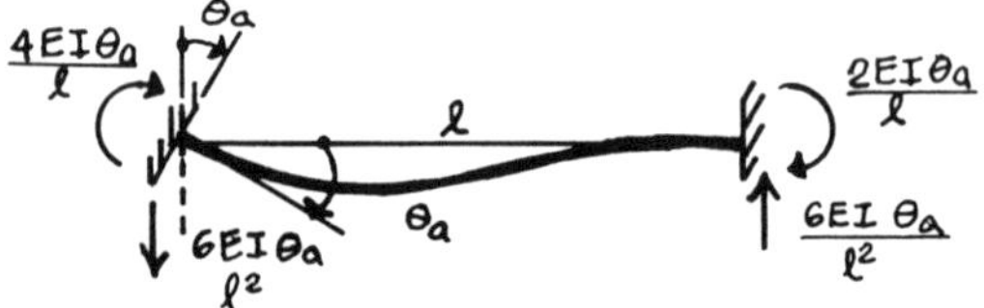

Figure 7.7

Using the fixed-end moment convention we see for prismatic beams

$$M_{ab}^{\theta_a} = +\frac{4EI\theta_a}{\ell} \quad \text{and} \tag{1}$$

$$M_{ba}^{\theta_a} = +\frac{2EI\theta_a}{\ell}. \tag{2}$$

Also the *stiffness* for a prismatic beam is found as:

$$K_{ab} = \frac{M_{ab}^{\theta_a}}{\theta_a} = \frac{4EI}{\ell}, \tag{3}$$

and the carry-over factor for a prismatic beam is found as:

$$\text{CF}|_{ab} = \frac{M_{ba}^{\theta_a}}{M_{ab}^{\theta_a}} = \frac{1}{2}. \tag{4}$$

It should be emphasized that, possibly contrary to intuition, when a clockwise moment is applied at pin a, the resulting moment at wall b is also clockwise. Flipping the beam around, we can easily determine the effects due to rotating "wall" b through an angle, θ_b.

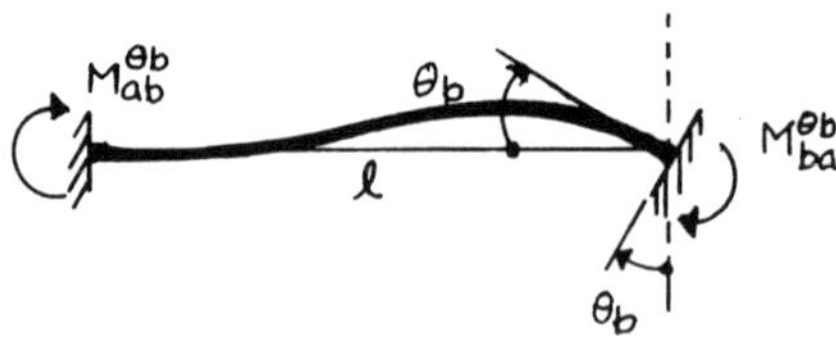

Figure 7.8

$$M_{ba}^{\theta_b} = +\frac{4EI\theta_b}{\ell} \quad \text{(C. W.)} \tag{5}$$

$$M_{ab}^{\theta_b} = +\frac{2EI\theta_b}{\ell} \quad \text{(C. W.)}. \tag{6}$$

Chord Rotations

Consider a beam fixed at both ends so that no angular rotation can take place (Figure 7.9). If one end is displaced vertically the beam will deform as shown; consequently, moments are generated throughout the beam. We are interested in only those moments at the ends. Using moment-area (and the bending moment sign convention) take M^{ρ}_{ba} and V^{ρ}_{b} as redundants. Moment-area implies $\Sigma\ \theta_b = 0$ and $\Sigma\ \delta_b = \rho\ell$, thus,

$$\sum \theta_b = 0 = \sum \left(\text{area of } \frac{M}{EI}\right)_a^b; \qquad \frac{M^{\rho}_{ba}\ell}{EI} + \frac{V^{\rho}_{b}\ell^2}{2EI} = 0, \tag{e}$$

$$\sum \delta_b = \rho\ell = -\sum t_{a/b}; \qquad \frac{M^{\rho}_{ba}\ell^2}{2EI} + \frac{V^{\rho}_{b}\ell^3}{3EI} = -\rho\ell. \tag{f}$$

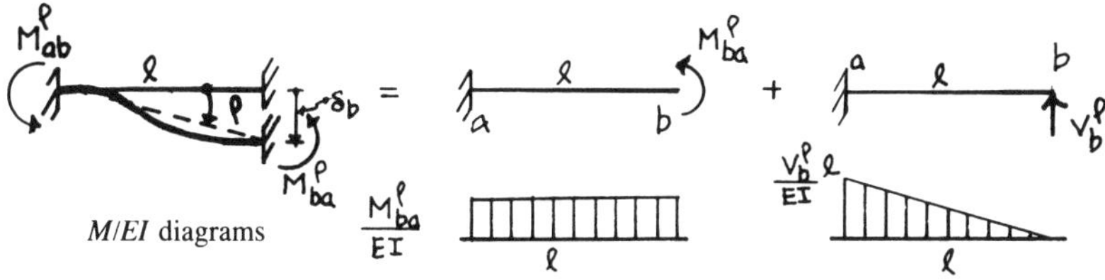

Figure 7.9

Solving (e) and (f) we find $M^{\rho}_{ba} = (+6\rho EI)/\ell$ (C. C. W.), and $V_b = (-12EI\rho)/\ell^2$; thus, from Figure 7.9, $M^{\rho}_{ab} = +M^{\rho}_{ba} + V^{\rho}_{b}\ell = (-6EI\rho)/\ell$ (C. C. W.). Reiterating with a fixed-end moment convention for prismatic beams:

$$M^{\rho}_{ba} = M^{\rho}_{ab} = \frac{-6EI\rho}{\ell} \text{ (C. C. W.)} \tag{7}$$

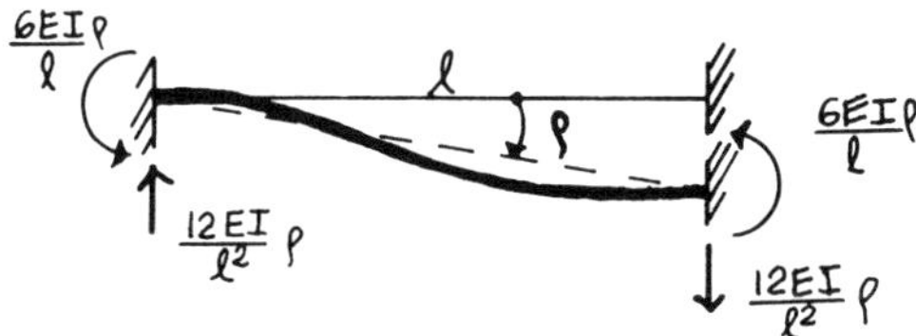

Figure 7.10

It should be noted that chord rotations occur when joint translations exist. These translations can be induced by frame sidesway (due to lateral loads, unsymmetrical vertical loads or lack of geometric frame symmetry), unequal support settlement or temperature changes. It is most important to realize when chord rotation may occur;

the example problems will help to clarify this point. Finally, putting the effects of load, joint rotation and chord rotation together, we find:

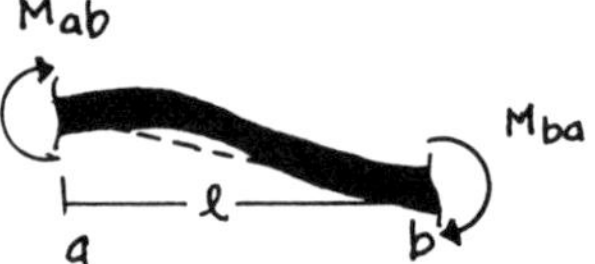

Figure 7.11

$$M_{ab} = M_{ab}^{FM} + M_{ab}^{\theta_a} + M_{ab}^{\theta_b} + M_{ab}^{\rho}$$

$$= M_{ab}^{FM} + \frac{4EI\theta_a}{\ell} + \frac{2EI\theta_b}{\ell} - \frac{6EI\rho}{\ell}$$

$$M_{ab} = M_{ab}^{FM} + \frac{EI}{\ell}\left[-6\rho + 4\theta_a + 2\theta_b\right], \qquad (8)$$

similarly,

$$M_{ba} = M_{ba}^{FM} + \frac{EI}{\ell}\left[-6\rho + 4\theta_b + 2\theta_a\right]. \qquad (9)$$

SIDESWAY

As previously mentioned, when a frame is subjected to wind loads or unsymmetric vertical loads, joint translation may take place. Of course all joints can't translate; at least one support must not move. In dealing with joint translation, two points should be kept in mind:

1) Axial change in member length is neglected,
2) The actual joint translations are extremely small compared to member lengths.

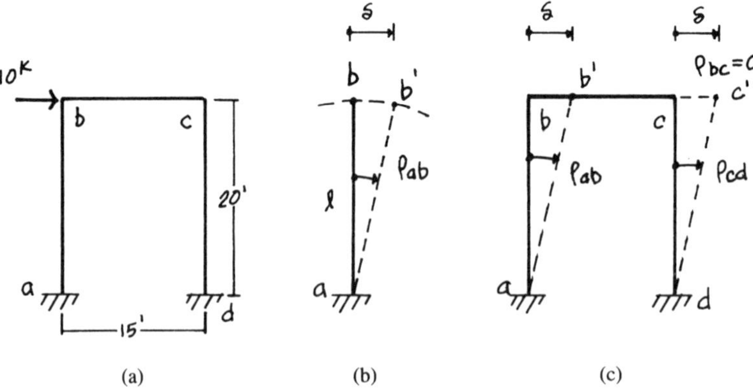

Figure 7.12

Consider the frame in Figure 7.12. Since a is fixed and member ab is considered not to change length, joint b moves in an arc to b' as shown in Figure 7.12b. However, the chord rotation ρ_{ab} is extremely small so that the arc length bb' is for all practical purposes the same as the horizontal translation δ. That is, $bb' = \rho_{ab}\ell$, the result being that b' is taken as being on the same line as bc in Figure 7.12c. Further, from point 1) we conclude that $bb' = cc' = \delta$. Finally, for this example we see $\rho_{ab} = \rho_{cd}$. (Both, we note, are positive, clockwise, rotations.)

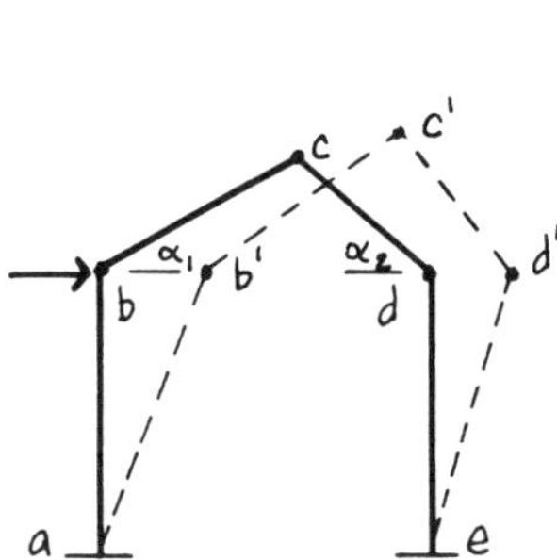

given frame and the final chord rotations (dotted)

Figure 7.13a

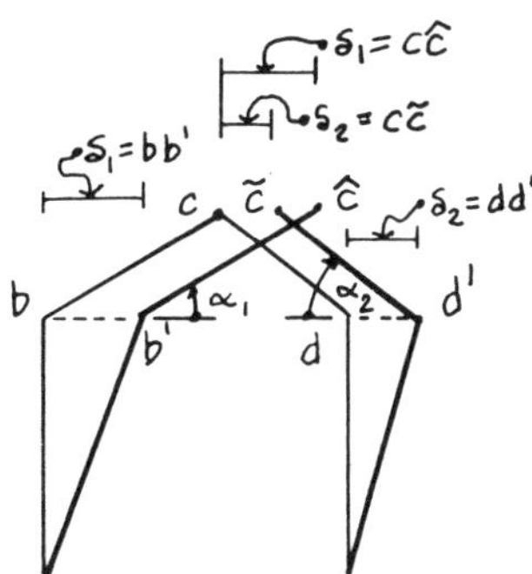

"cut" joint c; allow cb and cd to translate parallel to themselves amounts δ_1 and δ_2. (Note: their end points, $\tilde{c}$ and $\hat{c}$, are on the same horizontal line as point c.)

Figure 7.13b

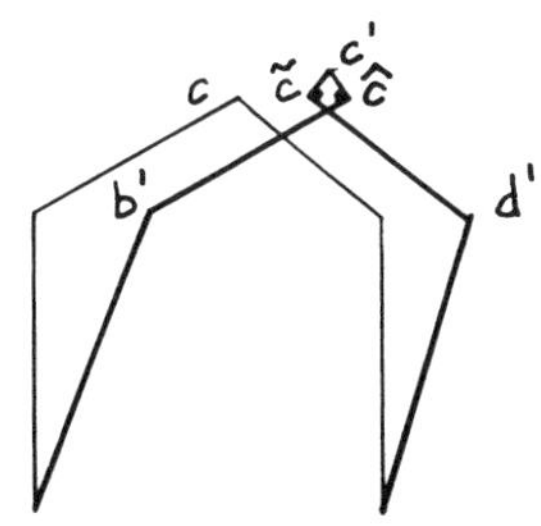

Since $\hat{c}$ and $\tilde{c}$ must really be the same point, c', we swing arcs (90° angles) to find their intersection, thus point c' is found.

Figure 7.13c

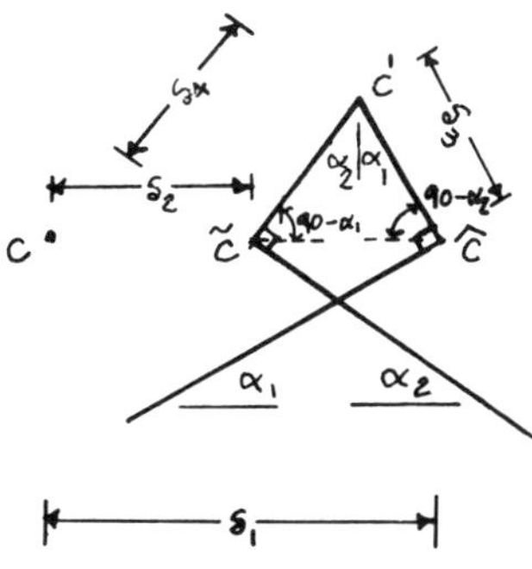

Blow-up of joint c'

Figure 7.13d

In encountering a frame with sloping roof members, as in Figure 7.13a, the problem of chord rotation is much more involved. The columns ab and de can rotate independently of each other; thus, we have two degrees of freedom from sidesway. The chord rotation of the roof members depends upon the column rotations. This interdependence is illustrated in Figure 7.13b, c, and d.

Referring to Figure 7.13d and using the law of sines we see:

$$\frac{\delta_1 - \delta_2}{\sin(\alpha_1 + \alpha_2)} = \frac{\delta_3}{\sin(90° - \alpha_2)} = \frac{\delta_4}{\sin(90° - \alpha_1)} \quad \text{or}$$

$$\frac{\delta_1 - \delta_2}{\sin(\alpha_1 + \alpha_2)} = \frac{\delta_3}{\cos \alpha_2} = \frac{\delta_4}{\cos \alpha_1}$$

$$\text{with } \rho_{bc} = \frac{\delta_3}{bc} \quad \text{and} \quad \rho_{dc} = \frac{\delta_4}{dc}$$

When chord rotations occur (via sidesway or settlement), additional equations of statics are formed by considering "end shear." For example, the frame in Figure 7.14a has three unknowns: θ_b, θ_c and ρ. Two equations are written by considering the equilibrium of joints b and c. The third equation is formed by summing horizontal forces on the total FBD (Figure 7.14b), i.e., $10^k = a_x + d_x$. By looking at the columns ab and cd separately we can phrase a_x and d_x in terms of the moments for columns ab and cd, respectively, i.e., $a_x = -(1/15')(M_{ab} + M_{ba})$ and $d_x = -(1/15')(M_{cd} + M_{dc})$. The slope-deflection relations allow us to put the moments in terms of the unknowns θ_b, θ_c and ρ. Thus the shears give a third independent equilibrium equation. This procedure is illustrated in Example 4.

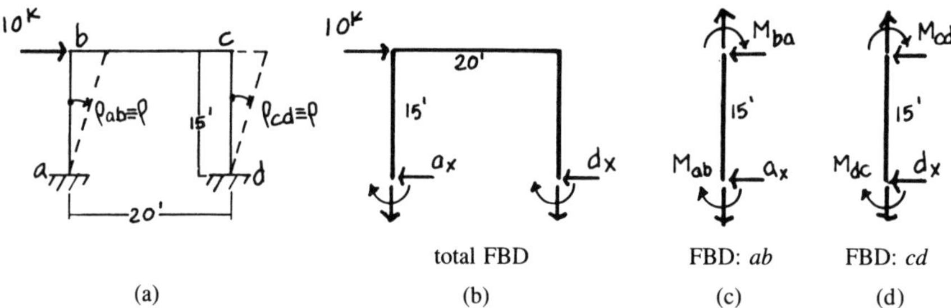

(a) (b) (c) (d)

Figure 7.14

With a two-story frame, a fourth equation is found by considering the end shears at the second floor level. Example 5 demonstrates the procedure for a two-story frame.

When sloping members are encountered it is usual to involve the end shears by taking moments. The point to take moments about is chosen in order to eliminate axial forces.

For the frame shown in Figure 7.15a and b, an equilibrium equation involving shears V_d and V_a (plus M_{ab}, M_{dc}) is formed by taking moments about point 0. The shears V_d and V_a are put in terms of the moments M_{cd}, M_{dc}, M_{ab} and M_{ba} by considering the members ab and cd in separate FBDs.

The frame shown in Figure 7.15c has two degrees of sidesway (see Figure 7.13). Thus, two shear equations are necessary. The first is formed by summing horizontal forces on the total FBD and proceeding as in a rectilinear frame. A second equation evolves by taking moments about point 0; this equation involves unknowns, V_a, V_d, M_{ab} and M_{dc}. The shears V_a and V_d are, again, put in terms of M_{cd}, M_{dc}, M_{ab} and M_{ba} by considering FBDs *ab* and *cd*. A few other illustrations are shown in Figure 7.16.

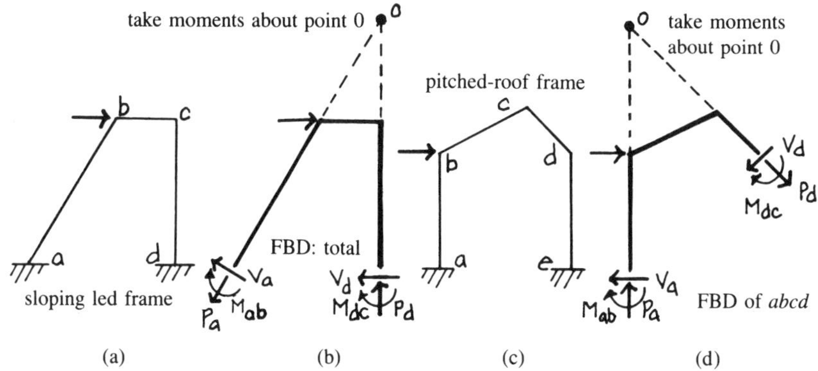

Figure 7.15

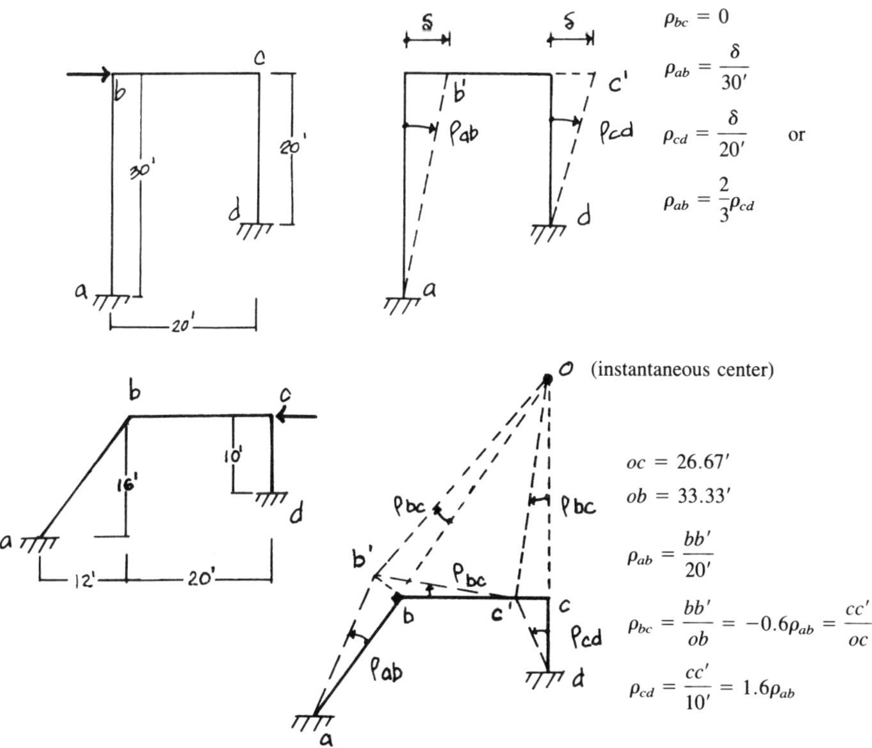

Figure 7.16

APPLICATION OF SLOPE-DEFLECTION

In applying slope-deflection, one breaks the beam or frame into a series of constituent members (or elements); a member is that part of the beam or frame between joints or supports (Figure 7.17). In general, each end of the member will involve a joint rotation, but keep in mind that abutting members share a joint rotation. Also, when a member is built-in at one end, the rotation is zero there. If we consider the beam shown in Figure 7.17, we first note that no chord rotations exist because no support settlement is specified, nor can the beam sidesway. Second, for this beam:

$\theta_a = 0$, wall at a; $\theta_b \neq 0$, shared by members, ab and bc;

$\theta_c \neq 0$, shared by members, bc and cd;

$\theta_d = 0$, wall at d,

thus we have two unknown rotations, θ_b and θ_c. All the joint moments are related to these two unknowns:

$$M_{ab} = M_{ab}^{FM} + \left(\frac{EI}{\ell}\right)_{ab}(-6\not\phi + 4\not\theta_a + 2\theta_b)$$

with M_{ab} determined strictly from the loads on ab,

$$M_{ba} = M_{ba}^{FM} + \left(\frac{EI}{\ell}\right)_{ab}(+4\theta_b)$$

$$M_{bc} = M_{bc}^{FM} + \left(\frac{EI}{\ell}\right)_{bc}(+4\theta_b + 2\theta_c)$$

$$M_{cb} = M_{cb}^{FM} + \left(\frac{EI}{\ell}\right)_{cb}(+4\theta_c + 2\theta_b)$$

$$M_{cd} = M_{cd}^{FM} + \left(\frac{EI}{\ell}\right)_{cd}(+4\theta_c)$$

$$M_{dc} = M_{dc}^{FM} + \left(\frac{EI}{\ell}\right)_{cd}(+2\theta_c).$$

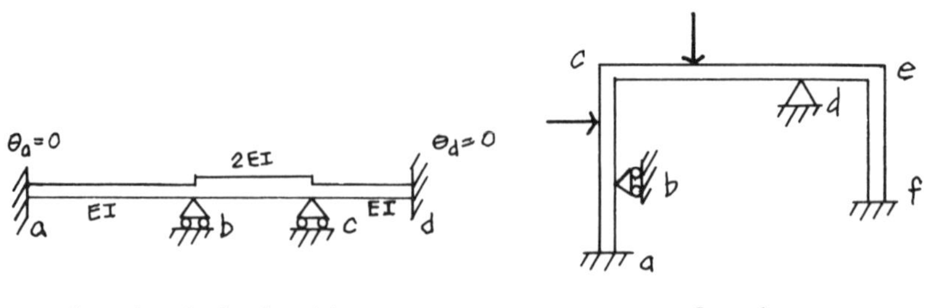

3 members (each prismatic)

5 members

Figure 7.17

Question: How to determine θ_b and θ_c?

Answer: Statics.

Considering the FBDs of joints b and c (Figure 7.18), the laws of equilibrium insist that:

$$M_{ba} + M_{bc} = 0 \text{ or } 0 = M_{ba}^{FM} + \left(\frac{EI}{\ell}\right)_{ab}(+4\theta_b) + M_{bc}^{FM} + \left(\frac{EI}{\ell}\right)_{bc}(+4\theta_b + 2\theta_c)$$

and

$$M_{cb} + M_{cd} = 0 \text{ or } 0 = M_{cb}^{FM} + \left(\frac{EI}{\ell}\right)_{cb}(+4\theta_c + 2\theta_b) + M_{cd}^{FM} + \left(\frac{EI}{\ell}\right)_{cd}(+4\theta_c).$$

Thereby two equations of equilibrium are generated for the two unknown rotations. Thus, slope-deflection comes down to solving sets of simultaneous equations for the unknown joint rotations and chord rotations, when they exist. The method is quite easy to set up but, as the number of unknowns increases, solving the simultaneous equations becomes a major problem unless a computer is at hand. When chord rotations also exist we'll see, by examples, that equations of statics in addition to those from joint consideration must be evolved.

Figure 7.18

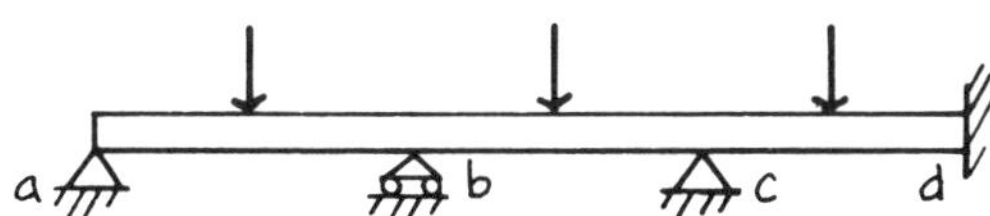

Figure 7.19

Pin Modification

When a pin or roller exists at an extreme end of a beam or frame, a modification of the slope deflection equations allows a short-cut resulting in the elimination of one unknown (the joint rotation at the pin or roller). For the beam in Figure 7.19, we know that the resulting moment at pin a must be zero (Note at roller b and pin c the moments are not zero). Thus, we write the slope-deflection equations (8), in particular

$$M_{ab} = 0 = M_{ab}^{FM} + \frac{EI}{\ell}(-6\rho_{ab} + 4\theta_a + 2\theta_b).$$

$$M_{ba} = M_{ba}^{FM} + \frac{EI}{\ell}(-6\rho_{ab} + 4\theta_b + 2\theta_a).$$

Solving the former equation for θ_a yields

$$\theta_a = \frac{-M_{ab}^{FM}\ell}{4EI} + 1.5\rho_{ab} - .5\theta_b$$

Thus, replacing θ_a in the M_{ba} expression we arrive at:

$$M_{ba} = M_{ba}^{FM} - .5M_{ab}^{FM} + \frac{3EI}{\ell}(\theta_b - \rho_{ab}). \tag{10}$$

Consequently, θ_a has been eliminated. The pin modification is used in Examples 3 and 5.

Symmetry Modification

It is not uncommon in engineering to encounter a symmetric beam; that is, one which is geometrically symmetric as well as symmetrically loaded. Consider the five-span beam in Figure 7.20. By symmetry we find $\theta_b = -\theta_e$ and $\theta_c = -\theta_d$. Therefore, only spans *ab*, *bc* and *cd* need be considered. Instead of four joint rotations to be found (θ_b, θ_c, θ_d, θ_e) the symmetry condition reduces the number to two (θ_b and θ_c) thus reducing the number of simultaneous equations to be solved also from four to two.

For even number spans a similar modification exists. The beam in Figure 7.21 illustrates the point. Here $\theta_c = 0$ and $\theta_b = \theta_d$. Thus, the four-span beam contracts to a two-span beam (Figure 7.22).

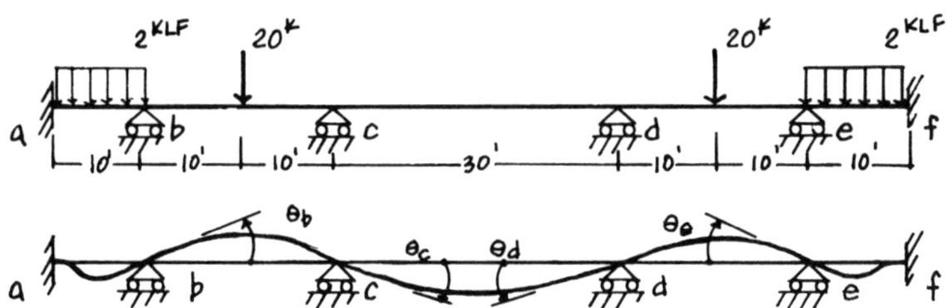

Figure 7.20

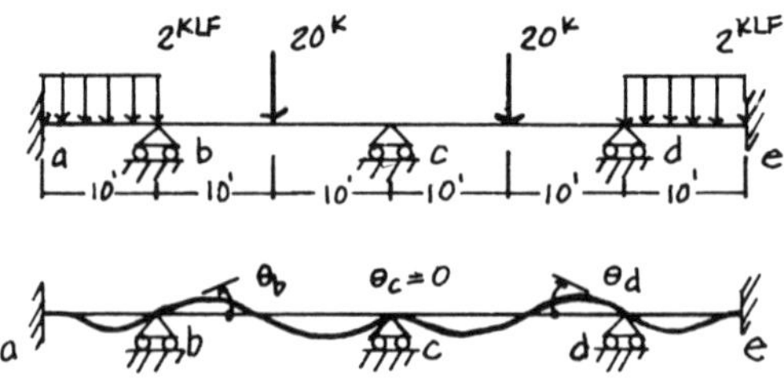

Figure 7.21

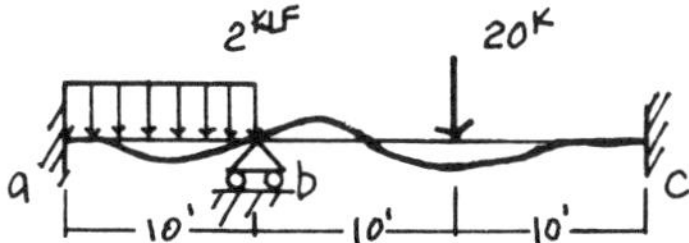

Figure 7.22

Antisymmetry Modification

A geometrically symmetric frame subjected only to wind loads (Figure 7.23) falls into this classification. Here we see $\theta_b = \theta_c$. Thus, the number of unknowns has been reduced by one. Other illustrations of antisymmetry are shown in Figure 7.24 and in Example 5.

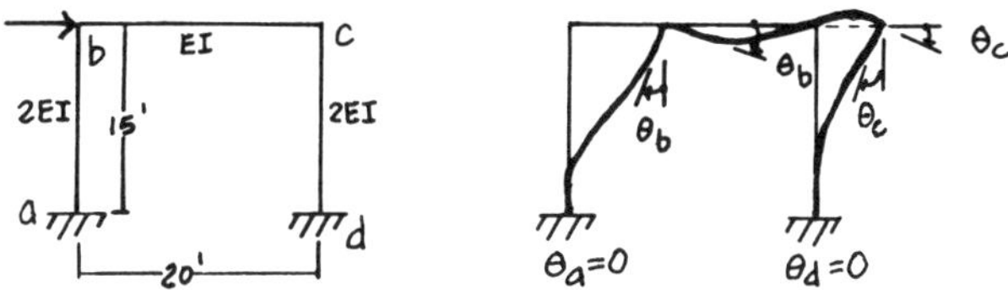

Figure 7.23

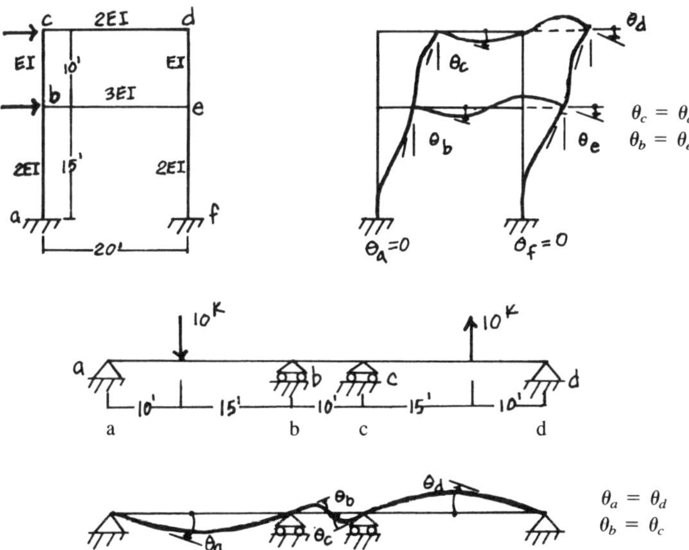

Figure 7.24

Other Methods

Related methods called the "three-moment equation" and "four-moment equation" can be taken as variations on slope-deflection (see Reference 2).

EXAMPLE #1: SLOPE-DEFLECTION: FRAME WITHOUT JOINT TRANSLATION

Given: Rigid-jointed frame as shown; EI values shown on frame.

Find: V & M diagrams; sketch the deflection curve.

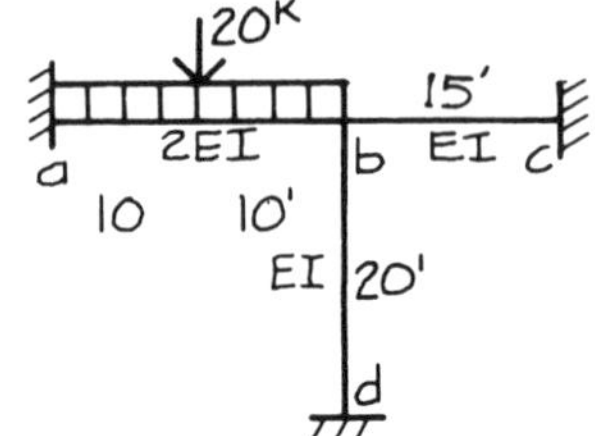

Chord rotations: Since all joints are restrained from translating all chord rotations are zero, $\rho_{ab}=\rho_{bc}=\rho_{bd}=0$

Joint rotations: Since joints a, c and d are built-in the rotations there are zero, ie, $\theta_a=\theta_c=\theta_d=0$. $\theta_b \neq 0$

Fixed-end moments: Since spans bc and bd are unloaded we have $M_{bc}^{FM}=M_{cb}^{FM}=M_{bd}^{FM}=M_{db}^{FM}=0$.
For span ab we can add the concentrated and distributed load effects; using the F.E.M. charts we have:

2KLF, 20K: M_{ab}^{FM}, M_{ba}^{FM} = 2KLF: $-\frac{(2)(20)^2}{12}$, $+\frac{(2)(20)^2}{12}$ + 20K (10', 10'): $-\frac{(20)(20)}{8}$, $+\frac{(20)(20)}{8}$

Adding we find $M_{ab}^{FM}=-116.67^{K\text{-}FT}$; $M_{ba}^{FM}=+116.67^{K\text{-}FT}$

$$M_{ab}=-116.67+(2EI/20)[-6\rho_{ab}+4\theta_a+2\theta_b]=-116.67+(EI/10)(+2\theta_b)$$
$$M_{ba}=+116.67+(2EI/20)[-6\rho_{ab}+4\theta_b+2\theta_a]=+116.67+(EI/10)(+4\theta_b)$$
$$M_{bc}=0+(EI/15)[-6\rho_{bc}+4\theta_b+2\theta_c]=(EI/15)(+4\theta_b)$$
$$M_{cb}=0+(EI/15)[-6\rho_{bc}+4\theta_c+2\theta_b]=(EI/15)(+2\theta_b)$$
$$M_{bd}=0+(EI/20)[-6\rho_{bd}+4\theta_b+2\theta_d]=(EI/20)(+4\theta_b)$$
$$M_{db}=0+(EI/20)[-6\rho_{bd}+4\theta_d+2\theta_b]=(EI/20)(+2\theta_b)$$

We have only one unknown, θ_b, which can be found by equilibrium of joint b.

$\Sigma M_{jt.\,b}=0$: $M_{ba}+M_{bc}+M_{bd}=0$

$116.67+(EI/10)(+4\theta_b)+(EI/15)(+4\theta_b)+(EI/20)(+4\theta_b)=0$

Thus $EI\theta_b=-134.62$ (C.C.W.)

FBD joint b: M_{bc}, M_{ba}, M_{bd}

The moment reactions are now calculated from slope-deflection relations.

$$M_{ab}=-116.67+(EI/10)(+2)(-134.62/EI)=-143.59^{K\text{-}FT}$$
$$M_{ba}=+116.67+(EI/10)(+4)(-134.62/EI)=+62.82^{K\text{-}FT}$$
$$M_{bc}=(EI/15)(+4)(-134.62/EI)=-35.90^{K\text{-}FT}$$

$M_{cb} = (EI/15)(+2)(-134.62/EI) = -17.95$ K-FT
$M_{bd} = (EI/20)(+4)(-134.62/EI) = -26.92$ K-FT
$M_{db} = (EI/20)(+2)(-134.62/EI) = -13.46$ K-FT

All signs are according to the F.E.M. convention. The correct moment picture of joint b is shown.

62.82
35.90
26.92
Joint b

The shear reactions are now found from statics.

$\Sigma M_a = 0$: $ba_y = \frac{1}{20}[62.82 - 143.59 + 20(10) + 40(10)] = +25.96^K$

$\Sigma F_Y = 0 \quad ab_y = 34.04^K$

$\Sigma F_X = 0 \quad ab_x = ba_x$

143.59
20K
62.82
10'
10'
ab_y
ab_y
ba_y
ba_x
FBD/ span ab

$\Sigma M_b = 0 \quad cb_y = -\frac{1}{15}[17.95 + 35.90] = -3.59^K$

$\Sigma F_Y = 0 \quad bc_y = +3.59^K$

$\Sigma F_X = 0 \quad bc_x = cb_x$

35.90
17.95
15'
bc_x
bc_y
cb_y
cb_x
FBD/ span bc

$\Sigma M_b = 0 \quad db_x = \frac{-1}{20}[13.46 + 26.92] = -2.02^K$

$\Sigma F_X = 0 \quad bd_x = +2.02^K$

By summing forces in vertical direction for the total structure we find $db_y = -29.56$ (↑)

bd_y
bd_x
26.92
20'
FBD/ span bd
db_x
13.46
db_y

143.59K-FT
20K
2KLF
ab_x
34.04K
θ_b
θ_b
17.95 (K-FT)
3.59K
cb_x
θ_b
2.02K
13.46K-FT
29.56K

Load diagram and deflection curve. (Note ab_x & cb_x remain undetermined)

V
a
b
34.04
14.04
-5.96
-25.96
b
c
+3.59

M
a
b
+96.81
-62.82
-143.59
b
c
+17.95
-35.90

M
b
d
-26.92
+13.46
V
b
d
+2.02
Δ

EXAMPLE #2: SLOPE-DEFLECTION: BEAM WITH SUPPORT SETTLEMENT

Given: Continuous beam as shown; support b settles 3"; $E = 30 \times 10^3$ KSI; $I = 499$ in^4.

Find: V & M diagrams; sketch the deflection curve.

3KLF

a b c d

15' 15' 10'

Chord rotations: From the sketch, right, we find:

$\rho_{ab} = +\frac{.25'}{15'} = .01667$ radians (C.W.)

$\rho_{bc} = -\frac{.25'}{15'} = -.01667$ radians (C.C.W)

$\rho_{cd} = 0$

ρ_{ab} ρ_{bc}

3"

Chord Rotations

Joint rotations:

$\theta_a = 0$; $\theta_d = 0$; $\theta_b \neq 0$; $\theta_c \neq 0$

Fixed-end moments:

$$M^{FM}_{ab} = M^{FM}_{bc} = -\frac{(3)(15)^2}{12} = -56.25^{K\text{-}FT}$$

$$M^{FM}_{ba} = M^{FM}_{cb} = +\frac{(3)(15)^2}{12} = +56.25^{K\text{-}FT}$$

$$M^{FM}_{cd} = -\frac{(3)(10)^2}{12} = -25.^{K\text{-}FT} = -M^{FM}_{dc}$$

Slope deflection relations:

$$M_{ab} = -56.25 + {}^{EI}/_{15'}[-6(+.01667) + 4\cancel{\theta_a} + 2\theta_b] = -56.25 + [-693.47 + 13{,}866.7\theta_b]$$

$$M_{ba} = +56.25 + {}^{EI}/_{15'}[-6(+.01667) + 4\theta_b + 2\cancel{\theta_a}] = +56.25 + [-693.47 + 27{,}733\,\theta_b]$$

$$M_{bc} = -56.25 + {}^{EI}/_{15'}[-6(-.01667) + 4\theta_b + 2\theta_c] = -56.25 + [+693.47 + 27{,}733\theta_b + 13{,}866.7\,\theta_c]$$

$$M_{cb} = +56.25 + [+693.47 + 27{,}733\,\theta_c + 13{,}866.7\theta_b]$$

$$M_{cd} = -25. + {}^{EI}/_{10'}[-6(0) + 4\theta_c + 2\cancel{\theta_d}] = -25 + [+41{,}600\,\theta_c]$$

$$M_{dc} = -25 + {}^{EI}/_{10}[-6(0) + 4\cancel{\theta_d} + 2\theta_c] = -25 + [+20{,}800\,\theta_c]$$

The two unknowns, θ_b and θ_c, can be found from equilibrium of joints b and c.

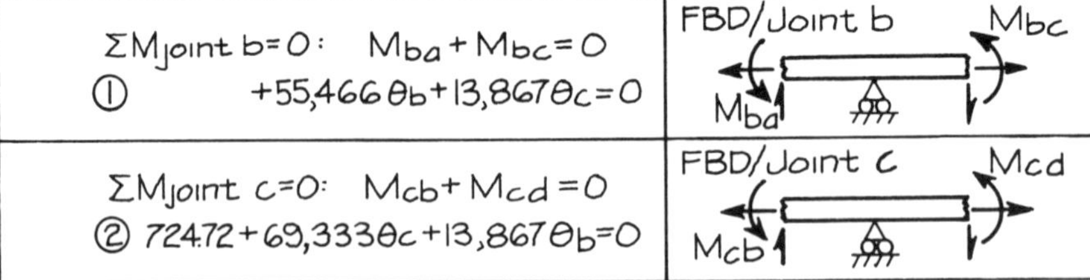

$\Sigma M_{joint\ b} = 0$: $M_{ba} + M_{bc} = 0$

① $+55{,}466\,\theta_b + 13{,}867\theta_c = 0$

$\Sigma M_{joint\ c} = 0$: $M_{cb} + M_{cd} = 0$

② $724.72 + 69{,}333\theta_c + 13{,}867\,\theta_b = 0$

Solving equations ① and ② simultaneously we find:

$\theta_b = +.002751$ radians (C.W.)

$\theta_c = -.011003$ radians (C.C.W.)

With θ_b and θ_c known the moments are calculated from the slope-deflection relations.

$M_{ab} = -56.25 + [-693.47 + 13{,}866.7(+.002751) = -711.59^{K\text{-}FT}$

$M_{ba} = +56.25 + [-693.47 + 27{,}733(+.002751) = -560.93^{K\text{-}FT}$

$M_{bc} = +560.93^{K\text{-}FT}$

$M_{cb} = +482.72^{K\text{-}FT}$

$M_{cd} = -482.72^{K\text{-}FT}$

$M_{dc} = -203.86^{K\text{-}FT}$

Taking FBDs of the individual spans the shear forces are found.

$\Sigma M_a = 0$: $V_{ba} = \frac{1}{15}[7.5(45) - 560.93 - 711.59]$ $V_{ba} = -62^K$ $\Sigma F_y = 0$: $V_{ab} = +107^K$	711.59^K-FT, 45^K, 560.93 K-FT, a, b, 15', V_{ab}, V_{ba} FBD span ab
$\Sigma M_b = 0$: $V_{cb} = \frac{1}{15}[7.5(45) + 482.72 + 560.93]$ $V_{cb} = +92^K$ $\Sigma F_y = 0$: $V_{bc} = -47^K$	560.93^K-FT, 45^K, 482.72 K-FT, b, c, 15', V_{ba}, V_{cb} FBD span bc
$\Sigma M_c = 0$: $V_{dc} = \frac{1}{10}[5(30) - 203.86 - 482.72]$ $V_{dc} = -54^K$ $\Sigma F_y = 0$ $V_{cd} = +84^K$	482.72^K-FT, 30^K, 203.86 K-FT, c, d, 10', V_{cd}, V_{dc} FBD span cd
712^K-FT, 3KLF, 204^K-FT, a, 15', b, 15', c, 10', d, 107^K, 62^K, 47^K, 92^K, 84^K, 54^K	Loading Diagram
+107^K, +62^K, +84^K, +54^K, a, -47, -92^K	V Diagram
+561, +204, a, b, c, d, -712, -483	M Diagram
$\theta_a = 0$, a, 3", b, c, d, $\theta_c = .011003$ radians, $\theta_d = 0$, $\theta_b = .002751$ radians	Deflection Curve

EXAMPLE #3: SLOPE-DEFLECTION: BEAM WITH PIN MODIFICATION

Given: Continuous beam as shown; EI constant

Find: V & M diagrams; sketch the deflection curve.

Chord rotations:

No settlement thus $\rho_{ab}=\rho_{bc}=\rho_{cd}=0$

Joint rotations:

$\theta_a=0;\ \theta_b\neq 0;\ \theta_c\neq 0$ and $\theta_d\neq 0$

Fixed-end moments:

$M_{ab}^{FM}=M_{cd}^{FM}=-M_{ba}^{FM}=-M_{dc}^{FM}=-\frac{(30)(20)}{8}=-75^{K\text{-}FT};\ M_{bc}^{FM}=M_{cb}^{FM}=0$

Since a pin exists at support d the moment there, M_{dc}, is zero. Using the pin modification for M_{cd} we get:

$M_{cd}=M_{cd}^{FM}-\frac{1}{2}M_{dc}^{FM}+\frac{3EI}{20}[\theta_c-\cancel{\rho_{cd}}]=-75-.5(75)+.15EI[\theta_c]$ or

$M_{cd}=-112.5+.15EI\theta_c$; and

$M_{ab}=M_{ab}^{FM}+\frac{EI}{20}[-6\cancel{\rho_{ab}}+4\cancel{\theta_a}+2\theta_b]=-75+.1EI\theta_b$;

$M_{ba}=M_{ba}^{FM}+\frac{EI}{20}[-6\cancel{\rho_{ab}}+4\theta_b+2\cancel{\theta_a}]=+75+.2EI\theta_b$; similarly

$M_{bc}=+.2EI\theta_b+.1EI\theta_c$

$M_{cb}=+.1EI\theta_b+.2EI\theta_c$

By using the pin modification only two unknowns (θ_b & θ_c) are to be found in order to calculate the moments; θ_d has been eliminated. Thus two equations of statics are written.

$\Sigma M_{joint\ b}=0:\quad M_{ba}+M_{bc}=0$

$75+.4EI\theta_b+.1EI\theta_c=0$ ①

FBD/Joint b

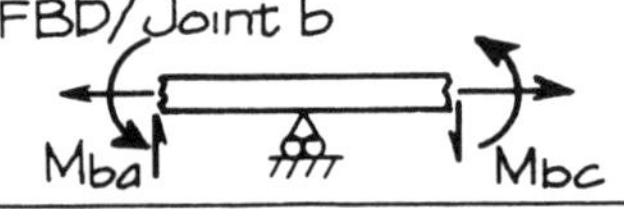

$\Sigma M_{joint\ c}=0:\quad M_{cb}+M_{cd}=0$

$-112.5+.1EI\theta_b+.35EI\theta_c=0$ ②

FBD/Joint c

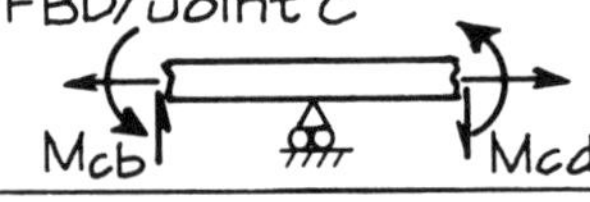

Solving equations ① and ② simultaneously yields:

$EI\theta_b=-288.46$ (c.c.w.)

$EI\theta_c=+403.85$ (c.w.)

Finding θ_b and θ_c allows the calculation of all the moments.

$M_{ab}=-75+.1(-288.46)=-103.85^{K\text{-}FT}$

$M_{ba}=+75+.2(-288.46)=+17.31^{K\text{-}FT}=-M_{bc}$

$M_{cb}=+.1(-288.46)+.2(+403.85)=+51.92^{K\text{-}FT}=-M_{cd}$

The shear forces are found by statics.

$\Sigma M_a = 0$: $V_{ba} = +10.7^K$ $\Sigma F_y = 0$: $V_{ab} = +19.3^K$	FBD/ Span ab 103.85, 30^K, 17.31, a, b, 10', 10', V_{ab}, V_{ba}
$\Sigma M_b = 0$: $V_{cb} = +1.7^K$ $\Sigma F_y = 0$: $V_{bc} = -1.7^K$	FBD/ Span bc 17.31, 51.92, b, c, 20', V_{bc}, V_{cb}
$\Sigma M_c = 0$: $V_{dc} = +12.4^K$ $\Sigma F_y = 0$: $V_{cd} = +17.6^K$	FBD/ Span cd 51.92, 30^K, c, d, 10', 10', V_{cd}, V_{dc}
104, a, 30^K, 10', 10', b, 20', c, 30^K, 10', 10', d, 19.3, 10.7, 1.7, 1.7, 17.6, 12.4	Loading Diagram
+19.3, a, -10.7, b, -1.7, +17.6, c, d, -12.4	V Diagram
+89, 0, -104, -17, -52, +124, 0	M Diagram
$\theta_a = 0$, θ_b, θ_c	Deflection Curve

EXAMPLE #4: SLOPE-DEFLECTION: FRAME WITH SIDESWAY

Given: Frame as shown; EI constant

Find: V & M diagrams; sketch the deflection curve.

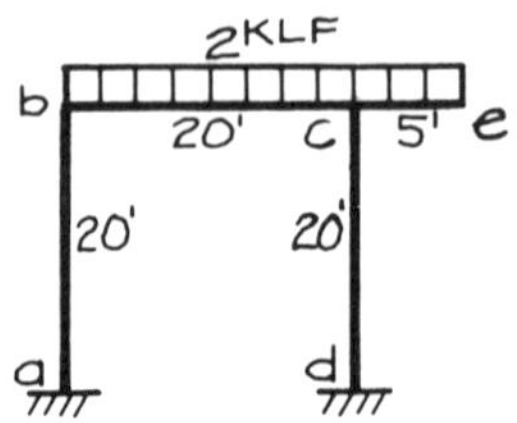

Joint rotations:

$\theta_a = 0$; $\theta_d = 0$

Chord rotations: From the sketch, right, we see the column rotations are the same.

$\rho_{ab} = \rho_{cd} = +\delta/20 \equiv +\rho$

(Note: clockwise rotations are positive)

$\rho_{bc} = 0$

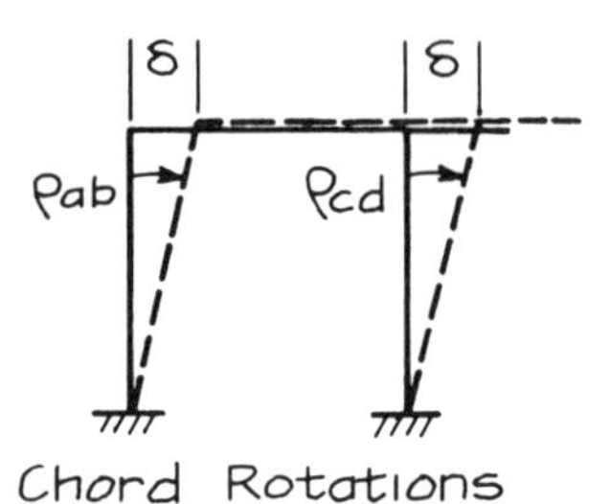

Chord Rotations

Fixed-end moments: First, what to do with the 5' cantilever, ce? Since ce is statically determinate it is best to sever ce and bring its contribution to joint c (sketch, right). Note: The 10^K and $25^{K\text{-}FT}$ "loads" at joint c do not affect the FEMs.

$M_{ab}^{FM} = M_{ba}^{FM} = M_{cd}^{FM} = M_{dc}^{FM} = 0$

$M_{bc}^{FM} = -M_{cb}^{FM} = \frac{-(2)(20)^2}{12} = -66.7^{K\text{-}FT}$

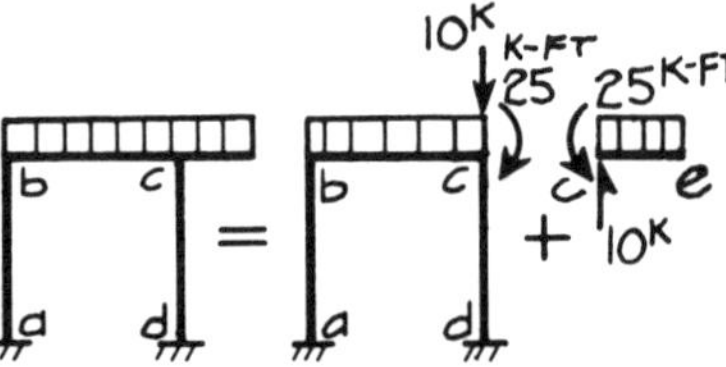

Cantilever Treatment

Slope-deflection relations

$M_{ab} = M_{ab}^{FM} + (EI/20)[-6\rho_{ab} + 4\theta_a + 2\theta_b] = [-.3\rho + .1\theta_b]EI$

$M_{ba} = (EI/20)[-6\rho_{ab} + 4\theta_b + 2\theta_a] = [-.3\rho + .2\theta_b]EI$, similarly

$M_{bc} = -66.7 + [+.2\theta_b + .1\theta_c]EI$

$M_{cb} = +66.7 + [+.2\theta_c + .1\theta_b]EI$

$M_{cd} = [-.3\rho + .2\theta_c]EI$

$M_{dc} = [-.3\rho + .1\theta_c]EI$

The slope-deflection relations involve three unknowns, θ_b, θ_c and ρ. Three equations of statics are necessary. Two come from considering joints b and c. The third equation comes from considering the equilibrium of the total frame and involves the base shears.

$\Sigma M_{jt\,b}=0$: $M_{ba}+M_{bc}=0$ or $-66.7+[+.4\theta_b+.1\theta_c-.3\rho]EI=0$ (1)	FBD/ Joint b b, M_{bc}, M_{ba}
$\Sigma M_{jt\,c}=0$: $M_{cb}+M_{cd}-25=0$ or $41.7+[+.4\theta_c+.1\theta_b-.3\rho]EI=0$ (2)	FBD/ Joint c M_{cb}, 10^K, $25^{K\text{-}FT}$, M_{cd}
$\Sigma F_x=0$: $a_x+d_x=0$ (3) where a_x and d_x can be stated in terms of the unknowns (θ_b, θ_c, and ρ) by considering members ab and cd (below right)	FBD/ Total Frame 2^{KLF}, 10^K, 25 K-FT, a_x, d_x
$\Sigma M_b=0$: $a_x=\frac{1}{20}[M_{ab}+M_{ba}]$ or $a_x=.05[-.6\rho+.3\theta_b]EI$ (3a)	FBD/ Span ab M_{ba}, b, 20', a, a_x, M_{ab}
$\Sigma M_c=0$: $d_x=\frac{1}{20}[M_{cd}+M_{dc}]$ or $d_x=.05[-.6\rho+.3\theta_c]EI$ (3b)	FBD/ Span cd M_{cd}, c, 20', d, d_x, M_{dc}

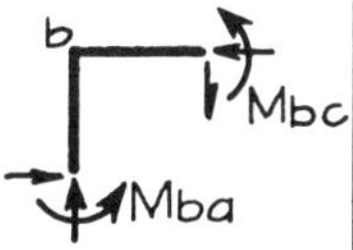

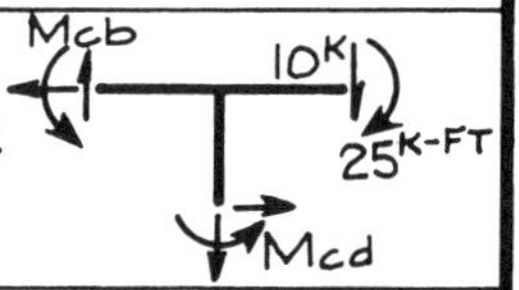

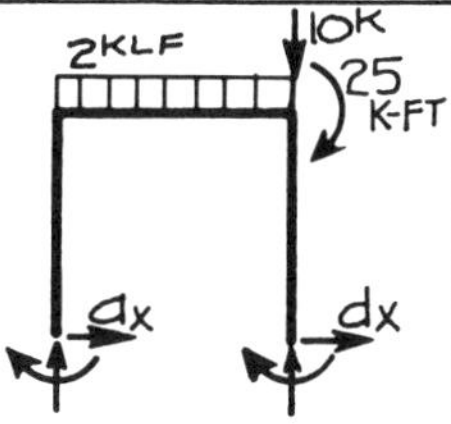

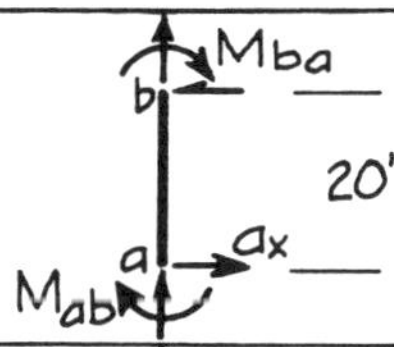

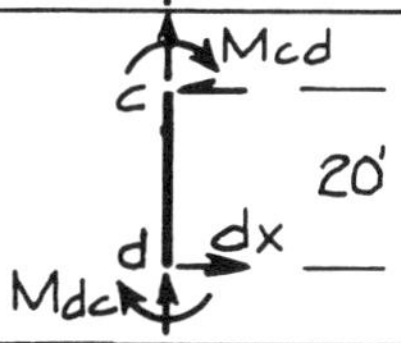

Putting (3a) and (3b) into (3) yields: $-1.2\rho+.3\theta_b+.3\theta_c=0$ (3c)
Solving (1), (2) and (3c) simultaneously yields:

$EI\theta_b=+216.38$ (C.W.)
$EI\theta_c=-144.94$ (C.C.W.)
$EI\rho=+17.86$ (C.W.)

Once θ_b, θ_c and ρ are found the moments are calculated from the slope-deflection relations. Also (3a) and (3b) give the horizontal base reactions.

$M_{ab}=[-.3(17.86)+.1(+216.38)]=+16.3^{K\text{-}FT}$ similarly
$M_{ba}=37.9^{K\text{-}FT}$
$M_{bc}=-37.9^{K\text{-}FT}$
$M_{cb}=+59.4^{K\text{-}FT}$
$M_{cd}=-34.4^{K\text{-}FT}$
$M_{dc}=-19.9^{K\text{-}FT}$
$a_x=+2.71$ (→)
$d_x=-2.71$ (←)

The vertical support reactions at a and d can now be found.

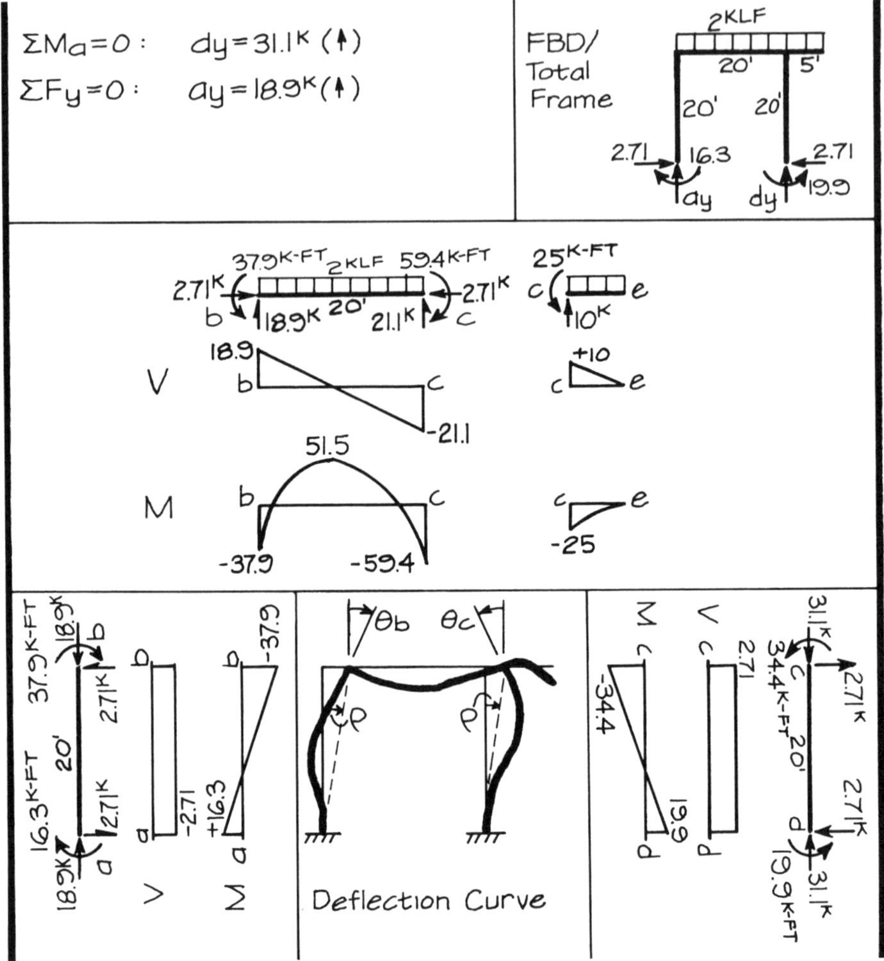
ΣMa=0 : dy=31.1K (↑)
ΣFy=0 : ay=18.9K (↑)
FBD/ Total Frame
2KLF
20'
5'
20'
20'
2.71
16.3
ay
dy
2.71
19.9
37.9K-FT
2KLF
59.4K-FT
2.71K
2.71K
b
c
18.9K
20'
21.1K
25K-FT
c
e
10K
V
18.9
-21.1
+10
M
51.5
-37.9
-59.4
-25
Deflection Curve
θb
θc

EXAMPLE #5: SLOPE DEFLECTION: FRAME WITH TWO DEGREES OF FREEDOM SIDESWAY/ ANTISYMMETRY MODIFICATION.

Given: Frame as shown; EI values as shown.

Find: V & M diagrams; P curve; sketch the deflection curve.

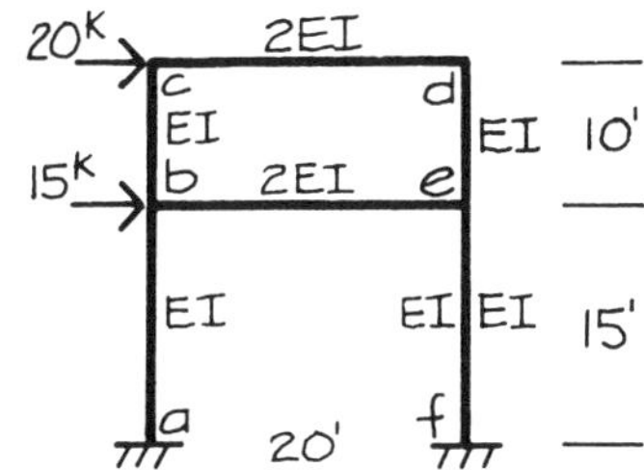

Chord rotations: From sketch,

$\rho_{ab} = \rho_{ef} = \delta_1/15' \equiv +\rho_1$

$\rho_{cb} = \rho_{de} = \delta_2/15' \equiv +\rho_2$

$\rho_{be} = \rho_{cd} = 0$

Joint rotations: From sketch, right below, we see

$\theta_b = \theta_e \neq 0$

$\theta_c = \theta_d \neq 0$

$\theta_a = \theta_f = 0$

Fixed-end moments: Since both loads are applied at joints, not within spans all FEMs are zero

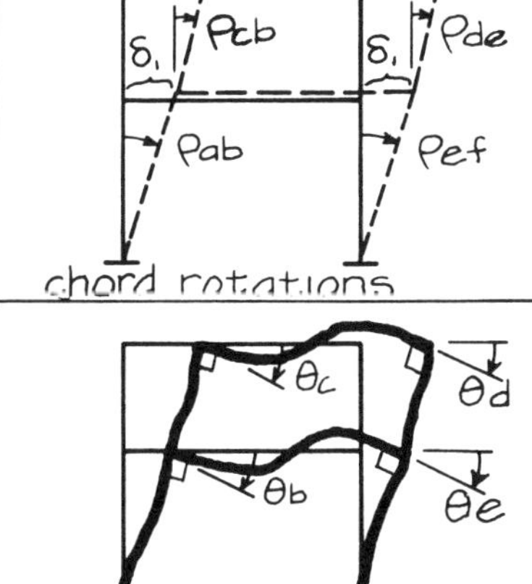

Slope-deflection relations

$M_{ab} = M_{fe} = (EI/15)[-6\rho_1 + 2\theta_b]$

$M_{ba} = M_{ef} = (EI/15)[-6\rho_1 + 4\theta_b]$

$M_{be} = M_{eb} = (2EI/20)[+4\theta_b + 2\theta_e] = (EI/10)(+6\theta_b)$

$M_{bc} = M_{ed} = (EI/10)[-6\rho_2 + 4\theta_b + 2\theta_c]$

$M_{cb} = M_{de} = (EI/10)[-6\rho_2 + 4\theta_c + 2\theta_b]$

$M_{cd} = M_{dc} = (2EI/20)[+4\theta_c + 2\theta_d] = (EI/10)[+6\theta_c]$

The S-D relations involve four unknowns ($\theta_b, \theta_c, \rho_1, \rho_2$). Four equations of equilibrium are required. Two come from considering the joints b and c; one comes from considering base shears and one from considering upper story shears.

$\Sigma M_{jt.\,b} = 0 \quad M_{ba} + M_{bc} + M_{be} = 0$

$EI[-.4\rho_1 - .6\rho_2 + 1.2667\theta_b + .2\theta_c] = 0$ ①

FBD Joint b

Mbc
15K
Mbe
Mba

$\Sigma M_{jt\,c} = 0$: $M_{cb} + M_{cd} = 0$ or

$EI[-.6\rho_2 + 1.0\theta_c + .2\theta_b] = 0$ ②

FBD Joint C

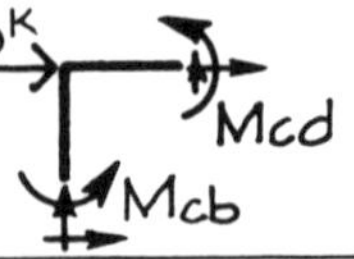

$\Sigma F_x = 0$: $a_x + f_x = 35^K$

due to antisymmetry $a_x = f_x$

$a_x = 17.5^K$ (←) ③

FBD total frame

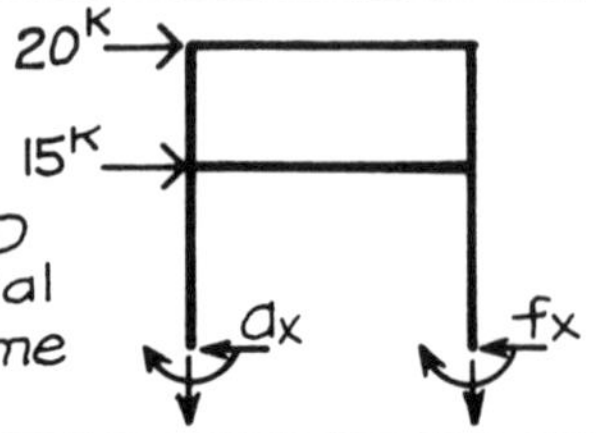

$\Sigma F_x = 0$ $bc_x + ed_x = 20^K$

due to antisymmetry $bc_x = ed_x$

$bc_x = 10^K$ (←) ④

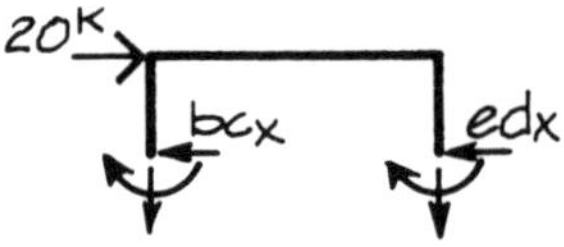

Using the FBDs of ab and bc the shear a_x and bc_x can be put in terms of moments and thus in terms of the unknowns.

For column ab

$\Sigma M_b = 0$: $15(17.5) + M_{ab} + M_{ba} = 0$ or

$-262.5 = [-.8\rho_1 + .4\theta_b]EI$ ③a

For column bc

$\Sigma M_c = 0$: $10(10) + M_{bc} + M_{cb} = 0$ or

④a $-100 = [-1.2\rho_2 + .6\theta_b + .6\theta_c]EI$

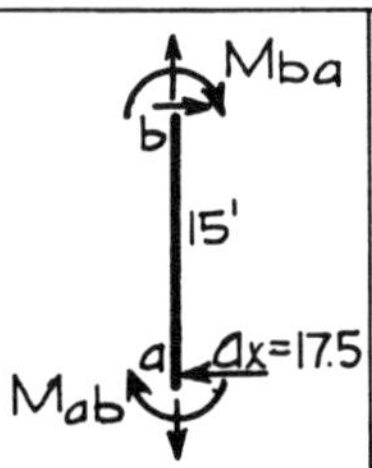

FBD: Column ab

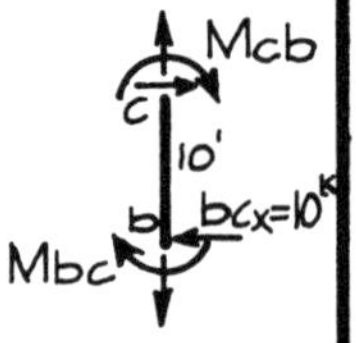

FBD: Column bc

Solving ①, ②, ③a and ④a yields: $EI\theta_b = +250$; $EI\theta_c = +107$; $EI\rho_1 = +453$; $EI\rho_2 = +263$

Putting these values into the S-D relations we calculate the moments:

$M_{ab} = -148^{K\text{-}FT}$ $(= M_{fe})$

$M_{ba} = -115$ $(= M_{ef})$

$M_{be} = +150$ $(= M_{eb})$

$M_{cb} = -64$ $(= M_{de})$

$M_{cd} = +64$ $(= M_{dc})$

$M_{bc} = -36$ $(= M_{ed})$

Once the moments are found the horizontal and vertical reactions at all joints can be obtained by statics.

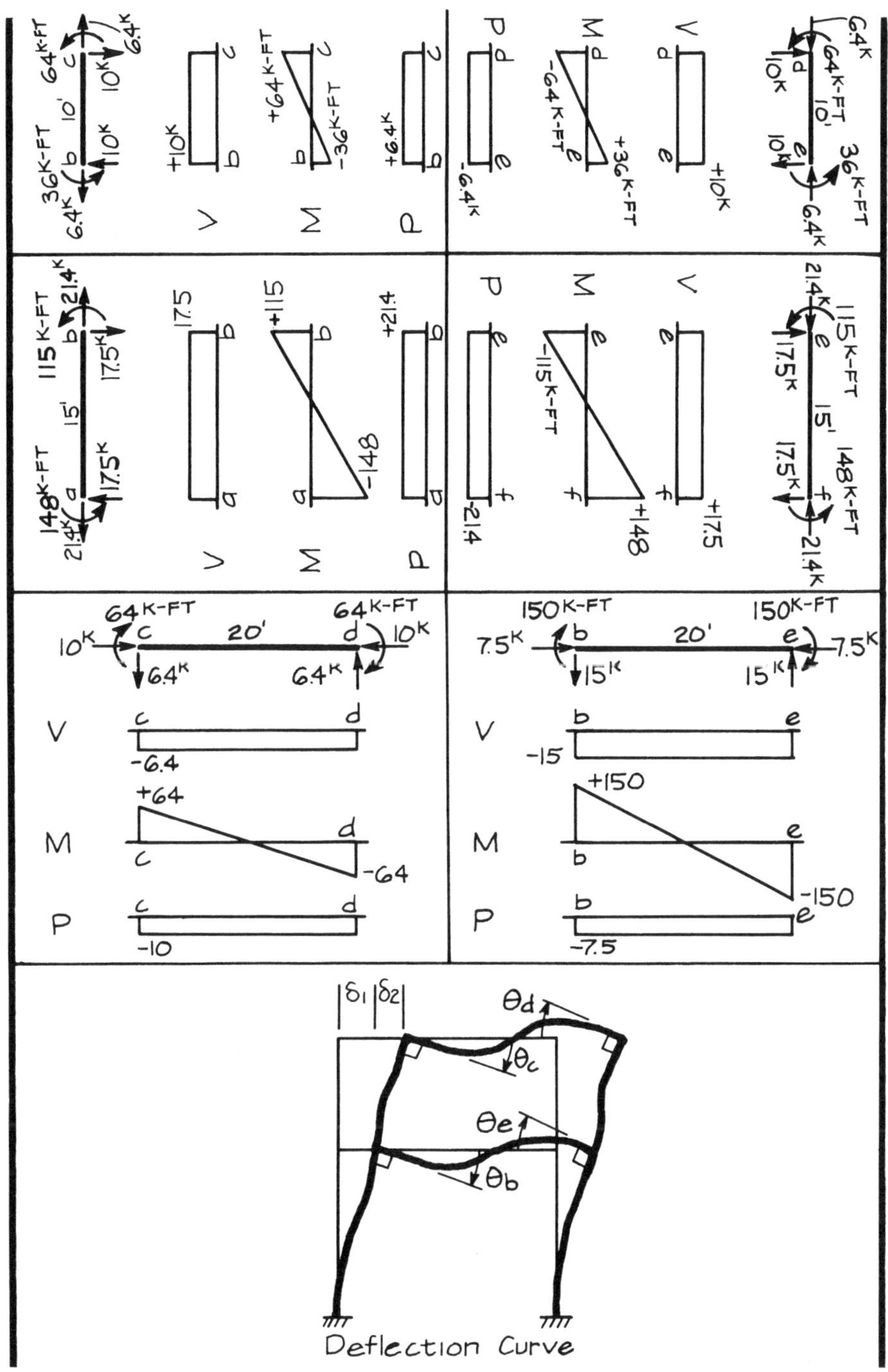
64K-FT
10K
6.4K
36K-FT
10'
V
M
P
+10K
+64K-FT
-36K-FT
+6.4K
-64K-FT
+36K-FT
-6.4K
115K-FT
21.4K
17.5K
15'
148K-FT
+115
-148
+21.4
-115K-FT
+148
+17.5
-21.4
64K-FT
10K
c
20'
d
6.4K
V
-6.4
+64
M
-64
P
-10
150K-FT
7.5K
b
20'
e
15K
V
-15
+150
M
-150
P
-7.5
δ1
δ2
θd
θc
θe
θb
Deflection Curve

PROBLEMS

Problems 1–5

For the beams shown find the V and M diagrams and sketch the deflection curves. Use slope deflection.

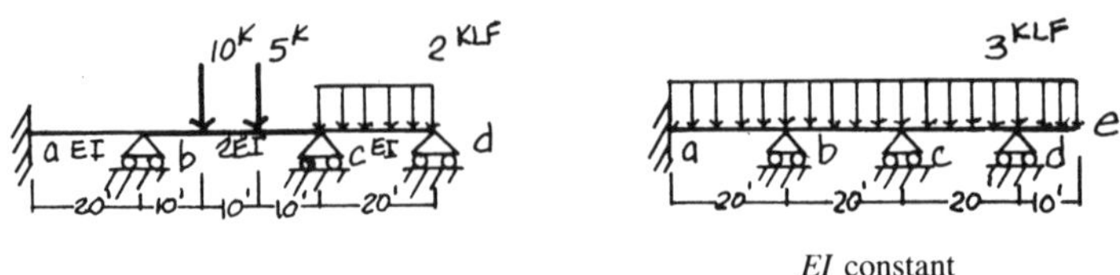

EI constant

beam 1

beam 2

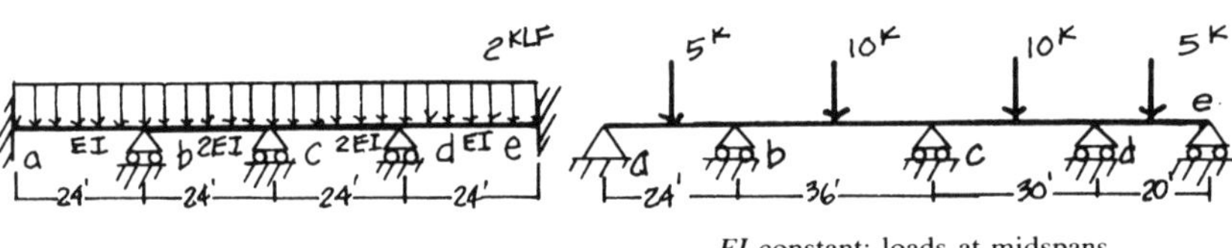

EI constant; loads at midspans

beam 3

beam 4

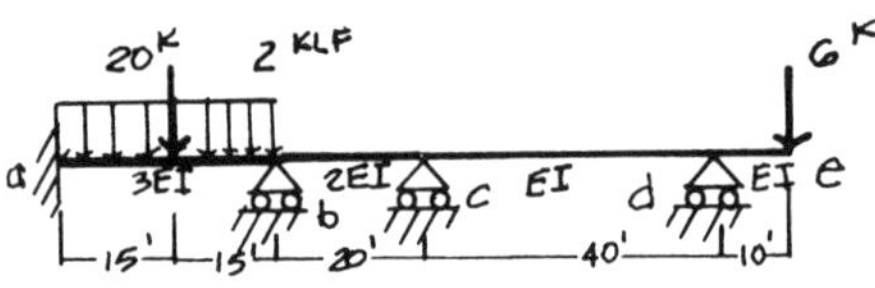

beam 5

Problems 6–8

Use $E = 30 \times 10^3$ ksi; $I = 1000$ in.^{4}

- Pr. 6: for beam 1 support b settles 2″; find V, M and deflection diagrams using slope deflection.
- Pr. 7: for beam 2 support c settles 1″; find V, M and deflection diagrams using slope deflection.
- Pr. 8: for beam 1 support b settles 1″, support c settles 2″; find V, M and deflection diagrams using slope deflection.

Problems 9–13

For the frames shown find the V and M diagrams and sketch the deflection curves. Use slope deflection.

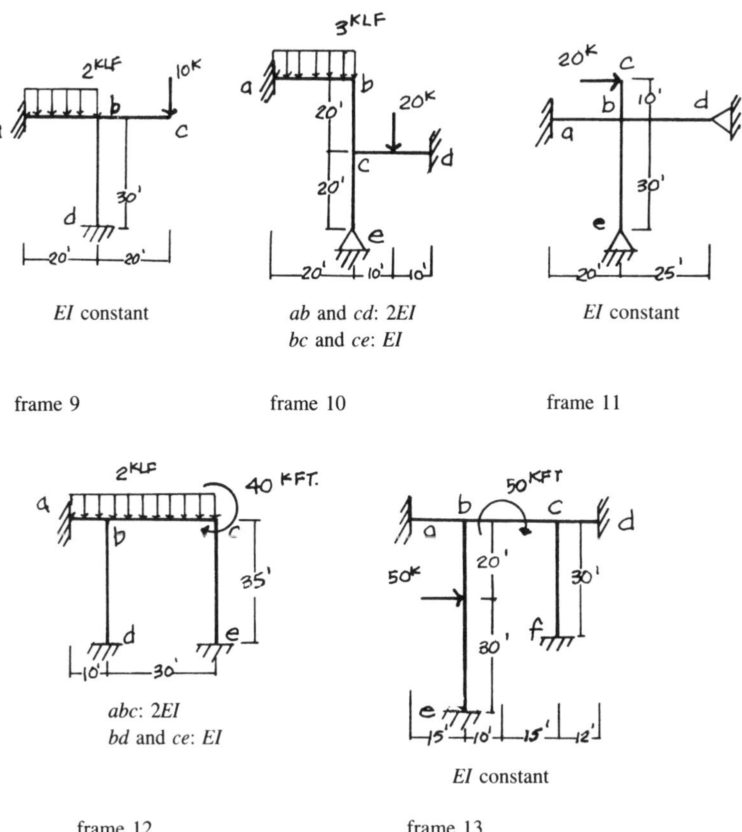

EI constant — frame 9

ab and *cd*: 2*EI*
bc and *ce*: *EI* — frame 10

EI constant — frame 11

abc: 2*EI*
bd and *ce*: *EI* — frame 12

EI constant — frame 13

Problems 14–25

For the frames shown find the V and M diagrams and sketch deflection curves. Use slope deflection.

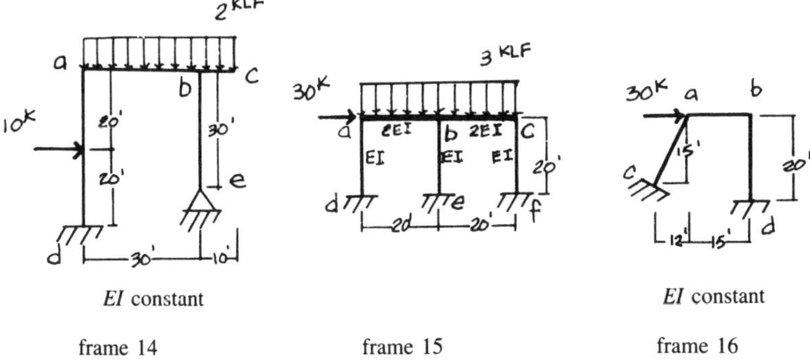

EI constant — frame 14

frame 15

EI constant — frame 16

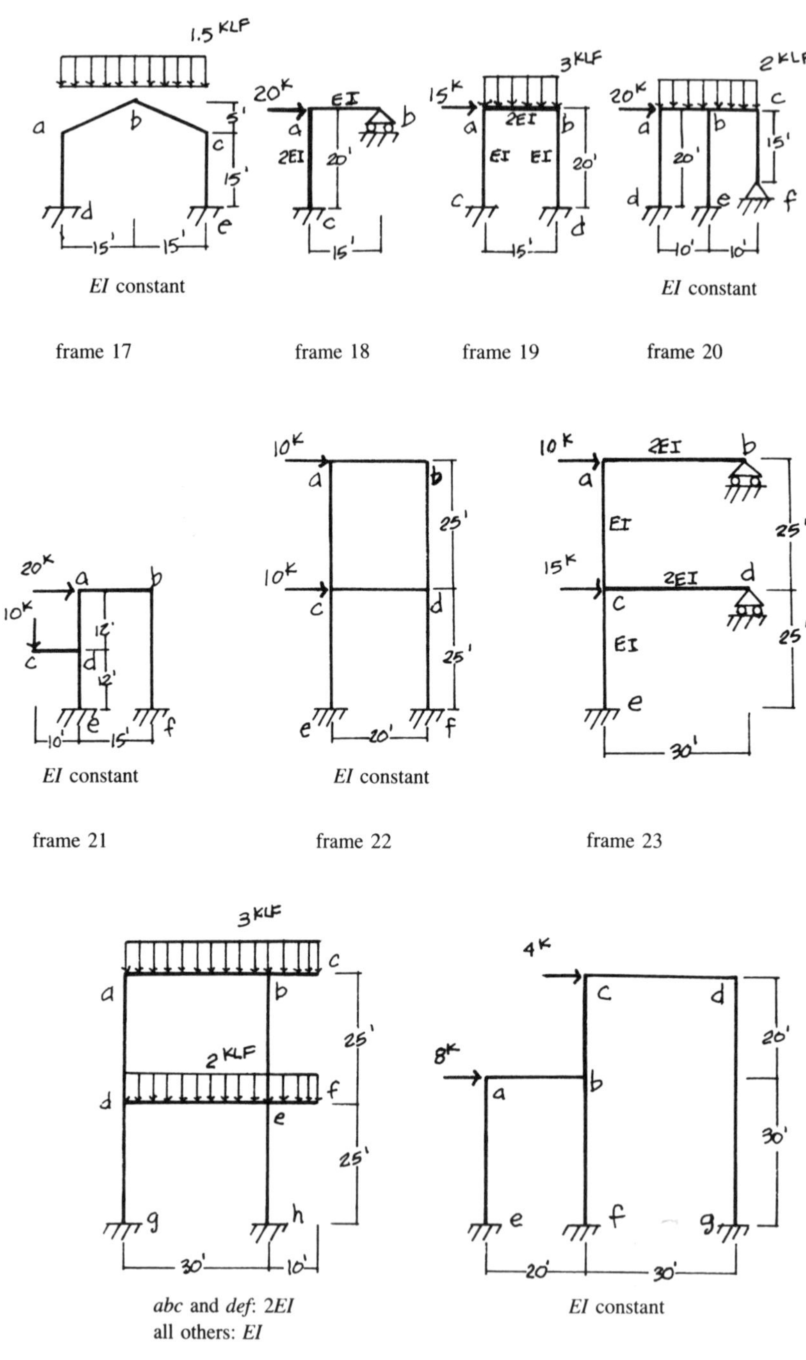

EI constant

frame 17

frame 18

frame 19

EI constant

frame 20

EI constant

frame 21

EI constant

frame 22

frame 23

abc and *def*: 2*EI*
all others: *EI*

frame 24

EI constant

frame 25

8

MOMENT DISTRIBUTION

INTRODUCTION

Moment distribution, developed in 1930 by eminent American engineer and educator Hardy Cross, is an extremely clever scheme used for solution of indeterminant beams and frames. Like slope-deflection, moment distribution is a stiffness method. In fact, many of the terms used in moment distribution have already been introduced in the slope-deflection chapter.

Fundamentally, the two methods are similar but the forms they take obscure their likeness. Moment distribution avoids—except for problems with multiple chord rotations—forming simultaneous equations, as was done in slope-deflection. Instead, moment distribution resorts to a process called successive corrections whereby a trial solution is proposed based on all fixed-end (no end rotations) members. Successive corrections are made which reconcile the errors in joint displacements and joint equilibrium. In slope-deflection, joint displacement compatability and joint equilibrium are explicitly stated; in moment distribution, these basic conditions are implicitly used. While slope-deflection, with its set of simultaneous equations, often requires computer aid, moment distribution lends itself nicely to hand computation.

The assumptions of moment distribution are the same as those in slope-deflection. Further, we will consider only prismatic members although the method can be extended to nonprismatic ones (Reference 1). Also, as in slope-deflection, short cuts will be introduced for several special cases.

COMPONENT FACTORS

Four of the five component factors we need for moment distribution have already been discussed in the slope-deflection chapter. The accompanying chart reviews these four factors. The fifth, the distribution factor, is defined as the proportion of the applied moment at a joint which is resisted by a specific member framing into the joint. Consider the frame with the rigid joint q, Figure 8.5.

Fixed-end moments: the moments at the walls for various loading conditions. The F.E.M. sign convention is used: C.W. on beam and C.C.W. on joint are positive.	Example 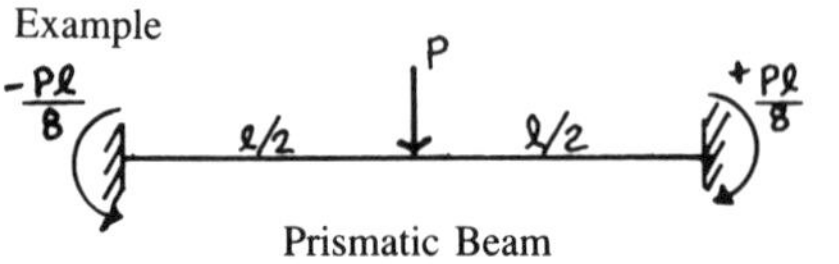 $-\frac{P\ell}{8}$ P $\ell/2$ $\ell/2$ $+\frac{P\ell}{8}$ Prismatic Beam **Figure 8.1**
Chord rotation moments: the moments at the walls due to relative displacements of ends. F.E.M. sign convention is used.	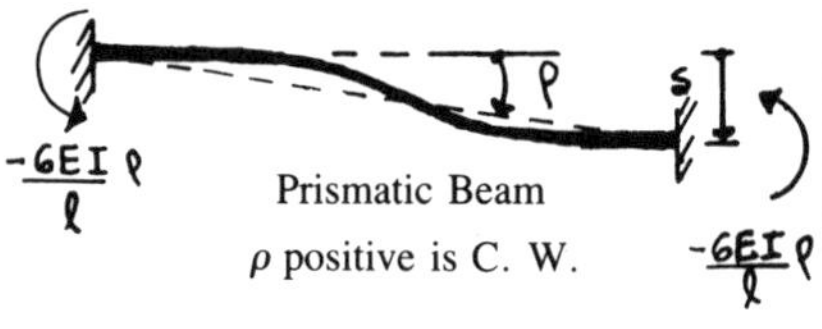 $-\frac{6EI}{\ell}\rho$ ρ δ $-\frac{6EI}{\ell}\rho$ Prismatic Beam ρ positive is C. W. **Figure 8.2**
Stiffness: the ratio of applied moment to corresponding rotation ($K_{ab} = M_b/\theta_b$). "Regular" stiffness is based on fixed/pin beam. F.E.M. sign convention is used.	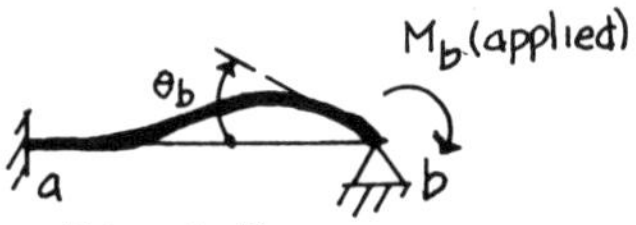 M_b (applied) θ_b a b Prismatic Beam $K_{ab} = \lvert M_b/\theta_b \rvert = 4EI/\ell$ **Figure 8.3**
Carry-over factor: the ratio of the reaction moment to the applied moment. "Regular" carry-over is based on fixed/pin beam. F.E.M. sign convention is used.	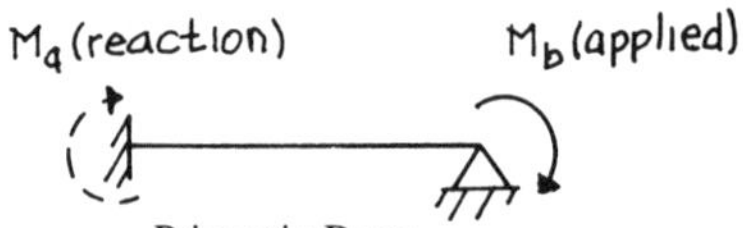 M_a (reaction) M_b (applied) Prismatic Beam $M_a = (1/2)M_b$. Both C. W.; carry-over factor is (1/2). **Figure 8.4**

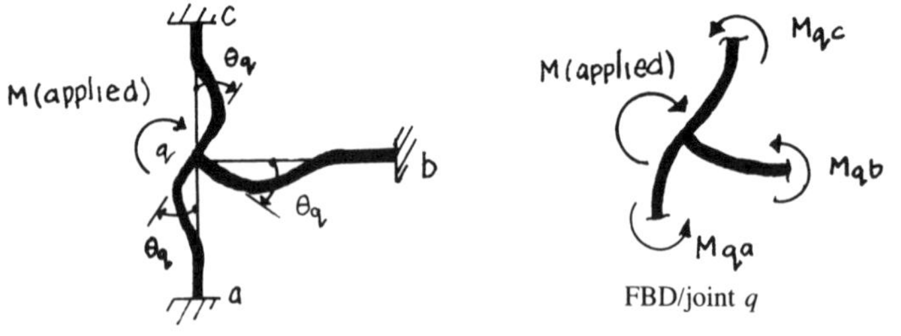

FBD/joint q

From statics $M = M_{qa} + M_{qb} + M_{qc}$.

Figure 8.5

By definition of stiffness

$$M_{qa} = K_{qa}\theta_q; \quad M_{qb} = K_{qb}\theta_q; \quad M_{qc} = K_{qc}\theta_q.$$

Thus,

$$M = K_{qa}\theta_q + K_{qb}\theta_q + K_{qc}\theta_q = (K_{qa} + K_{qb} + K_{qc})\theta_q = \left(\sum K_q\right)\theta_q \qquad \text{or}$$

$$\theta_q = \frac{M}{\Sigma K_q} \qquad \therefore$$

$$M_{qa} = K_{qa}\left(\frac{M}{\Sigma K_q}\right) = M\left(\frac{K_{qa}}{\Sigma K_q}\right).$$

Similarly,

$$M_{qb} = M\left(\frac{K_{qb}}{\Sigma K_q}\right) \text{ and } M_{qc} = M\left(\frac{K_{qc}}{\Sigma K_q}\right).$$

Thus we see that the proportion of the applied moment each member sees is related to its stiffness and inversely related to the total joint stiffness. That is, the distribution factor for, say, member *qa* is

$$DF\,\big|_{qa} = \frac{K_{qa}}{\Sigma K_q}.$$

Note the sum of distribution factors at a joint must be unity. Using this fact we investigate two specific cases. The distribution factor for the *beam* at a wall is zero since all the moment is absorbed by the *support* thus none by the *beam*. For the *beam* at a pin or roller, the distribution factor is unity. Here the *support* sees none of the moment, meaning all the moment goes to the *beam*.

The above derivation of the distribution factor illustrates an important concept dealing with the structural behavior of statically indeterminate structures. In general, for statically indeterminate structures, when it is possible for several members to resist a given load, the stiffer member will contribute the greatest resistance. Here the idea of "stiffness" is extended to axial load stiffness, shear stiffness as well as bending stiffness. From a design point of view, this concept means that by designing certain members very stiff (compared to others) the designer can determine the fundamental paths the load will take on its way to the foundation. This is true only for statically indeterminate structures. For statically determinate structures, deformation and thus stiffness has nothing to do with load distribution. The equations of equilibrium do not involve stiffness terms.

METHOD RATIONALE

Let's look at a two-span indeterminate beam problem—a simple problem that, hopefully, will clearly illustrate the method rationale. First, a warning: While all the interacting activity may make moment distribution appear complicated, we'll see that the clever bookkeeping scheme reduces the complexity tremendously.

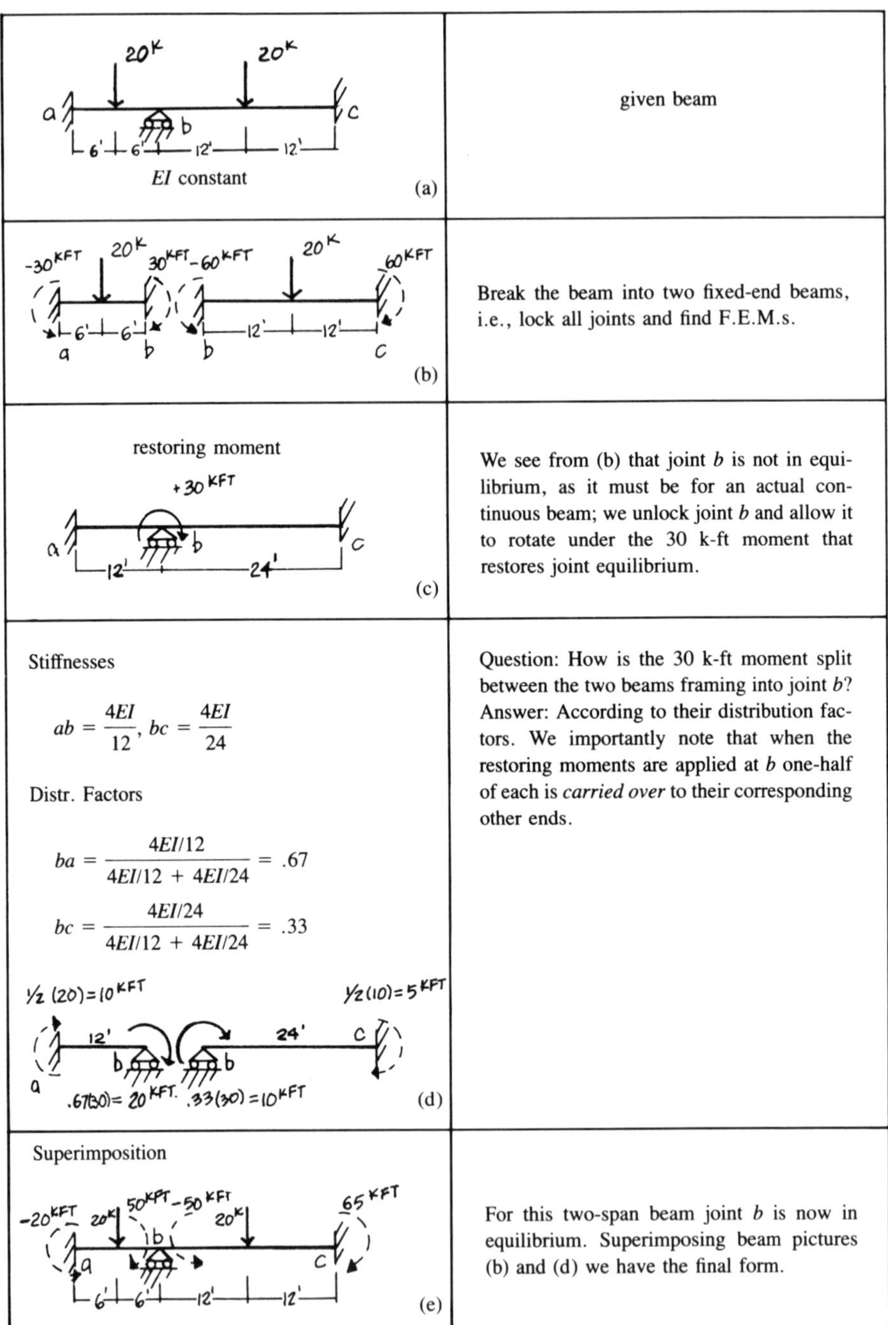

Figure 8.6 Rationale of M-D for a two-span beam

Note that if, instead, we had three spans, with internal *b* and *c*, that when a restoring moment is applied to joint *b* the carry-over moment to joint *c* will cause *c* once again to be out of balance; we balance *c* and carry over to *b*, and etc. Each cycle generally reduces the magnitude of the restoring moment needed, thus convergence to a solution results. Example 1, with its running commentary, illustrates; Example 2 treats a frame; and Example 3 investigates a beam with settlement.

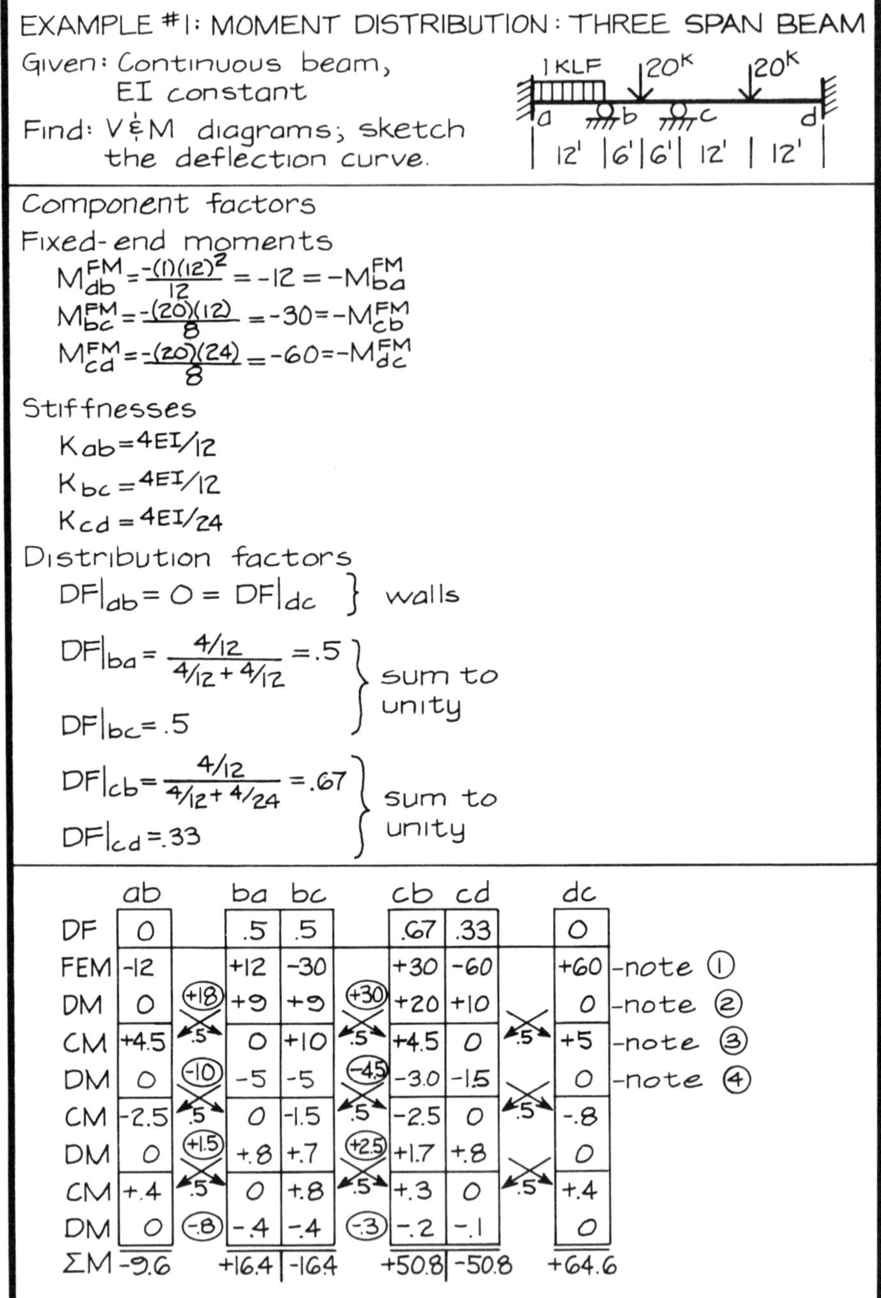

EXAMPLE #1: MOMENT DISTRIBUTION: THREE SPAN BEAM

Given: Continuous beam, EI constant

Find: V & M diagrams; sketch the deflection curve.

Component factors

Fixed-end moments

$$M_{db}^{FM} = \frac{-(1)(12)^2}{12} = -12 = -M_{ba}^{FM}$$

$$M_{bc}^{FM} = \frac{-(20)(12)}{8} = -30 = -M_{cb}^{FM}$$

$$M_{cd}^{FM} = \frac{-(20)(24)}{8} = -60 = -M_{dc}^{FM}$$

Stiffnesses

$K_{ab} = 4EI/12$

$K_{bc} = 4EI/12$

$K_{cd} = 4EI/24$

Distribution factors

$DF|_{ab} = 0 = DF|_{dc}$ } walls

$DF|_{ba} = \dfrac{4/12}{4/12 + 4/12} = .5$

$DF|_{bc} = .5$

} sum to unity

$DF|_{cb} = \dfrac{4/12}{4/12 + 4/24} = .67$

$DF|_{cd} = .33$

} sum to unity

	ab		ba	bc		cb	cd		dc	
DF	0		.5	.5		.67	.33		0	
FEM	-12		+12	-30		+30	-60		+60	-note ①
DM	0	(+18)	+9	+9	(+30)	+20	+10		0	-note ②
CM	+4.5	.5	0	+10	.5	+4.5	0	.5	+5	-note ③
DM	0	(-10)	-5	-5	(-4.5)	-3.0	-1.5		0	-note ④
CM	-2.5	.5	0	-1.5	.5	-2.5	0	.5	-.8	
DM	0	(+1.5)	+.8	+.7	(+2.5)	+1.7	+.8		0	
CM	+.4	.5	0	+.8	.5	+.3	0	.5	+.4	
DM	0	(-.8)	-.4	-.4	(-.3)	-.2	-.1		0	
ΣM	-9.6		+16.4	-16.4		+50.8	-50.8		+64.6	

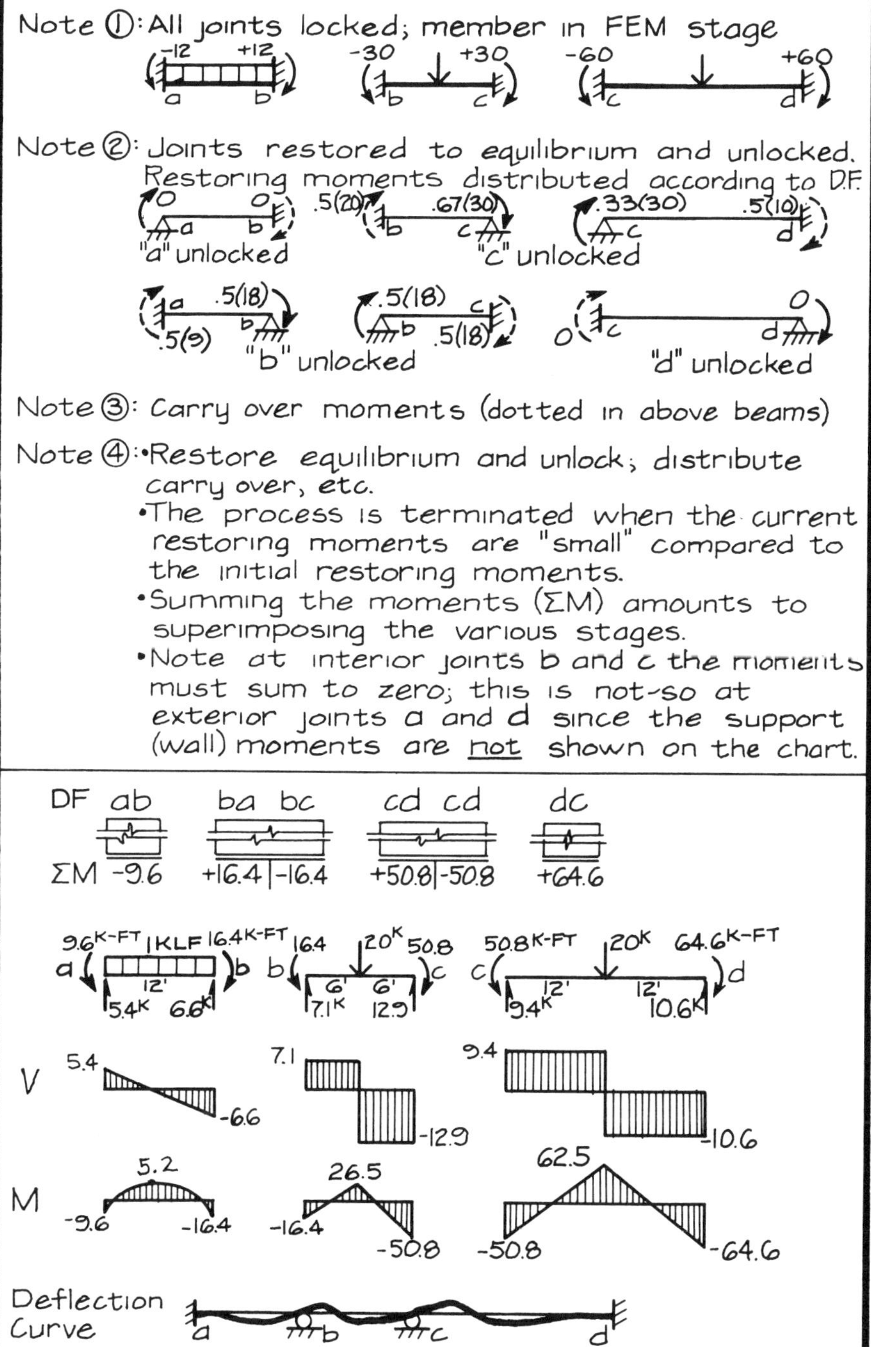
Note ①: All joints locked; member in FEM stage
-12 +12 -30 +30 -60 +60
a b b c c d
Note ②: Joints restored to equilibrium and unlocked. Restoring moments distributed according to D.F.
0 0 .5(20) .67(30) .33(30) .5(10)
"a" unlocked
"c" unlocked
.5(18) .5(9) .5(18) .5(18) 0 0
"b" unlocked
"d" unlocked
Note ③: Carry over moments (dotted in above beams)
Note ④: •Restore equilibrium and unlock; distribute carry over, etc.
•The process is terminated when the current restoring moments are "small" compared to the initial restoring moments.
•Summing the moments (ΣM) amounts to superimposing the various stages.
•Note at interior joints b and c the moments must sum to zero; this is not-so at exterior joints a and d since the support (wall) moments are not shown on the chart.
DF ab ba bc cd cd dc
ΣM -9.6 +16.4 -16.4 +50.8 -50.8 +64.6
9.6K-FT 1KLF 16.4K-FT 16.4 20K 50.8 50.8K-FT 20K 64.6K-FT
a b b c c d
12' 6' 6' 12' 12'
5.4K 6.6K 7.1K 12.9 9.4K 10.6K
V
5.4 -6.6 7.1 -12.9 9.4 -10.6
M
5.2 -9.6 -16.4 26.5 -16.4 -50.8 62.5 -50.8 -64.6
Deflection Curve
a b c d

EXAMPLE #2: MOMENT DISTRIBUTION
FRAME WITHOUT JOINT TRANSLATION
(Same as ex.#1 of slope deflection chapter)

Given: Rigid-jointed frame; EI values shown on frame.

Find: V and M diagrams; sketch the deflection curve.

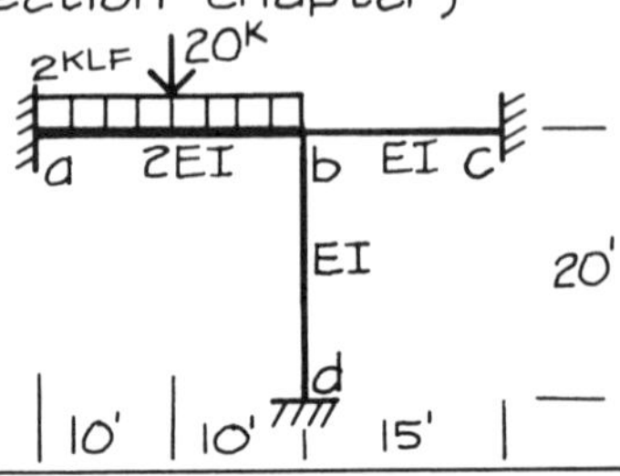

Component factors

Fixed-end moments

$M_{ab}^{FM} = \frac{-(2)(20)^2}{12} - \frac{(20)(20)}{8} = -116.7$

$M_{ba}^{FM} = +116.7$

$M_{bc}^{FM} = M_{bd}^{FM} = M_{cb}^{FM} = M_{db}^{FM} = 0$

Stiffnesses

$K_{ab} = 4(2EI)/20 \sim .4$

$K_{bc} = 4(EI)/15 \sim .27$

$K_{bd} = 4(EI)/20 \sim .2$

Distribution factors

$DF|_{ba} = \frac{.4}{.4+.27+.2} = .46$

$DF|_{bc} = \frac{.27}{.87} = .31$

$DF|_{bd} = \frac{.2}{.87} = .23$

$DF|_{ab} = DF|_{cb} = DF|_{db} = 0$

carry-over (.5)

	ab	ba	bd	bc	cb	db
DF	0	.46	.23	.31	0	0
FM	-116.7	+116.7	0	0	0	0
DM	0	-53.7	-26.8	-36.2	0	0
CM	-26.9	0	0	0	-18.1	-13.4
DM	0	0	0	0	0	0
ΣM	-143.6	+63.0	-26.8	-36.2	-18.1	-13.4

Note: The resultant moment values compare favorably with those found by slope deflection.

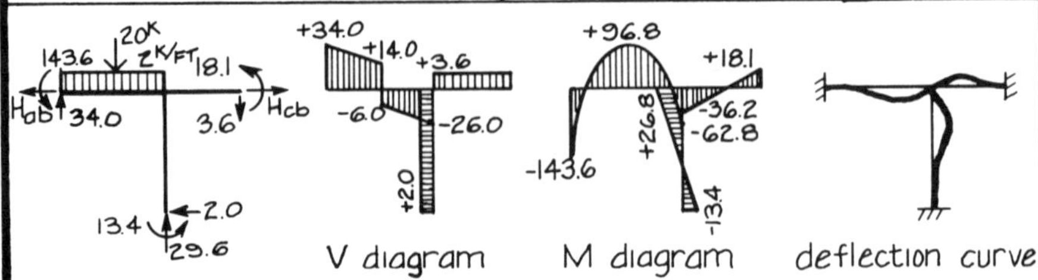

EXAMPLE #3: MOMENT DISTRIBUTION: BEAM WITH SETTLEMENT

(Same as ex. #2 of slope deflection chapter)

Given: Continuous beam with constant EI; $E = 30\times10^3$ KSI, $I = 499$ in^4; support "b" settles 3".

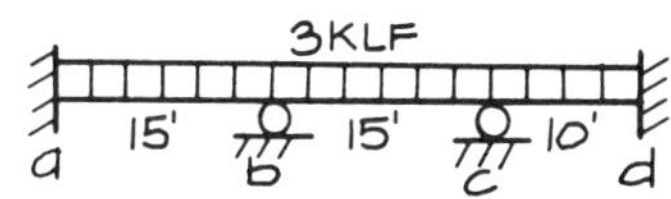

Find: V and M diagrams; sketch the deflection curve.

Component factors

Fixed-end moments

The moments due to chord rotations are treated exactly like F.E.M.s due to loads.

ρ_{ab} ρ_{bc} 3"

$\rho_{ab} = +\frac{.25'}{15'} = +.01667$ $\qquad EI = 1.04\times10^5$ K-FT^2

$\rho_{bc} = -\frac{.25'}{15'} = -.01667$

$\rho_{cd} = 0$

$M_{ab}^{FM} = \frac{-(3)(15)^3}{12} - \frac{6EI\rho_{ab}}{\ell_{ab}} = -56.3 - 693.5 \approx -750^{K\text{-}FT}$

$M_{ba}^{FM} = +56.3 - 693.5 \approx -637^{K\text{-}FT}$

$M_{bc}^{FM} = -56.3 + 693.5 \approx +637^{K\text{-}FT}$

$M_{cb}^{FM} = +56.3 + 693.5 \approx +750^{K\text{-}FT}$

$M_{cd}^{FM} = -25 = -M_{dc}^{FM}$

Stiffnesses

$K_{ab} = 4EI/15'$; $K_{bc} = 4EI/15'$; $K_{cd} = 4EI/10'$

Distribution factors

$DF|_{ba} = \frac{.27}{.27+.27} = .5$

$DF|_{bc} = .5$

$DF|_{cb} = \frac{.27}{.27+.4} = .4$

$DF|_{cd} = .6$

$DF|_{ab} = DF|_{bc} = 0$

	ab	(Joint b)	ba	bc	(Joint c)	cb	cd	dc
DF	0		.5	.5		.4	.6	0
FM	-750		-637	+637	(-725)	+750	-25	+25
DM	0	(0)	0	0		-290	-435	0
CM	0	(+145)	0	-145	(0)	0	0	-218
DM	0		+73	+72		0	0	0
CM	+36	(0)	0	0	(-36)	+36	0	0
DM	0		0	0		-14	-22	0
CM	0	(+7)	0	-7	(0)	0	0	-11
DM	0		+3	+4		0	0	0
CM	+2	(0)	0	0	(-2)	+2	0	0
DM	0		0	0		-1	-1	0
ΣM	-712		-561	+561		+483	-483	-204

712 K-FT 3KLF 204 K-FT
107 K 109 K 176 K 54 K

deflection curve

+107 K +62 K +84 K +54 K
-47 K -92 K
V diagram

+561 K-FT +204 K-FT
-712 K-FT -483 K-FT
M diagram

SIDESWAY

Sidesway and its effects are handled in a fashion similar to slope deflection. First, all chord rotations are identified by investigating the joint translations of the structure (see the sidesway section in the slope-deflection chapter). These chord rotations (with magnitudes unknown) give rise to moments ($6EI\rho/\ell$); the chord rotation moments are treated exactly like fixed-end moments, i.e., they are distributed and carried over, etc. The magnitudes of the chord rotations are found via statics usually using the balance of shear forces for each story. Example 4 illustrates the process for a one-story frame; Example 7 treats a frame of two stories.

MODIFICATIONS

Pin Modification

When an exterior pin or roller exists in a structure, the resulting moment at the pin or roller is, of course, zero. This predetermined fact allows a modification in the M-D process which causes a more rapid convergence. Consider the beam shown (Figure 8.7). Here we note that member *ab* is terminated with an exterior pin, *a*. In our first cycle we still lock all joints, including *a* and *b*. Then we unlock; unlock joint *a* and carry over to *b* as usual; unlock joint *b* but instead of treating joint *a* as a fixed support as usual, we treat joint *a* as a pin, Figure 8.8.

Figure 8.7

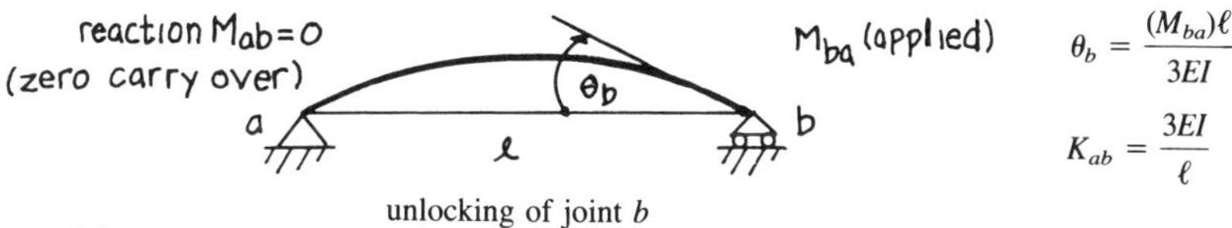

$$\theta_b = \frac{(M_{ba})\ell}{3EI}$$

$$K_{ab} = \frac{3EI}{\ell}$$

Figure 8.8

With joint *a* treated as a pin, two important modifications occur: there is no carry-over to joint *a* from joint *b* and the stiffness of the member is reduced: $K_{ab}|\text{pin} = 3EI/\ell$ (prismatic members). As a recapitulation of the pin modification, for only those members that have an exterior pin or roller reduce the stiffness of the member to $3EI/\ell$ (this of course alters distribution factors), carry over once from the exterior pin or roller to the interior joint but do not carry over at all from the interior to the pin or roller. Example 5 illustrates the pin modification. Note in Example 5 how the restoring moment at c rapidly diminishes when the carry-over from d is stopped; one cycle later this effect is felt at joint b (-16 to -1.5).

Symmetry Modification

When a structure with an odd number of spans is geometrically symmetric and symmetrically loaded, a modification of great time-saving can be used. Consider the three-span beam in Figure 8.9. By inspection it is clear $|M_{ab}| = |M_{dc}|$; $|M_{ba}| = |M_{cd}|$ and most importantly $|M_{bc}| = |M_{cb}|$. After fixing all joints, we unlock them. Consider, in particular, member bc: using symmetry we can unlock both joints at the same time and consider M_{bc} and M_{cb} as applied loads. Calculating the stiffness in the usual way, we find $K_{bc}|\text{symm} = 2EI/\ell$ (prismatic members). Further, no carry-over takes place from joint b to joint c or from joint c to joint b.

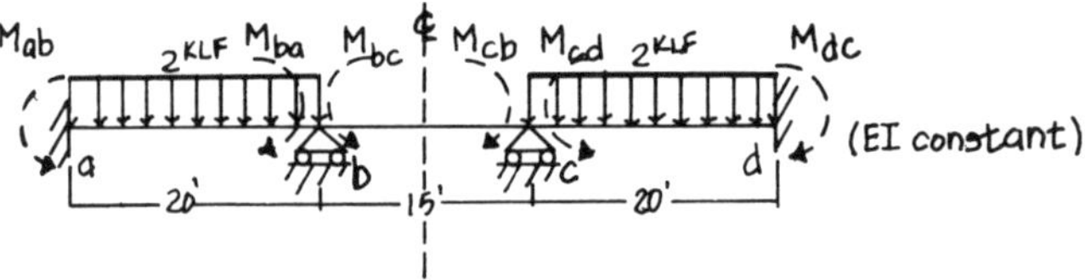

Figure 8.9

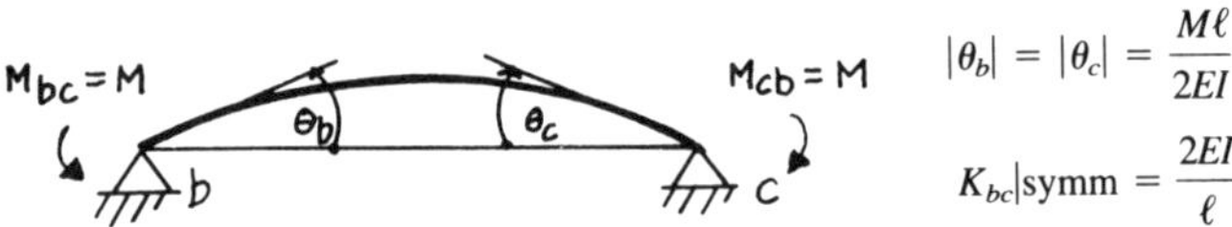

Figure 8.10

It is most important to realize that the stiffness is modified only for those members that have equal end moments. For the above illustration $|M_{ab}| \neq |M_{ba}|$ therefore the stiffness for ab is not modified.

Recapping, for symmetric structures of odd number spans, the "linking" spans have their stiffnesses modified to $2EI/\ell$; the chart is symmetrically reduced and no carry-over takes place across the line of symmetry. Example 6 illustrates. For symmetric problems of even-number spans, no stiffness or carry-over modification is used. However, due to symmetry, only half of the structure need be dealt with and generally the rotation is zero over the mid-support (see the symmetry section in the slope-deflection chapter).

Antisymmetry Modification

When a structure is geometrically symmetric but antisymmetrically loaded a modification can be used in the moment distribution process. Consider the beam shown (Figure 8.11). Here span *bc* is the "linking" member. Again the stiffness of *bc* is altered—and only *bc*. We find (Figure 8.12), $K_{bc}|\text{antisymm} = 6EI/\ell$ (prismatic member). No carry-over takes place across the centerline and only one-half of the structure need be analyzed. The chart is similar to that for the symmetry modification.

For an antisymmetric frame (Figure 8.13), the stiffness of *bc* (only) is modified and only *ab*, *ba* and *bc* need be charted. There is no carry-over to or from *bc*. See Example 17.

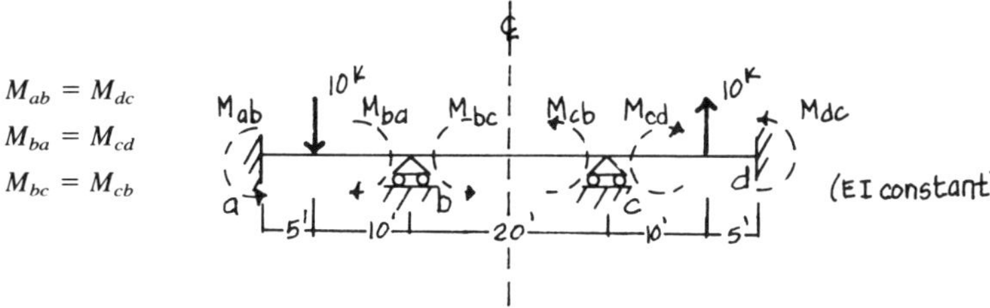

Figure 8.11

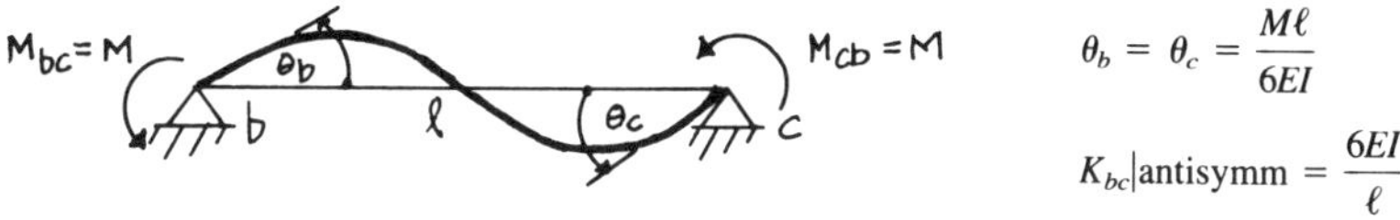

Figure 8.12

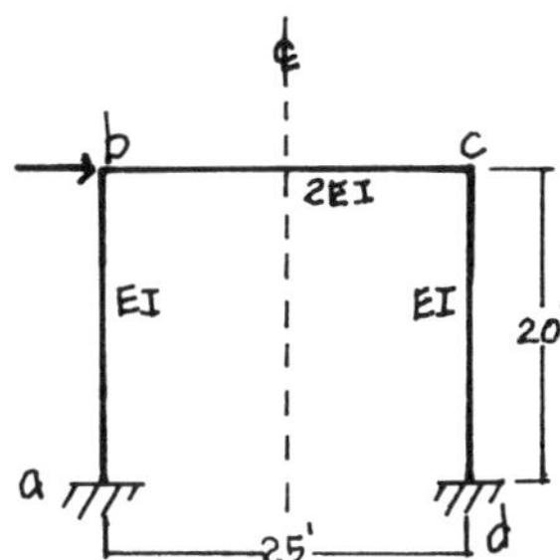

Figure 8.13

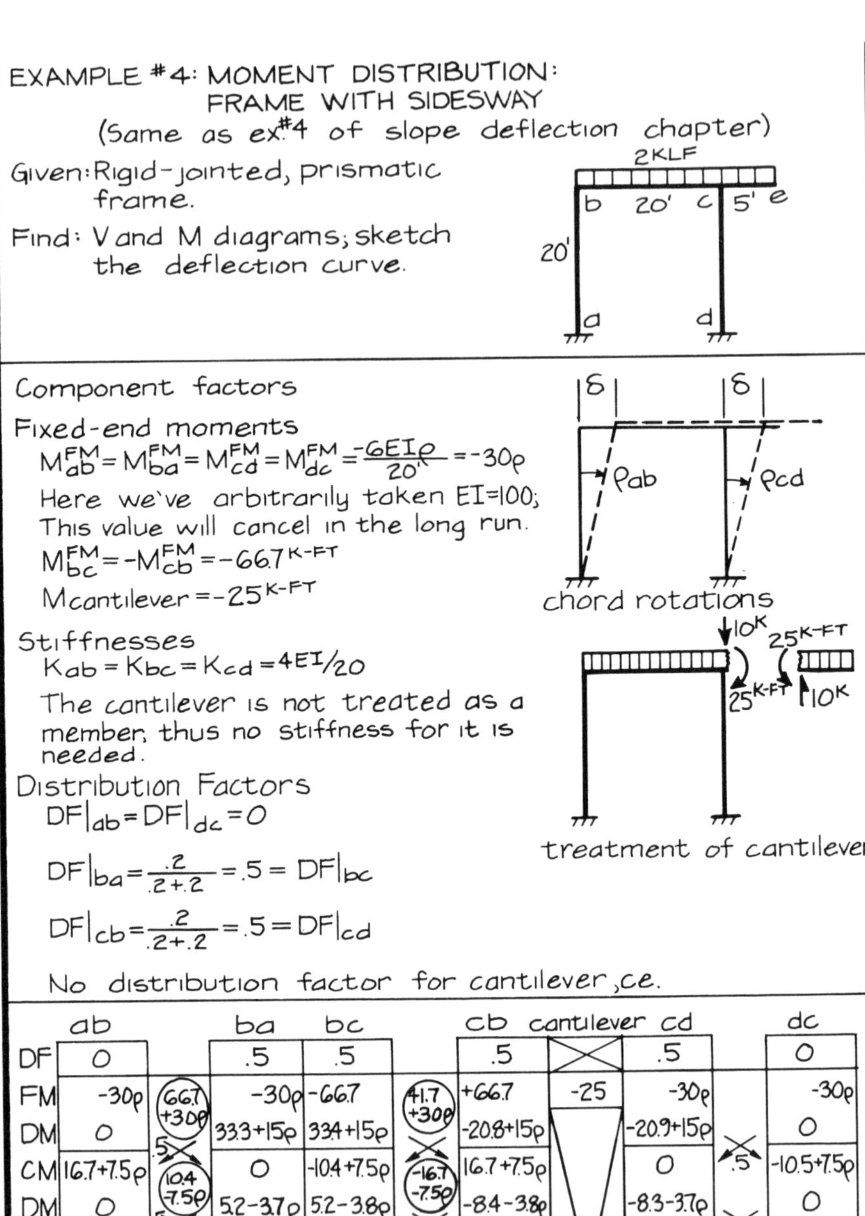

EXAMPLE #4: MOMENT DISTRIBUTION:
FRAME WITH SIDESWAY

(Same as ex.#4 of slope deflection chapter)

Given: Rigid-jointed, prismatic frame.

Find: V and M diagrams; sketch the deflection curve.

Component factors

Fixed-end moments

$M^{FM}_{ab} = M^{FM}_{ba} = M^{FM}_{cd} = M^{FM}_{dc} = \frac{-6EI\rho}{20} = -30\rho$

Here we've arbitrarily taken EI=100; This value will cancel in the long run.

$M^{FM}_{bc} = -M^{FM}_{cb} = -66.7^{K\text{-}FT}$

$M_{cantilever} = -25^{K\text{-}FT}$

chord rotations

Stiffnesses

$K_{ab} = K_{bc} = K_{cd} = 4EI/20$

The cantilever is not treated as a member, thus no stiffness for it is needed.

treatment of cantilever

Distribution Factors

$DF|_{ab} = DF|_{dc} = 0$

$DF|_{ba} = \frac{.2}{.2+.2} = .5 = DF|_{bc}$

$DF|_{cb} = \frac{.2}{.2+.2} = .5 = DF|_{cd}$

No distribution factor for cantilever, ce.

	ab	b bal.	ba	bc	c bal.	cb	cantilever	cd	dc
DF	0		.5	.5		.5		.5	0
FM	-30ρ	66.7 +30ρ	-30ρ	-66.7	41.7 +30ρ	+66.7	-25	-30ρ	-30ρ
DM	0		33.3+15ρ	33.4+15ρ		-20.8+15ρ		-20.9+15ρ	0
CM	16.7+7.5ρ	10.4 -7.5ρ	0	-10.4+7.5ρ	-16.7 -7.5ρ	16.7+7.5ρ		0	-10.5+7.5ρ
DM	0		5.2-3.7ρ	5.2-3.8ρ		-8.4-3.8ρ		-8.3-3.7ρ	0
CM	2.6-1.9ρ	4.24 +1.9ρ	0	-4.2-1.9ρ	-2.6 +1.9ρ	2.6-1.9ρ		0	-4.2-1.9ρ
DM	0		2.1+.9ρ	2.1+1.0ρ		-1.3+1.0ρ		-1.3+.9ρ	0
CM	1.0+.5ρ	.6 -.5ρ	0	-.6+.5ρ	-1.0 -.5ρ	1.0+.5ρ		0	-.6+.5ρ
DM	0		.3-.2ρ	+.3-.3ρ		-.5-.3ρ		-.5-.2ρ	0
CM	.2-.1ρ	.3 +.2ρ	0	-.3-.2ρ	-.2 +.2ρ	+.2-.2ρ		0	-.2-.1ρ
DM	0		.1+.1ρ	+.2+.1ρ		-.1+.1ρ		-.1+.1ρ	0
ΣM	20.5 -24.0ρ		41.0 -17.9ρ	-41.0 +17.9ρ		56.1 +17.9ρ	-25	-31.1 -17.9ρ	-15.5 -24.0ρ

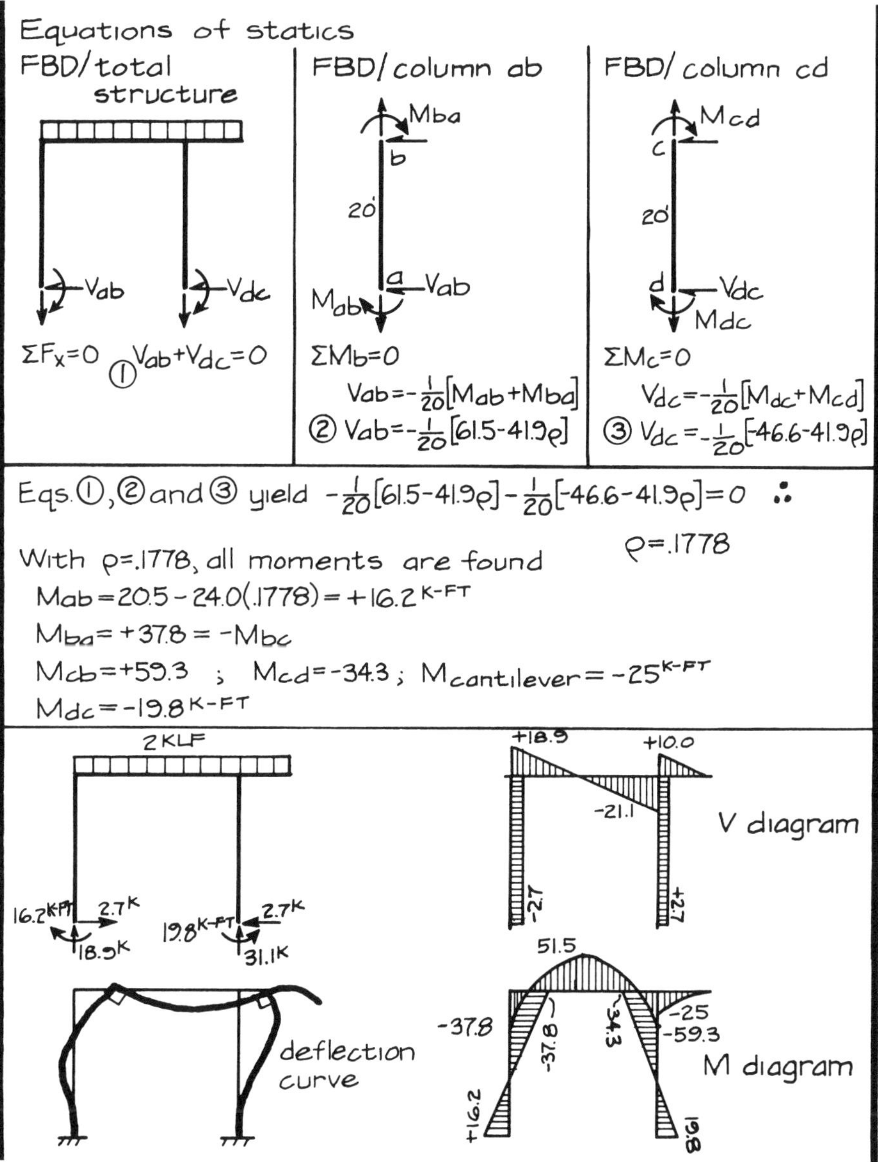

Equations of statics
FBD/total structure
FBD/column ab
FBD/column cd
Mba
b
20'
a
Vab
Mab
Mcd
c
d
Vdc
Mdc
ΣFx=0 ① Vab+Vdc=0
ΣMb=0
Vab=-1/20[Mab+Mba]
② Vab=-1/20[61.5-41.9ρ]
ΣMc=0
Vdc=-1/20[Mdc+Mcd]
③ Vdc=-1/20[-46.6-41.9ρ]
Eqs. ①, ② and ③ yield -1/20[61.5-41.9ρ]-1/20[-46.6-41.9ρ]=0 ∴
ρ=.1778
With ρ=.1778, all moments are found
Mab=20.5-24.0(.1778)=+16.2 K-FT
Mba=+37.8=-Mbc
Mcb=+59.3 ; Mcd=-34.3 ; Mcantilever=-25 K-FT
Mdc=-19.8 K-FT
2KLF
16.2 K-FT
2.7K
18.9K
19.8 K-FT
2.7K
31.1K
deflection curve
+18.9
+10.0
-21.1
-2.7
+2.7
V diagram
51.5
-37.8
-37.8
-34.3
-25
-59.3
+16.2
19.8
M diagram

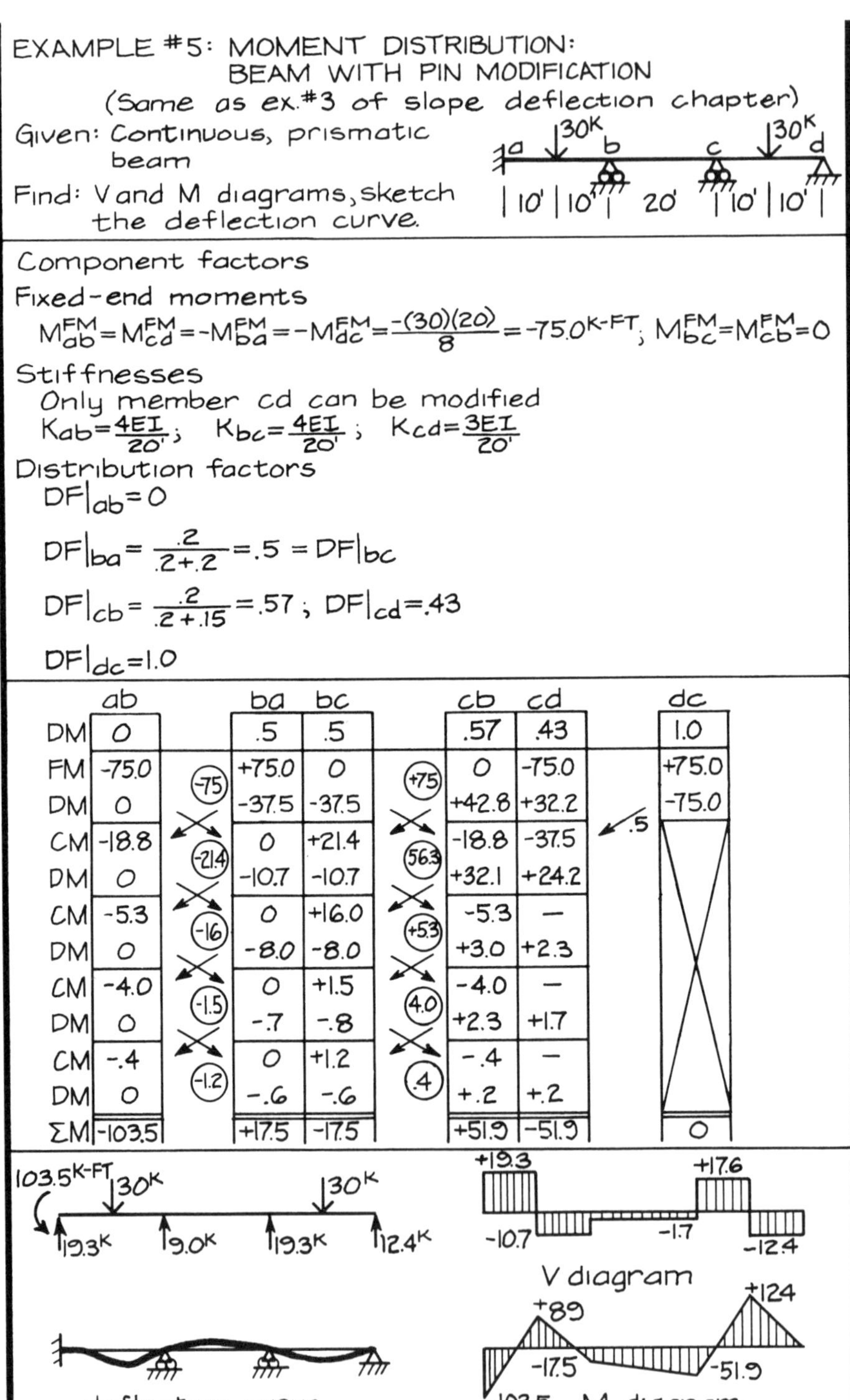

EXAMPLE #5: MOMENT DISTRIBUTION: BEAM WITH PIN MODIFICATION

(Same as ex. #3 of slope deflection chapter)

Given: Continuous, prismatic beam

Find: V and M diagrams, sketch the deflection curve.

Component factors

Fixed-end moments

$M^{FM}_{ab} = M^{FM}_{cd} = -M^{FM}_{ba} = -M^{FM}_{dc} = \frac{-(30)(20)}{8} = -75.0^{K\text{-}FT}$; $M^{FM}_{bc} = M^{FM}_{cb} = 0$

Stiffnesses

Only member cd can be modified

$K_{ab} = \frac{4EI}{20'}$; $K_{bc} = \frac{4EI}{20'}$; $K_{cd} = \frac{3EI}{20'}$

Distribution factors

$DF|_{ab} = 0$

$DF|_{ba} = \frac{.2}{.2+.2} = .5 = DF|_{bc}$

$DF|_{cb} = \frac{.2}{.2+.15} = .57$; $DF|_{cd} = .43$

$DF|_{dc} = 1.0$

	ab		ba	bc		cb	cd		dc
DM	0		.5	.5		.57	.43		1.0
FM	-75.0	(-75)	+75.0	0	(+75)	0	-75.0		+75.0
DM	0		-37.5	-37.5		+42.8	+32.2	.5	-75.0
CM	-18.8	(-21.4)	0	+21.4	(56.3)	-18.8	-37.5		
DM	0		-10.7	-10.7		+32.1	+24.2		
CM	-5.3	(-16)	0	+16.0	(+5.3)	-5.3	—		
DM	0		-8.0	-8.0		+3.0	+2.3		
CM	-4.0	(-1.5)	0	+1.5	(4.0)	-4.0	—		
DM	0		-.7	-.8		+2.3	+1.7		
CM	-.4	(-1.2)	0	+1.2	(.4)	-.4	—		
DM	0		-.6	-.6		+.2	+.2		
ΣM	-103.5		+17.5	-17.5		+51.9	-51.9		0

EXAMPLE #6: MOMENT DISTRIBUTION: BEAM ODD SPAN SYMMETRY

Given: Continuous beam; EI values as shown.

Find: V and M diagrams, sketch the deflection curve.

Component factors

Fixed-end moments

$M_{ab}^{FM} = -M_{ba}^{FM} = -66.7^{K\text{-}FT} = -M_{fe}^{FM} = M_{ef}^{FM} = \frac{-(2)(20)^2}{12}$

$M_{bc}^{FM} = -M_{cd}^{FM} = -37.5^{K\text{-}FT} = -M_{ed}^{FM} = M_{de}^{FM} = \frac{-(2)(15)^2}{12}$

$M_{cd}^{FM} = -M_{dc} = -62.5^{K\text{-}FT} = \frac{-(25)(20)}{8}$

Stiffnesses

$K_{ab} = 4(2EI)/20$; $K_{bc} = 4EI/15$; $K_{cd}\big|_{symm} = 2EI/20$ (only span cd is modified)

Distribution factors (only half of the beam is considered)

$DF|_{ab} = 0$

$DF|_{ba} = \frac{.4}{.4+.27} = .6$; $DF|_{bc} = .4$

$DF|_{cb} = \frac{.27}{.27+.7} = .73$; $DF|_{cd} = .27$

ab		ba	bc		cb	cd
0		.6	.4		.73	.27
-66.7	+29.2	+66.7	-37.5	+25	+37.5	-62.5
0		-17.5	-11.7		+18.3	+6.7
-8.7	9.2	0	+9.2	5.8	-5.8	—
0		-5.5	-3.7		+4.2	+1.6
-2.8	2.1	0	+2.1	+1.9	-1.9	—
0		-1.3	-.8		+1.4	+.5
-.7	.7	0	+.7	+.4	-.4	—
0		-.4	-.3		+.3	+.1
-.2	.2	0	+.2	+.1	-.1	—
0		-.1	-.1		+.1	0
-79.1		+41.9	-41.9		+53.6	-53.6

No carry over to or from cd

symmetrical about ℄

EXAMPLE #7: MOMENT DISTRIBUTION: FRAME WITH SIDESWAY – TWO DEGREES OF FREEDOM: ANTISYMMETRY MODIFICATION

(Same as ex#5 of slope deflection chapter)

Given: Two story rigid-jointed frame; EI as shown

Find: V and M diagrams; sketch the deflection curve.

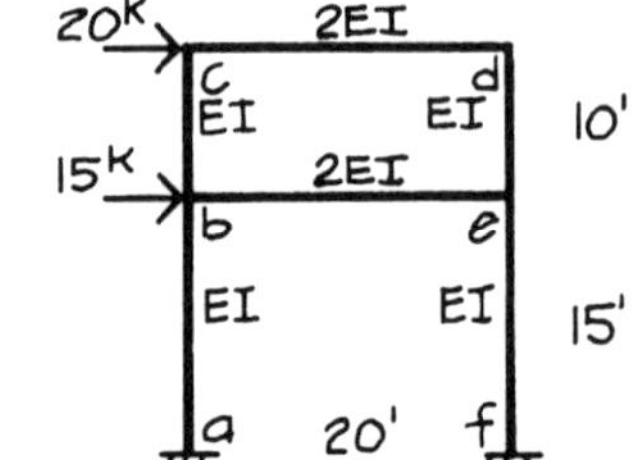

Component factors

Fixed-end moments

Loads cause no F.E.M.s, all due to sidesway.

$M^{FM}_{ab} = M^{FM}_{ba} = M^{FM}_{fe} = M^{FM}_{ef} = \frac{-6EI\rho_{ab}}{15'} = -40\rho_1$

Here we're taking, arbitrarily, EI=100; this value will cancel out.

$M^{FM}_{bc} = M^{FM}_{cb} = M^{FM}_{de} = M^{FM}_{ed} = \frac{-6EI\rho_{bc}}{10'} = -60\rho_2$

$M^{FM}_{be} = M^{FM}_{eb} = M^{FM}_{cd} = M^{FM}_{dc} = 0$

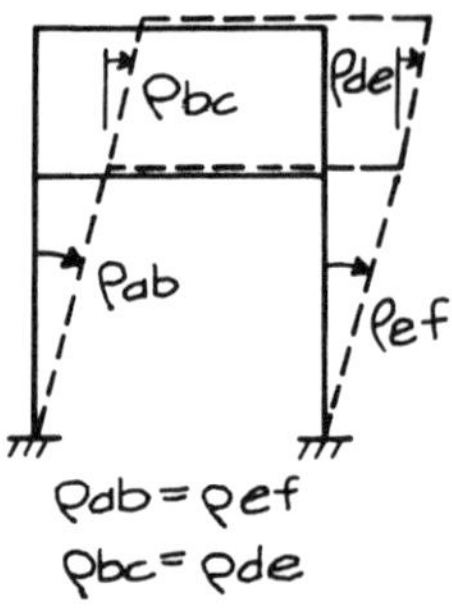

$\rho_{ab} = \rho_{ef}$

$\rho_{bc} = \rho_{de}$

Stiffnesses

Spans be and cd are the "linking" spans, therefore

$K_{be}\big|_{\substack{anti\\symm}} = \frac{6(2EI)}{20'}$; $K_{cd}\big|_{\substack{anti\\symm}} = \frac{6(2EI)}{20'}$

$K_{ab} = \frac{4EI}{15'}$; $K_{cb} = \frac{4EI}{10'}$

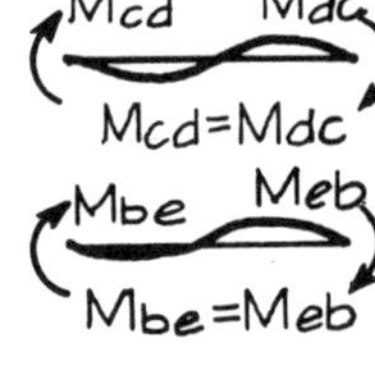

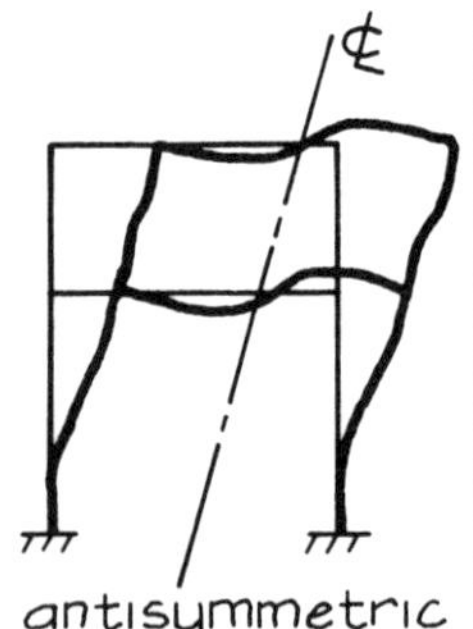

Distribution factors

$DF|_{ab} = 0$

$DF|_{ba} = \frac{.27}{.27+.4+.6} = .21$

$DF|_{be} = \frac{.6}{.27+.4+.6} = .47$

$DF|_{bc} = \frac{.4}{.27+.4+.6} = .32$

$DF|_{cb} = \frac{.4}{.4+.6} = .4$

$DF|_{cd} = \frac{.6}{.4+.6} = .6$

(no carryover to or from be, cd)

	ab		ba	be	bc		cb	cd
DF	0		.21	.47	.32		.4	.6
FM	$-40\rho_1$	$40\rho_1$ $+60\rho_2$	$-40\rho_1$	0	$-60\rho_2$	$+60\rho_2$	$-60\rho_2$	0
DM	0		$8.4\rho_1+12.6\rho_2$	$18.8\rho_1+28.2\rho_2$	$12.8\rho_1+19.2\rho_2$		$+24\rho_2$	$+36\rho_2$
CM	$4.2\rho_1+6.3\rho_2$	$-12\rho_2$	0	—	$+12\rho_2$	$-6.4\rho_1$ $-9.6\rho_2$	$6.4\rho_1+9.6\rho_2$	—
DM	0		$-2.5\rho_2$	$-5.6\rho_2$	$-3.9\rho_2$		$-2.6\rho_1-3.8\rho_2$	$-3.8\rho_1-5.8\rho_2$
CM	$-1.3\rho_2$	$+1.3\rho_1$ $+1.9\rho_2$	0	—	$-1.3\rho_1-1.9\rho_2$	$+1.9\rho_2$	$-1.9\rho_2$	—
DM	0		$.3\rho_1+.4\rho_2$	$.6\rho_1+.9\rho_2$	$.4\rho_1+.6\rho_2$		$.8\rho_2$	$1.1\rho_2$
CM	$.2\rho_1+.2\rho_2$	$-.4\rho_2$	0	—	$.4\rho_2$	$-.2\rho_1$ $-.3\rho_2$	$+.2\rho_1+.3\rho_2$	—
DM	0		$-.1\rho_2$	$-.2\rho_2$	$-.1\rho_2$		$-.1\rho_1-.1\rho_2$	$-.1\rho_1-.2\rho_2$
ΣM	$-35.6\rho_1$ $+5.2\rho_2$		$-31.3\rho_1$ $+10.4\rho_2$	$19.4\rho_1$ $+23.3\rho_2$	$11.9\rho_1$ $-33.7\rho_2$		$3.9\rho_1$ $-31.1\rho_2$	$-3.9\rho_1$ $+31.1\rho_2$

FBD/total structure

20^K, 15^K, V_{ab}, V_{ed}

FBD/ column ab

M_{ba}, b, 15', $V_{ab}=17.5^K$, a, M_{ab}

From total structure

$\Sigma F_x = 0$: $V_{ab}+V_{ed}=35^K$

due to antisymm: $V_{ab}=V_{ed}$ ∴ $V_{ab}=17.5^K$

From column ab

$\Sigma M_b=0$: $15(17.5)+M_{ab}+M_{ba}=0$

or using moment dist. results

$262.5=-(35.6\rho_1+5.2\rho_2)-(31.3\rho_1+10.4\rho_2)$

$262.5=66.9\rho_1-15.6\rho_2$ ①

FBD/ upper story

20^K, V_{bc}, V_{ed}

FBD/ column bc

M_{cb}, c, 10', $V_{bc}=10^K$, b, M_{bc}

From upper story

$\Sigma F_x=0$: $V_{bc}+V_{ed}=20^K$

due to antisymm: $V_{bc}=V_{ed}$

$V_{bc}=10^K$

From column bc

$\Sigma M_c=0$: $10(10)+M_{bc}+M_{cb}=0$

or using M-D results

$100=-(11.9\rho_1-33.7\rho_2)-(3.9\rho_1-31.1\rho_2)$

$100=-15.8\rho_1+64.8\rho_2$ ②

Solving ① and ② simultaneously yields:

$\rho_1=4.54$ or $EI\rho_{ab}=454$

$\rho_2=2.65$ or $EI\rho_{bc}=265$

The end-moments are calculated:

$M_{ab}=-35.6(4.54)+5.2(2.65)=-148^{K\text{-}FT}$ ($M_{FE}=-148$)

$M_{ba}=-115^{K\text{-}FT}$; $M_{be}=+150^{K\text{-}FT}$; $M_{bc}=-35^{K\text{-}FT}$

($M_{ef}=-115^{K\text{-}FT}$); ($M_{eb}=+150^{K\text{-}FT}$); ($M_{ed}=-35^{K\text{-}FT}$)

$M_{cb}=-65^{K\text{-}FT}$; $M_{ce}=+65^{K\text{-}FT}$

($M_{de}=-65^{K\text{-}FT}$); ($M_{ec}=+65^{K\text{-}FT}$)

The V and M are identical (except for some round-off differences) to those shown in ex.#5 of the slope deflection chapter. The deflection diagram is also shown there.

PROBLEMS

Problems 1–25

Do problems in slope-deflection chapter (Ch. 7, pages 168–170) by moment distribution.

9

MATRIX ANALYSIS OF STRUCTURES: THE STIFFNESS METHOD

INTRODUCTION

Matrix structural analysis is not a new theory of analysis but rather a bookkeeping procedure of classical methods. It is a technique for operating upon and solving the simultaneous equations which result from the classical analysis of structures. Reorganization of structural problems results.

Matrix structural analysis is computer oriented. In fact, if there were no digital computers, there would be no real need for matrix algebra because the use of matrices in hand calculations is very cumbersome. Indeed, in the example problems in this chapter it will seem that the matrix approach is a form of overkill. For relatively simple problems done by hand the matrix approach *is* overkill; but for complex problems done with computer aid the method shines.

There are two primary matrix structural approaches: the flexibility or force method and the stiffness or displacement method. Each method has its roots in classical approaches. Say, for statically indeterminate problems using moment-area, we formed a series of statically determinate problems by choosing, for instance, redundant support forces. Each of the resulting statically determinate structures was more flexible than the original structure; by geometric compatibility we were able to find the redundant forces. This procedure in matrix force is the flexibility or force method; support forces (or moments) are the unknowns.

The stiffness or displacement method is based on slope-deflection concepts; structures are made more rigid or stiffer by using fixed-end principles, then joints are released bringing forth elastic deformation terms. Finally, by joint equilibrium the unknown displacements are found.

Most commercially available structural analysis programs are based on the force method. Setting up and operating upon the basic matrices by computer gives an edge to the stiffness method. The possibility of choosing redundants that will render the structure unstable is also a handicap of the flexibility method in computer application. In the stiffness method there are no redundants to choose.

Matrix structural analysis has many levels of sophistication; in fact, entire books are written on just this subject. As the sophistication increases, the procedures become more mechanical; notation and bookkeeping become more detailed and precise, all as a prelude to computer interaction. Our intent here is to keep the level basic; we shall use only the stiffness method to deal with two-dimensional beams, frames, and trusses.

The stiffness method handles statically determinate, statically indeterminate, and sidesway problems, all with the same approach. Extensions can be made for settlement, temperature changes, elastic foundation, prestrains, and nonprismatic members (see Reference 4).

A short review of matrix algebra can be found in the appendix.

We shall introduce the stiffness method via the slope-deflection method for beams, then extend it to frames and trusses. Complete theoretical derivations can be found in Reference 4.

BEAMS

In Chapter 7, we found the basic slope-deflection Equations (8) and (9) for the end moments:

$$M_{ab} = M_{ab}^{FM} + \left\{\frac{EI}{L}\left[-6\left(\frac{\delta_a - \delta_b}{L}\right) + 4\theta_a + 2\theta_b\right]\right\} \tag{1}$$

$$M_{ba} = M_{ba}^{FM} + \left\{\frac{EI}{L}\left[-6\left(\frac{\delta_a - \delta_b}{L}\right) + 4\theta_b + 2\theta_a\right]\right\} \tag{2}$$

Here we substituted for $\rho = (\delta_a - \delta_b)/L$ (see Figure 7.1c). In these equations we have used the sign convention that clockwise is positive for the end-moments (M_{ab}, M_{ba}), the end rotations (θ_a, θ_b) and the fixed-end moment *reactions* (M_{ab}^{FM}, M_{ba}^{FM}).

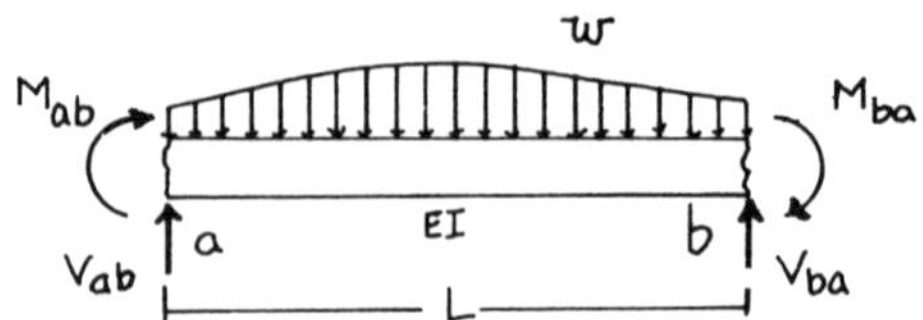

Figure 9.1

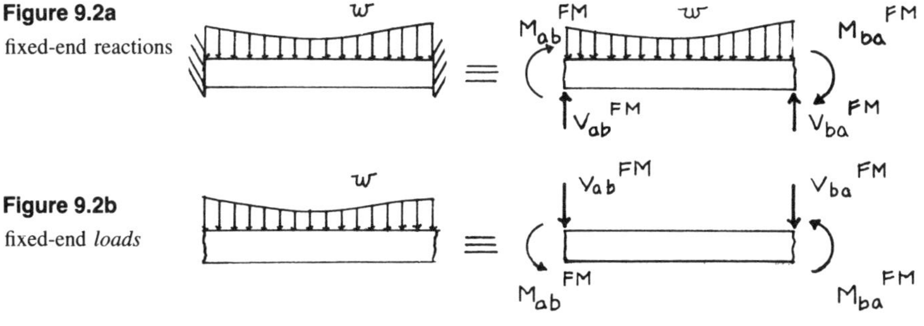

Figure 9.2a
fixed-end reactions

Figure 9.2b
fixed-end *loads*

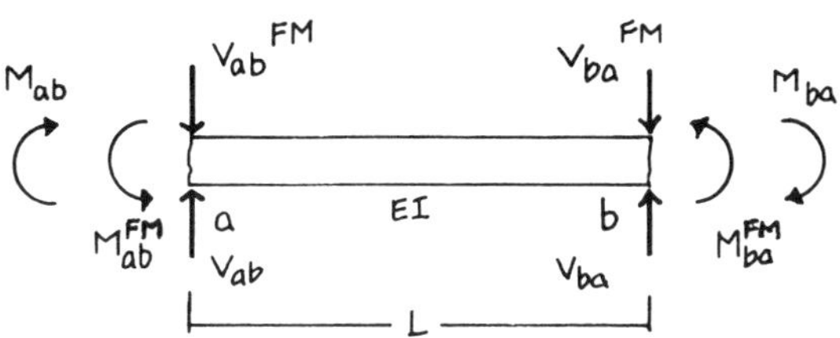

Figure 9.3

Let's look at the fixed-end reactions terms. Our sign convention takes clockwise fixed-end moment *reactions* as positive; we will take upward as positive for fixed-end shear *reactions*. In Figure 9.2a we see the reactions to the loading; in Figure 9.2b we see that the loading is *replaced* by fixed-end terms; here the "equivalent loads" are in the opposite directions as the reactions. Now consider the beam element of Figure 9.1 with fixed-end loads as shown in Figure 9.3. By taking moments about point b we find

$$(V_{ab} - V_{ab}^{FM})L + (M_{ab} - M_{ab}^{FM}) + (M_{ba} - M_{ba}^{FM}) = 0 \qquad \text{or}$$

$$V_{ab} = V_{ab}^{FM} - \frac{M_{ab} - M_{ab}^{FM}}{L} - \frac{M_{ba} - M_{ba}^{FM}}{L}.$$

Using Equations 1 and 2 in the above equation yields:

$$V_{ab} = V_{ab}^{FM} + \left\{\frac{EI}{L^2}\left[\frac{12}{L}(\delta_a - \delta_b) - 6\theta_a - 6\theta_b\right]\right\}. \tag{3}$$

In a similar way we find

$$V_{ba} = V_{ba}^{FM} + \left\{\frac{EI}{L^2}\left[-12(\delta_a - \delta_b) + 6\theta_a + 6\theta_b\right]\right\}. \tag{4}$$

We shall now show that the elastic terms (in brackets) in Equations 1–4 are stiffness influence coefficients where *a stiffness influence coefficient, s_{ij},* is a joint reaction, r_i, due to a unit end displacement, d_j, with all other joint displacements kept zero.*

For the beam element in Figure 9.1, there are four possible joint displacements; we shall arbitrarily take them in this order: $d_1 = \delta_a$ (the vertical translation at left end), $d_2 = \theta_a$ (the rotation at left end), $d_3 = \delta_b$ (the vertical translation at right end), $d_4 = \theta_b$ (the rotation at right end). We must take the reactions in a corresponding order—that is, translations are associated with shears and rotations are associated with moments; $r_1 = V_{ab}$, $r_2 = M_{ab}$, $r_3 = V_{ba}$, $r_4 = M_{ba}$. Upward and clockwise are taken as positive.

Refer to Chapter 7 for values of the end moments (in Figure 9.4) due to the unit displacements; by statics the shear values easily can be found.

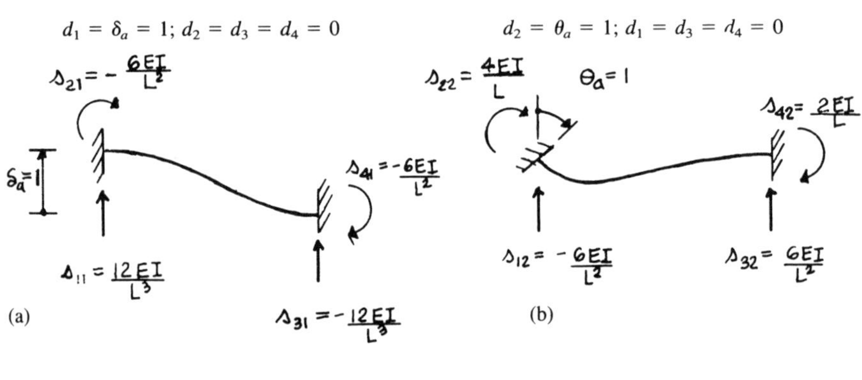

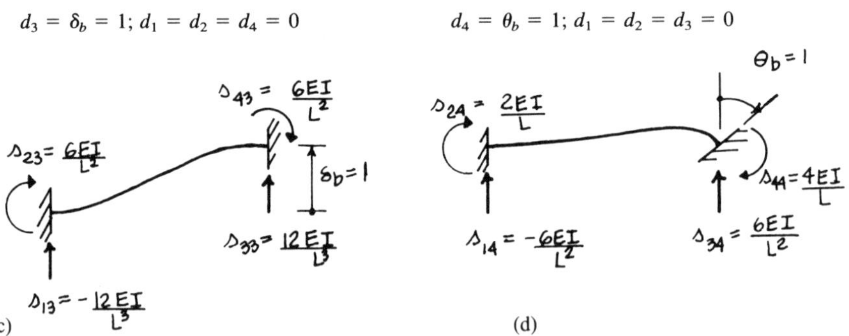

Figure 9.4

**Note:* The character s used in Chapter 9 is equivalent to the lower-case script "s" used in the Figure illustrations and Examples.

If we add to the fixed-end shear all the elastic terms (from Figure 9.4) associated directly with the shear V_{ab} and recall that in forming the chart we took in order $\delta_a = 1$, $\theta_a = 1$, $\delta_b = 1$, $\theta_b = 1$, we arrive at

$$V_{ab} = V_{ab}^{FM} + \frac{12EI}{L^3}\delta_a - \frac{6EI}{L^2}\theta_a - \frac{12EI}{L^3}\delta_b - \frac{6EI}{L^2}\theta_b, \tag{5a}$$

which exactly matches Equation 3.

We could rewrite the above as

$$V_{ab} = V_{ab}^{FM} + k_{11}d_1 + k_{12}d_2 + k_{13}d_3 + k_{14}d_4 \tag{5b}$$

In a similar way, we find

$$M_{ab} = M_{ab}^{FM} - \frac{6EI}{L^2}\delta_a + \frac{4EI}{L}\theta_a + \frac{6EI}{L^2}\delta_b + \frac{2EI}{L}\theta_b, \tag{6a}$$

which exactly matches Equation 1, or, rewriting,

$$M_{ab} = M_{ab}^{FM} + k_{21}d_1 + k_{22}d_2 + k_{23}d_3 + k_{24}d_4. \tag{6b}$$

$$V_{ba} = V_{ba}^{FM} - \frac{12EI}{L^3}\delta_a + \frac{6EI}{L^2}\theta_a + \frac{12EI}{L^3}\delta_b + \frac{6EI}{L^2}\theta_b, \tag{7a}$$

which matches Equation 4; rewriting we get

$$V_{ba} = V_{ba}^{FM} + k_{31}d_1 + k_{32}d_2 + k_{33}d_3 + k_{34}d_4. \tag{7b}$$

$$M_{ba} = M_{ba}^{FM} - \frac{6EI}{L^2}\delta_a + \frac{2EI}{L}\theta_a + \frac{6EI}{L^2}\delta_b + \frac{4EI}{L}\theta_b, \tag{8a}$$

which matches Equation 2; or

$$M_{ba} = M_{ba}^{FM} + k_{41}d_1 + k_{42}d_2 + k_{43}d_3 + k_{44}d_4 \tag{8b}$$

We can write Equations 5–8 in matrix form:

$$\begin{bmatrix} V_{ab} \equiv r_1 \\ M_{ab} \equiv r_2 \\ V_{ba} \equiv r_3 \\ M_{ba} \equiv r_4 \end{bmatrix} = \begin{bmatrix} V_{ab}^{FM} \\ M_{ab}^{FM} \\ V_{ba}^{FM} \\ M_{ba}^{FM} \end{bmatrix} + EI \begin{bmatrix} 12/L^3 & -6/L^2 & -12/L^3 & -6/L^2 \\ -6/L^2 & 4/L & 6/L^2 & 2/L \\ -12/L^3 & 6/L^2 & 12/L^3 & 6/L^2 \\ -6/L^2 & 2/L & 6/L^2 & 4/L \end{bmatrix} \begin{bmatrix} \delta_a \equiv d_1 \\ \theta_a \equiv d_2 \\ \delta_b \equiv d_3 \\ \theta_b \equiv d_4 \end{bmatrix}$$

or

$$\bar{r}_{\textcircled{i}} = \bar{r}_{\textcircled{i}}^{FM} + \bar{\bar{k}}_{\textcircled{i}} \qquad \bar{d}_{\textcircled{i}} \tag{9}$$

We call $\bar{\bar{k}}_{\textcircled{i}}$ the member stiffness matrix for the element $\textcircled{i}$; $\bar{r}_{\textcircled{i}}^{FM}$ means the fixed-end *reactions* to the applied loading.

Looking again at the member stiffness matrix (Figure 9.5) (valid for all beam elements) we see that:

1) All diagonal terms are positive. This merely means that a translation (or rotation) and the force (or moment) necessary to cause the translation (or rotation) must both be in the same direction.
2) The member stiffness matrix is symmetric. This is an affirmation of the Maxwell reciprocal theorem, which is treated in Chapter 11.
3) The first row is made up of all the elastic terms associated with the first reaction, V_{ab} (or r_1); the second row is made up of all the elastic terms associated with the second reaction, M_{ab} (or r_2); etc.
4) The first column is made up of all the elastic terms associated with the first displacement, δ_a (or d_1); the second column is associated with the second displacement, θ_a (or d_2); etc.

$$
EI\begin{bmatrix} 12/L^3 & -6/L^2 & -12/L^3 & -6/L^2 \\ -6/L^2 & 4/L & 6/L^2 & 2/L \\ -12/L^3 & 6/L^2 & 12/L^3 & 6/L^2 \\ -6/L^2 & 2/L & 6/L^2 & 4/L \end{bmatrix}
\begin{matrix} \text{assoc. w. } r_1 = V_{ab} \\ \text{assoc. w. } r_2 = M_{ab} \\ \text{assoc. w. } r_3 = V_{ba} \\ \text{assoc. w. } r_4 = M_{ba} \end{matrix}
$$

$$
\begin{matrix} \text{assoc. w.} & \text{assoc. w.} & \text{assoc. w.} & \text{assoc. w.} \\ d_1 = \delta_a & d_2 = \theta_a & d_3 = \delta_b & d_4 = \theta_b \end{matrix}
$$

Figure 9.5 The member stiffness matrix for beams.

Well, great. Equation 9 ($\overline{r}_{ⓘ} = \overline{r}^{FM}_{ⓘ} + \overline{\overline{k}}_{ⓘ}\, \overline{d}_{ⓘ}$) tells us in matrix form how to find the member joint shears and moments in terms of the stiffness influence coefficients once the member joint displacements are found. But how do we find those displacements? Again, let's return to the slope-deflection scheme. Consider the three-spans continuous beam shown in Figure 9.6. In the slope-deflection procedure we would sum moments at joints b and c; namely,

$$M_{ba} + M_{bc} = 0 \qquad \text{and} \qquad M_{cb} + M_{cd} = 0,$$

or in terms of the displacements:

$$\left[M_{ba}^{FM} + \frac{E_1I_1}{L_1}(4\theta_b)\right] + \left[M_{bc}^{FM} + \frac{E_2I_2}{L_2}(4\theta_b + 2\theta_c)\right] = 0, \tag{10}$$

$$\left[M_{cb}^{FM} + \frac{E_2I_2}{L_2}(4\theta_c + 2\theta_b)\right] + \left[M_{cd}^{FM} + \frac{E_3I_3}{L_3}(4\theta_c)\right] = 0. \tag{11}$$

$\theta_a = 0$
$\theta_d = 0$
$\delta_a = \delta_b = \delta_c = \delta_d = 0$
$\theta_b \neq 0$
$\theta_c \neq 0$

Figure 9.6

These two equilibrium equations, 10 and 11, can be put in a general matrix form involving the stiffness influence coefficients. The process goes as follows: first, we recognize the unknown displacements (here θ_b and θ_c); next, we look at the separated members and affect the displacement $d_1 = 1$, holding all the other joint displacements at zero (here we take $d_1 = \theta_b = 1$); then we take $d_2 = 1$ while holding the other joint displacements at zero. When forming these disturbances, we look only for the stiffness influence coefficients *directly* associated with the unknown displacements (for our example we seek only stiffness terms associated with M_b and M_c—see Figure 9.7). Thus the relevant stiffness terms for our example (Figure 9.7) are

$$\bar{\bar{S}} = \begin{bmatrix} 4E_1I_1/L_1 + 4E_2I_2/L_2 & 2E_2I_2/L_2 \\ 2E_2I_2/L_2 & 4E_2I_2/L_2 + 4E_3I_3/L_3 \end{bmatrix} \begin{matrix} M_b \\ M_c \end{matrix}$$

$$\qquad\theta_b \qquad\qquad\qquad \theta_c$$

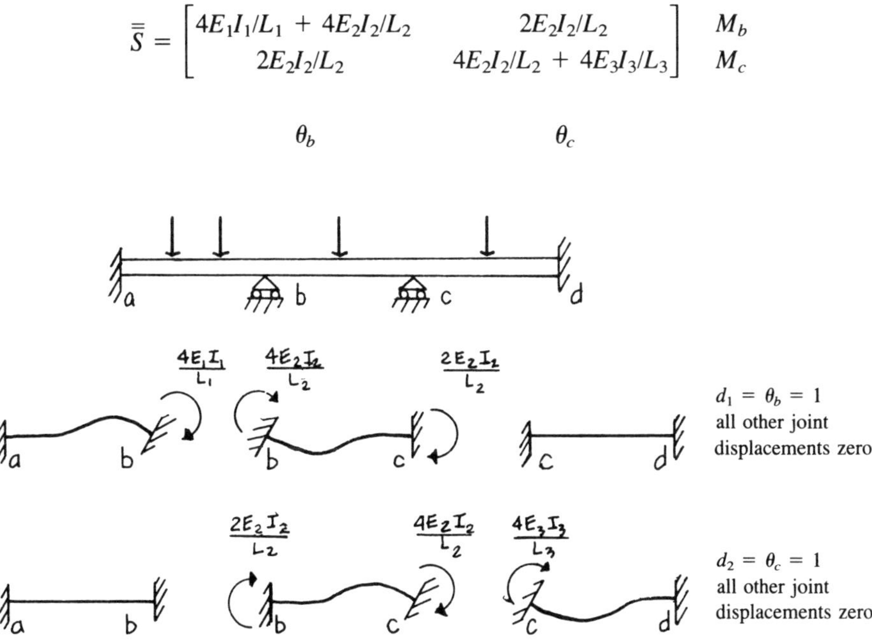

Figure 9.7

We see that Equations 10 and 11 can be written in matrix form as follows:

$$\begin{bmatrix} M_b \equiv 0 \\ M_c \equiv 0 \end{bmatrix} = \begin{bmatrix} M_{ba}^{FM} + M_{bc}^{FM} \\ M_{cb}^{FM} + M_{cd}^{FM} \end{bmatrix}$$

$$+ \begin{bmatrix} 4E_1I_1/L_1 + 4E_2I_2/L_2 & 2E_2I_2/L_2 \\ 2E_2I_2/L_2 & 4E_2I_2/L_2 + 4E_3I_3/L_3 \end{bmatrix} \begin{bmatrix} \theta_b \\ \theta_c \end{bmatrix}, \tag{12}$$

or

$$\overline{R} \equiv 0 = \overline{R}^{FM} + \overline{\overline{S}}\overline{D}. \tag{13}$$

The set of equilibrium Equations 13 is a general equation and is valid for all structures. The key to setting up the fundamental matrix Equations 9 and 13 is identifying the unknown displacement; the ordering of all the matrices is driven by the ordering of the $\overline{D}$ matrix. In Equation 12, θ_b and θ_c are the unknowns; we choose the rotation θ_b first, thus $\overline{R}$, which is the *external* reaction matrix, will have the external moment at b first, then M_c. Note that in all cases if θ_b is the unknown displacement, the corresponding *external* support moment at b must be zero—no support or joint can violate this condition. The $\overline{R}^{FM}$ vector consists of the external fixed-end reaction terms—here the fixed-end moments at b and c. Looking at the *structure stiffness matrix*, $\overline{\overline{S}}$ (Figure 9.5), we see it is square, a 2×2, corresponding to two unknown displacements; it, too, is always symmetric and will always have positive diagonal terms. The columns are associated with the unknown displacements while the rows correlate the corresponding reactions. A note on our notation: we have used lower case letters for member, or element, terms and capital letters for entire structure terms.

The size and individual terms of the structure stiffness matrix, of course, vary from problem to problem. This matrix can always be formed by making the displacement perturbations in a way similar to that used above. These perturbations give a good physical interpretation to the stiffness terms, but this can be a laborious process for large structures. There is a mechanical, systematic way of forming the structure stiffness matrix.

To form $\overline{\overline{S}}$, we first write out all the *member* stiffness matrices—they all have the same form as in Figure 9.5. For the beam in Figure 9.6, we would have three such matrices.

We can now form the *structure* stiffness matrix from the member stiffness matrices by collecting all the terms associated with the unknown displacements *and* the corresponding reactions and adding terms associated with the same joint. In Figure 9.8 we pick out the terms associated with θ_b, θ_c and M_b, M_c and add all terms associated purely with joint b ($k_{44\textcircled{1}} + k_{22\textcircled{2}}$) and add all terms associated purely with joint c ($k_{44\textcircled{2}} + k_{22\textcircled{3}}$), (Figure 9.9).

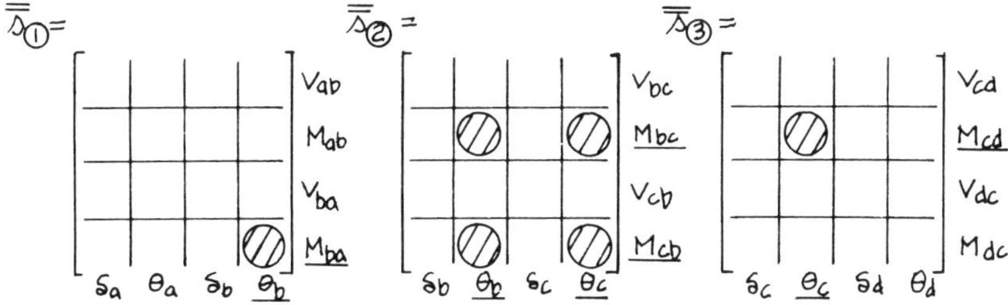

Figure 9.8

$$\bar{\bar{S}} = \begin{bmatrix} k_{44①} + k_{22②} & k_{24②} \\ k_{42②} & k_{44②} + k_{22③} \end{bmatrix} \begin{matrix} M_b \\ M_c \end{matrix}$$

$$\begin{matrix} \theta_b & \theta_c \end{matrix}$$

Figure 9.9

This process lends itself easily to the computer; once the member stiffness matrices are stored, the addition and compilation of the "right" stiffness terms is easily done.

To reiterate the basic process, we have two basic matrix equations:

$$0 = \bar{R}^{FM} + \bar{\bar{S}}\bar{D} \qquad \text{or}$$

$$\bar{D} = -\bar{\bar{S}}^{-1}\bar{R}^{FM} \qquad \text{and} \tag{14}$$

$$\bar{r}_{ⓘ} = \bar{r}^{FM}_{ⓘ} + \bar{\bar{k}}_{ⓘ}\bar{d}_i. \tag{15}$$

Equation 14 gives the unknown joint displacements in terms of the elastic terms (actually the inverse of the structure stiffness matrix) and the associated loads (the negatives of the fixed-end reactions).

The process works in this order:

1) Identify and order the unknown structure displacements, $\bar{D}$;
2) form the member stiffness matrices, $\bar{\bar{k}}_{ⓘ}$;
3) form the structure stiffness matrix, $\bar{\bar{S}}$, by the perturbation idea or by selective addition of the member stiffness terms;
4) invert the structure stiffness matrix using one of various techniques, (co-factor, Gaussian elimination techniques, programmable calculator or computer sub-routine);
5) calculate all the fixed-end reactions, $\bar{R}^{FM}$, and pick out those associated with the unknown displacements (recall that moments are associated with rotations; shears are associated with translations);

6) carry out Equation 14 to find the unknown structure displacements, $\bar{D}$;
7) once the structure displacement matrix, $\bar{D}$, is found, the member reactions, $\bar{r}_{(i)}$, can be found from Equation 15. Here it is very important to identify which structure displacements in $\bar{D}$ go to which member displacements, $\bar{d}_{(i)}$.

Let's look at an example. In Example 1 we see that the most important step is to identify (and order) the unknown displacements; all structure matrices depend upon the size and order of the structure displacement matrix, $\bar{D}$. We should also note that the degree of statical indeterminateness has nothing at all to do with the stiffness method; statically determinate problems can be treated as well.

In Example 1 we see the relative ease in handling more than one load. The major chore is forming (and inverting) $\bar{\bar{S}}$ which is strictly a function of geometry, support conditions, and elastic properties; $\bar{\bar{S}}$ does not depend upon the loadings. Example 1 illustrates the handling of loads applied at the joints. These loads do affect the *structure* fixed-end terms but do not affect the *member* fixed-end terms. The effect of joint loads upon the member reactions is through the member displacements.

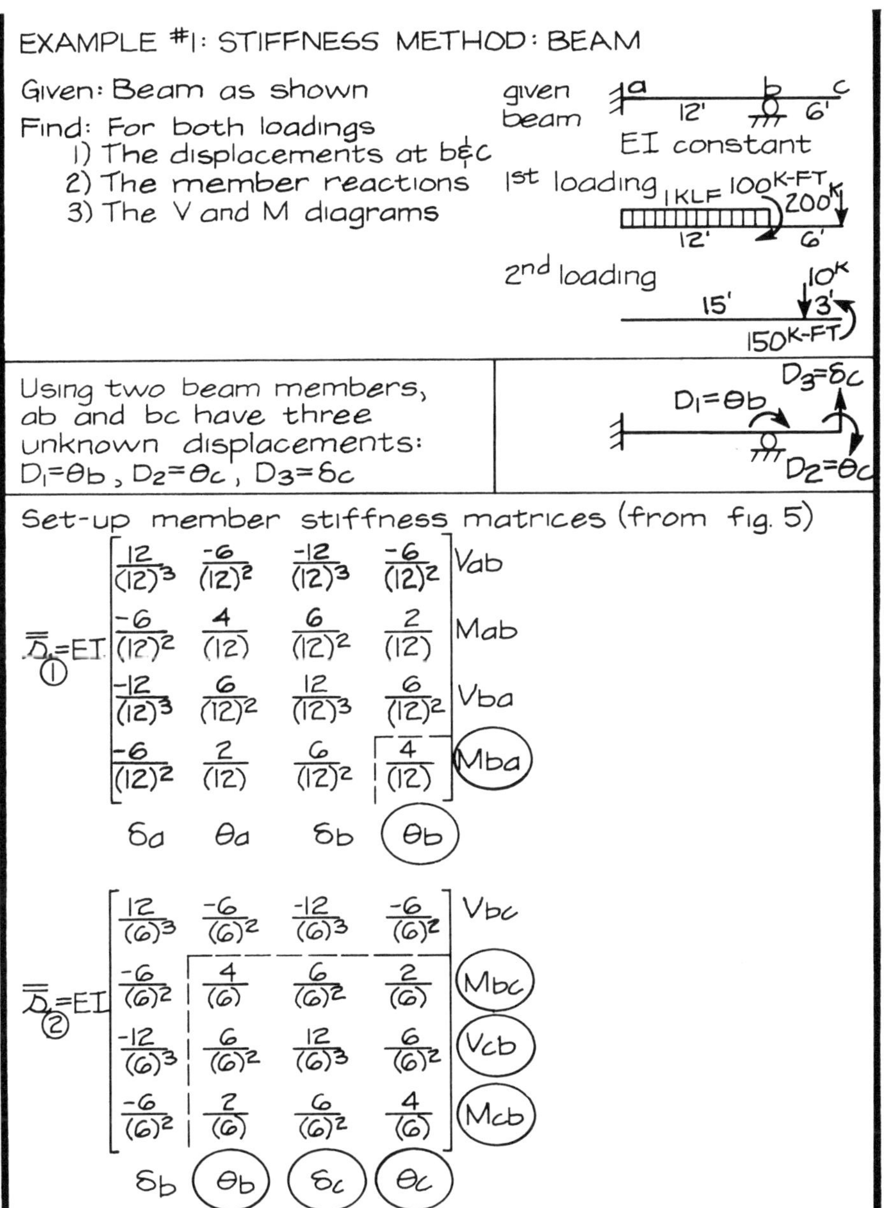

$$\bar{\bar{D}}_1 = EI\begin{bmatrix} \frac{12}{(12)^3} & \frac{-6}{(12)^2} & \frac{-12}{(12)^3} & \frac{-6}{(12)^2} \\ \frac{-6}{(12)^2} & \frac{4}{(12)} & \frac{6}{(12)^2} & \frac{2}{(12)} \\ \frac{-12}{(12)^3} & \frac{6}{(12)^2} & \frac{12}{(12)^3} & \frac{6}{(12)^2} \\ \frac{-6}{(12)^2} & \frac{2}{(12)} & \frac{6}{(12)^2} & \frac{4}{(12)} \end{bmatrix} \begin{matrix} V_{ab} \\ M_{ab} \\ V_{ba} \\ M_{ba} \end{matrix}$$

$$\delta_a \quad \theta_a \quad \delta_b \quad \theta_b$$

$$\bar{\bar{D}}_2 = EI\begin{bmatrix} \frac{12}{(6)^3} & \frac{-6}{(6)^2} & \frac{-12}{(6)^3} & \frac{-6}{(6)^2} \\ \frac{-6}{(6)^2} & \frac{4}{(6)} & \frac{6}{(6)^2} & \frac{2}{(6)} \\ \frac{-12}{(6)^3} & \frac{6}{(6)^2} & \frac{12}{(6)^3} & \frac{6}{(6)^2} \\ \frac{-6}{(6)^2} & \frac{2}{(6)} & \frac{6}{(6)^2} & \frac{4}{(6)} \end{bmatrix} \begin{matrix} V_{bc} \\ M_{bc} \\ V_{cb} \\ M_{cb} \end{matrix}$$

$$\delta_b \quad \theta_b \quad \delta_c \quad \theta_c$$

Form structure stiffness matrix from member stiffness matrices by choosing terms associated with $(\theta_b, \theta_c, \delta_c)$ and (M_b, M_c, V_c).

$$\bar{\bar{S}} = EI \begin{bmatrix} \left(\frac{4}{12}+\frac{4}{6}\right) & \frac{2}{6} & \frac{6}{(6)^2} \\ \frac{2}{6} & \frac{4}{6} & \frac{6}{(6)^2} \\ \frac{6}{(6)^2} & \frac{6}{(6)^2} & \frac{12}{(6)^3} \end{bmatrix} \begin{matrix} M_b \\ M_c \\ V_c \end{matrix} \equiv \frac{EI}{18} \begin{bmatrix} 18 & 6 & 3 \\ 6 & 12 & 3 \\ 3 & 3 & 1 \end{bmatrix} \begin{matrix} M_b \\ M_c \\ V_c \end{matrix}$$

$$\qquad\quad \theta_b \quad \theta_c \quad \delta_c \qquad\qquad\qquad \theta_b \quad \theta_c \quad \delta_c$$

OR

We could form the structure stiffness matrix by making a series of displacement perturbations and picking out the influence coefficients associated with M_b, M_c and V_c.

$D_1 \equiv \theta_b = 1$; $D_2 \equiv \theta_c = 0$; $D_3 \equiv \delta_c = 0$	$D_2 = \theta_c = 1$; $D_1 \equiv \theta_b = 0$; $D_3 = \delta_c = 0$	$D_3 = \delta_c = 1$; $D_1 \equiv \theta_b = 0$; $D_2 \equiv \theta_c = 0$
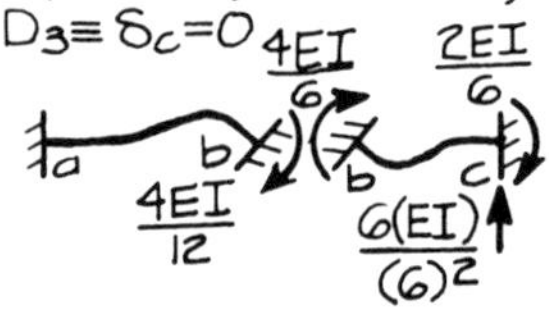	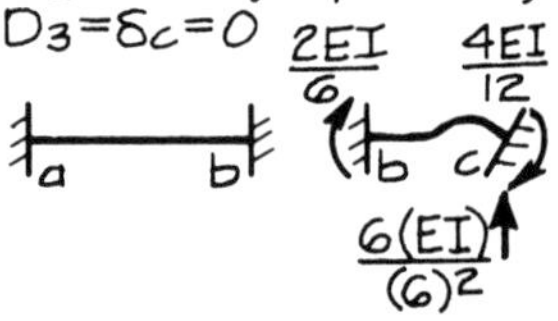	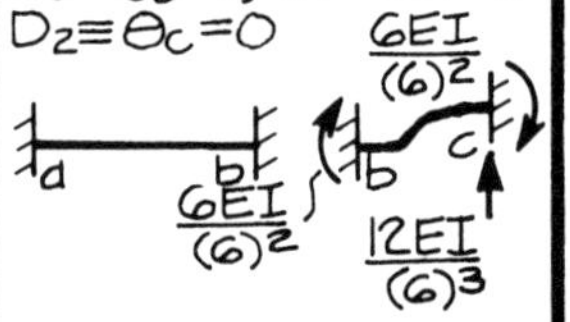
$S_{11} = \frac{4EI}{12} + \frac{4EI}{6}$	$S_{12} = \frac{2EI}{6}$	$S_{13} = \frac{6EI}{(6)^2}$
$S_{21} = \frac{2EI}{6}$	$S_{22} = \frac{4EI}{6}$	$S_{23} = \frac{6EI}{(6)^2}$
$S_{31} = \frac{6EI}{(6)^2}$	$S_{32} = \frac{6EI}{(6)^2}$	$S_{33} = \frac{12EI}{(6)^3}$

Invert $\bar{\bar{S}}$ using standard inversion methods; check $\bar{\bar{S}}^{-1}$

$$\bar{\bar{S}}^{-1} = \frac{1}{EI}\begin{bmatrix} 3 & 3 & -18 \\ 3 & 9 & -36 \\ -18 & -36 & 180 \end{bmatrix}; \quad \text{Does } \bar{\bar{S}}\bar{\bar{S}}^{-1} = \bar{\bar{I}}\ ? \quad \frac{EI}{18}\begin{bmatrix} 18 & 6 & 3 \\ 6 & 12 & 3 \\ 3 & 3 & 1 \end{bmatrix} \frac{1}{EI}\begin{bmatrix} 3 & 3 & -18 \\ 3 & 9 & -36 \\ -18 & -36 & 180 \end{bmatrix} = \begin{bmatrix} 1 & 0 & 0 \\ 0 & 1 & 0 \\ 0 & 0 & 1 \end{bmatrix} \text{ Yes}$$

Now consider loading #1: Note that the $100^{K\text{-}FT}$ and 200^K loads applied at joints b and c; we wish to find the external reactions at b and c, thus the directions of the $100^{K\text{-}FT}$ and 200^K loads must be reversed to be reactions. We also form fixed-end members ab and bc.

The $\bar{R}^{FM}$ vector ordering must follow the $\bar{D}$ vector ordering,

$$\bar{D} = \begin{bmatrix} \theta_b \\ \theta_c \\ \delta_c \end{bmatrix} \Rightarrow \bar{R}^{FM} = \begin{bmatrix} M_b^{FM} \\ M_c^{FM} \\ V_c^{FM} \end{bmatrix} = \begin{bmatrix} 12-100 \\ 0 \\ 0+200 \end{bmatrix} = \begin{bmatrix} -88^{K\text{-}FT} \\ 0 \\ 200^K \end{bmatrix}$$

Note: ↻ is positive and ↑ is positive

Therefore for loading #1 the joint displacements are found:

$$\bar{D} = -\bar{\bar{S}}^{-1}\bar{R}^{FM} = \frac{-1}{EI}\begin{bmatrix} 3 & 3 & -18 \\ 3 & 9 & -36 \\ -18 & -36 & 180 \end{bmatrix}\begin{bmatrix} -88 \\ 0 \\ 200 \end{bmatrix} = \begin{bmatrix} 3864 \\ 7464 \\ -37584 \end{bmatrix}\frac{1}{EI}$$

Thus for loading #1: $\theta_b = \frac{3864}{EI}$ (↻); $\theta_c = \frac{7464}{EI}$ (↺); $\delta_c = \frac{-37584}{EI}$ (↓)

To find the member reactions we use $\bar{r}_{(i)} = \bar{r}^{FM}_{(i)} + \bar{\bar{\mathcal{D}}}_{(i)}\bar{d}_{(i)}$

For member ① = ab

$$\bar{d}_{①} = \begin{bmatrix} \delta_a \equiv 0 \\ \theta_a \equiv 0 \\ \delta_b \equiv 0 \\ \theta_b \equiv \frac{3864}{EI} \end{bmatrix}; \quad \bar{r}^{FM}_{①} = \begin{bmatrix} V^{FM}_{ab} \equiv 6 \\ M^{FM}_{ab} \equiv -12 \\ V^{FM}_{ba} \equiv 6 \\ M^{FM}_{ba} \equiv 12 \end{bmatrix}$$

For member ② = bc

$$\bar{d}_{②} = \begin{bmatrix} \delta_b \equiv 0 \\ \theta_b \equiv \frac{3864}{EI} \\ \delta_c \equiv \frac{-37584}{EI} \\ \theta_c = \frac{7464}{EI} \end{bmatrix}; \quad \bar{r}^{FM}_{②} = \begin{bmatrix} V^{FM}_{bc} \equiv 0 \\ M^{FM}_{bc} \equiv 0 \\ V^{FM}_{cb} \equiv 0 \\ M^{FM}_{cb} = 0 \end{bmatrix}$$

Note: The joint loads ($100^{K\text{-}FT}$ and 200^K) do not directly affect the fixed-end terms.

$$\bar{r}_{①} = \bar{r}^{FM}_{①} + \bar{\bar{\mathcal{D}}}_{①}\bar{d}_{①} = \begin{bmatrix} 6 \\ -12 \\ 6 \\ 12 \end{bmatrix} + \frac{EI}{12}\begin{bmatrix} \frac{1}{12} & -\frac{1}{2} & -\frac{1}{12} & -\frac{1}{2} \\ -\frac{1}{2} & 4 & \frac{1}{2} & 2 \\ -\frac{1}{12} & \frac{1}{2} & \frac{1}{12} & \frac{1}{2} \\ -\frac{1}{2} & 2 & \frac{1}{2} & 4 \end{bmatrix}\frac{1}{EI}\begin{bmatrix} 0 \\ 0 \\ 0 \\ 3864 \end{bmatrix} = \begin{bmatrix} -155^K \\ 632^{K\text{-}FT} \\ 167^K \\ 1300^{K\text{-}FT} \end{bmatrix}$$

$632^{K\text{-}FT}$ 1KLF $1300^{K\text{-}FT}$
a 12' b
155^K 167^K

$$\bar{r}_{②} = \bar{r}^{FM}_{②} + \bar{\bar{\mathcal{D}}}_{②}\bar{d}_{②} = \begin{bmatrix} 0 \\ 0 \\ 0 \\ 0 \end{bmatrix} + \frac{EI}{6}\begin{bmatrix} \frac{1}{3} & -1 & -\frac{1}{3} & -1 \\ -1 & 4 & 1 & 2 \\ -\frac{1}{3} & 1 & \frac{1}{3} & 1 \\ -1 & 2 & 1 & 4 \end{bmatrix}\frac{1}{EI}\begin{bmatrix} 0 \\ 3864 \\ -37584 \\ 7464 \end{bmatrix} = \begin{bmatrix} 200^K \\ -1200^{K\text{-}FT} \\ -200^K \\ 0 \end{bmatrix}$$

$1200^{K\text{-}FT}$ 0
b c
200^K 200^K

For loading #2: $\bar{\bar{S}}^{-1}$ remains the same as previously calculated. Consider fixed-end beams and again the joint <u>load.</u> Directions must be reversed in order to become joint <u>reactions.</u>

0 0
a 12' b
0 0

$7.5^{K\text{-}FT}$ 10^K $7.5^{K\text{-}FT}$ $150^{K\text{-}FT}$
3' 3'
5^K 5^K

$$\bar{R}^{FM} = \begin{bmatrix} M_b = -7.5^{K\text{-}FT} \\ M_c = 157.5^{K\text{-}FT} \\ V_c = 5^K \end{bmatrix}$$

For loading #2 the joint displacements are found:

$$\bar{D} = -\bar{\bar{S}}^{-1}\bar{R}^{FM} = -\frac{1}{EI}\begin{bmatrix} 3 & 3 & -18 \\ 3 & 9 & -36 \\ -18 & -36 & 180 \end{bmatrix}\begin{bmatrix} -7.5 \\ 157.5 \\ 5 \end{bmatrix} = \begin{bmatrix} -360 \\ -1215 \\ 4635 \end{bmatrix}\frac{1}{EI}$$

Thus for loading #2: $\theta_b = \frac{-360}{EI}$ (↶); $\theta_c = \frac{-1215}{EI}$ (↶); $\delta_c = \frac{4635}{EI}$ (↑)

To find the member reactions we use $\bar{r}_{(i)} = \bar{r}_{(i)}^{FM} + \bar{\bar{S}}_{(i)}\bar{d}_{(i)}$

For member ① = ab

$$\bar{d}_{(1)} = \begin{bmatrix} \delta_a \equiv 0 \\ \theta_a \equiv 0 \\ \delta_b \equiv 0 \\ \theta_b \equiv \frac{-360}{EI} \end{bmatrix}; \quad \bar{r}_{(1)}^{FM} = \begin{bmatrix} V_{ab}^{FM} \equiv 0 \\ M_{ab}^{FM} \equiv 0 \\ V_{ba}^{FM} \equiv 0 \\ M_{ba}^{FM} \equiv 0 \end{bmatrix}$$

For member ② = bc

$$\bar{d}_{(2)} = \begin{bmatrix} \delta_b \equiv 0 \\ \theta_b \equiv -360 \\ \delta_c \equiv 4635 \\ \theta_c \equiv -1215 \end{bmatrix}\frac{1}{EI}; \quad \bar{r}_{(2)}^{FM} = \begin{bmatrix} V_{bc}^{FM} \equiv 5 \\ M_{bc}^{FM} \equiv -7.5 \\ V_{cb}^{FM} \equiv 5 \\ M_{cb}^{FM} \equiv 7.5 \end{bmatrix}$$

$$\bar{r}_{(1)} = \bar{r}_{(1)}^{FM} + \bar{\bar{S}}_{(1)}\bar{d}_{(1)} = \begin{bmatrix} 0 \\ 0 \\ 0 \\ 0 \end{bmatrix} + \frac{EI}{12}\begin{bmatrix} \frac{1}{12} & -\frac{1}{2} & \frac{-1}{12} & \frac{-1}{2} \\ -\frac{1}{2} & 4 & \frac{1}{2} & 2 \\ -\frac{1}{12} & \frac{1}{2} & \frac{1}{12} & \frac{1}{2} \\ -\frac{1}{2} & 2 & \frac{1}{2} & 4 \end{bmatrix}\begin{bmatrix} 0 \\ 0 \\ 0 \\ -\frac{360}{EI} \end{bmatrix} = \begin{bmatrix} 15^K \\ -60^{K\text{-}FT} \\ -15^K \\ -120^{K\text{-}FT} \end{bmatrix}$$

60K-FT, 120K-FT, a, 12', b, 15K, 15K

$$\bar{r}_{(2)} = \bar{r}_{(2)}^{FM} + \bar{\bar{S}}_{(2)}\bar{d}_{(2)} = \begin{bmatrix} 5 \\ -7.5 \\ 5 \\ 7.5 \end{bmatrix} + \frac{EI}{6}\begin{bmatrix} \frac{1}{3} & -1 & \frac{-1}{3} & -1 \\ -1 & 4 & 1 & 2 \\ \frac{-1}{3} & 1 & \frac{1}{3} & 1 \\ -1 & 2 & 1 & 4 \end{bmatrix}\begin{bmatrix} 0 \\ -360 \\ 4635 \\ -1215 \end{bmatrix}\frac{1}{EI} = \begin{bmatrix} 10^K \\ 120^{K\text{-}FT} \\ 0 \\ -150^{K\text{-}FT} \end{bmatrix}$$

120K-FT, 10K, 3', 3', b, 10K, c, 150K-FT

Now the shear and bending moment diagrams are drawn

1KLF, 100K-FT, 200K, 12', 6'

V: 200K, 0, −155K, −167K

M: 632K-FT, −1200K-FT, −1300K-FT

Loading #1

10K, 12', 3', 3', 150K-FT

V: 15K, 10K, 0

M: 150K-FT, 120K-FT, −60K-FT

Loading #2

FRAMES

The same general process as used for beams can be used as well for frames—that is, the structure displacements are found by Equation 14 and the member reactions are evaluated by Equation 15. For frames there are, however, two "wrinkles" beam problems don't have. First, and easily handled, are axial elongation terms. True, in the classical "hand methods" we neglected them. However, most structural computer programs do include them—as we shall. The second complication is due to corners in frames which cause grief with respect to adding stiffness terms from adjoining members.

First the stiffness influence coefficients: we shall order the displacements (and reactions) as shown in Figure 9.10 with $d_1 = \delta_{ax}$, $d_2 = \delta_{ay}$, $d_3 = \theta_a$, $d_4 = \delta_{bx}$, $d_5 = \delta_{by}$, $d_6 = \theta_b$. The only two displacements not previously treated in the beam section are the axial displacements d_1 and d_4 (Figure 9.11). Thus, the frame member stiffness matrix based on its own member, *xy* axes, is as shown in Figure 9.12.

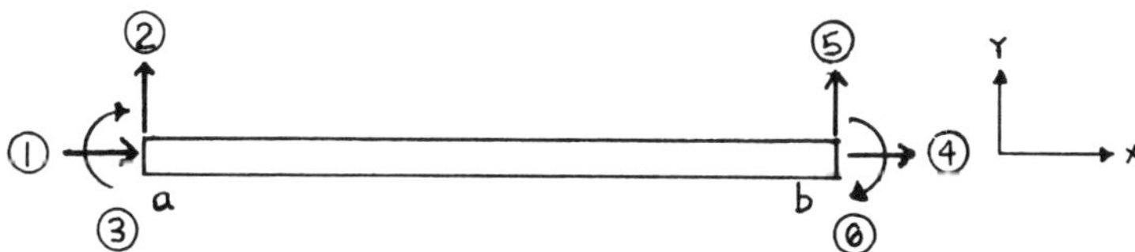

Figure 9.10

(a) $d_1 = \delta_{ax} = 1;\ d_2 = d_3 = d_4 = d_5 = d_6 = 0$

$\Delta_{11} = \frac{EA}{L}$ $\quad \Delta_{41} = -\frac{EA}{L}$ $\quad \delta_{ax} = 1$

$\Delta_{21} = \Delta_{31} = \Delta_{51} = \Delta_{61} = 0$

(b) $d_4 = \delta_{bx} = 1;\ d_1 = d_2 = d_3 = d_5 = d_6 = 0$

$\Delta_{14} = -\frac{EA}{L}$ $\quad \Delta_{44} = \frac{EA}{L}$ $\quad \delta_{bx} = 1$

$\Delta_{24} = \Delta_{34} = \Delta_{54} = \Delta_{64} = 0$

Figure 9.11

$$
\bar{\bar{k}} = \begin{bmatrix}
EA/L & 0 & 0 & -EA/L & 0 & 0 \\
0 & 12EI/L^3 & -6EI/L^2 & 0 & -12EI/L^3 & -6EI/L^2 \\
0 & -6EI/L^2 & 4EI/L & 0 & 6EI/L^2 & 2EI/L \\
-EA/L & 0 & 0 & EA/L & 0 & 0 \\
0 & -12EI/L^3 & 6EI/L^2 & 0 & 12EI/L^3 & 6EI/L^2 \\
0 & -6EI/L^2 & 2EI/L & 0 & 6EI/L^2 & 4EI/L
\end{bmatrix}
\begin{matrix} P_{ab} \\ V_{ab} \\ M_{ab} \\ P_{ba} \\ V_{ba} \\ M_{ba} \end{matrix}
$$

$$
\begin{matrix} \delta_{ax} & \delta_{ay} & \theta_a & \delta_{bx} & \delta_{by} & \theta_b \end{matrix}
$$

Figure 9.12 Member stiffness matrix for frames (in local coordinates).

In the beam section we saw we could form the structure stiffness matrix by merely adding, from the various member stiffness matrices, terms common to intersecting joints. These common terms represent, in fact, moment and shear influence coefficients. In beams, of course, shear forces of the separate elements are always vertical. In frames, shear and axial force directions vary with the inclination of the member; shears and axial forces are aligned with the *local* coordinate system (xy). Looking at Figure 9.13, we see some decomposition will have to be done in order to add common terms at intersecting joints. We will accomplish this by forming the stiffness influence coefficients with respect to the fixed *global coordinate system*, $\underline{\bar{X}}\bar{Y}$. As we will see, the stiffness influence coefficients are not vectors, thus they do not decompose like vectors. A stiffness influence coefficient is a force or moment per unit displacement. In changing coordinate systems the forces (not moments) decompose *and* the unit displacements decompose as well.

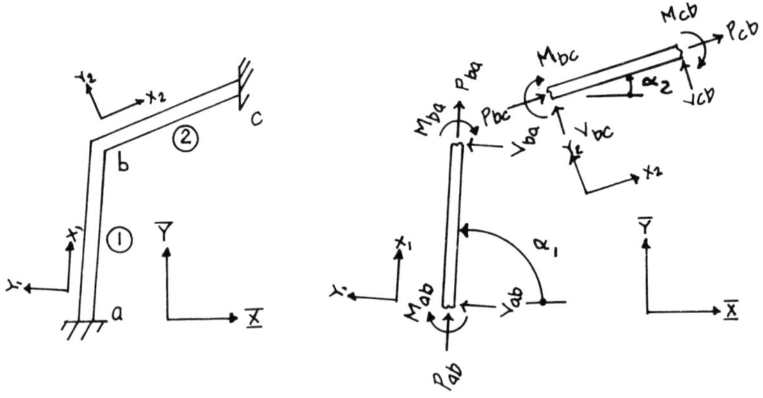

Figure 9.13a

In Figure 9.14, an inclined member is taken; the angle α is measured *from* the global $\underline{\bar{X}}$ axis *to* the local x axis with counterclockwise taken as positive. We then cause a unit $\delta_{a\underline{\bar{X}}}$ disturbance. The global $\delta_{a\underline{\bar{X}}}$ decomposes to local δ_{ax} and δ_{ay} displacements.

$$\delta_{a\underline{X}} = 1; \quad \delta_{a\overline{Y}} = \delta_{b\underline{X}} = \delta_{b\overline{Y}} = \theta_a = \theta_b = 0$$

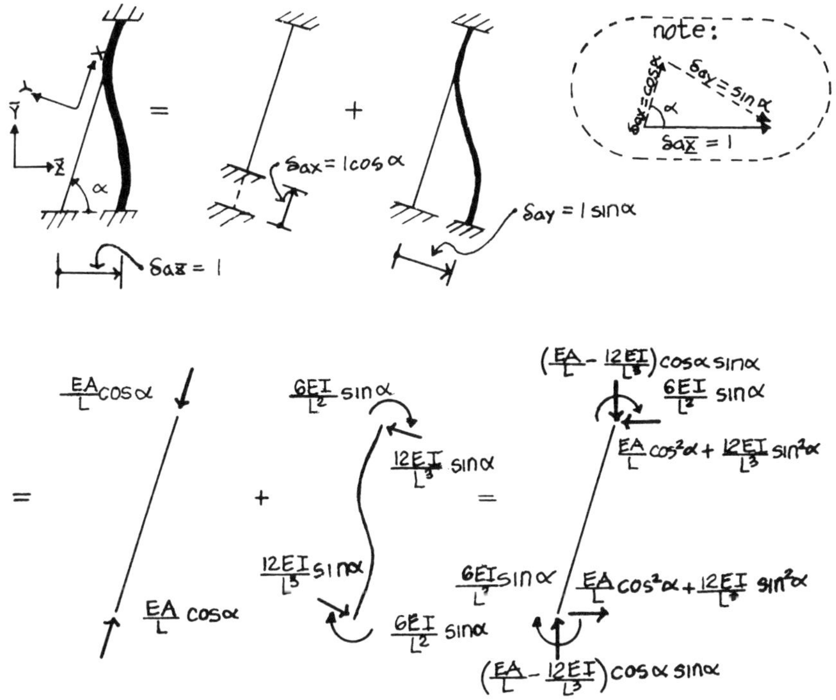

$$k_{11G} = \frac{EA}{L}\cos^2\alpha + \frac{12EI}{L^3}\sin^2\alpha; \quad k_{21G} = \left(\frac{EA}{L} - \frac{12EI}{L^3}\right)\cos\alpha\sin\alpha;$$

$$k_{31G} = \frac{6EI}{L^2}\sin\alpha;$$

$$k_{41G} = -\frac{EA}{L}\cos^2\alpha - \frac{12EI}{L^3}\sin^2\alpha; \quad k_{51G} = \left(-\frac{EA}{L} + \frac{12EI}{L^3}\right)\cos\alpha\sin\alpha;$$

$$k_{61G} = \frac{6EI}{L^2}\sin\alpha$$

Figure 9.14

$$\delta_{a\bar{Y}} = 1; \quad \delta_{a\underline{X}} = \delta_{b\underline{X}} = \delta_{b\bar{Y}} = \theta_a = \theta_b = 0$$

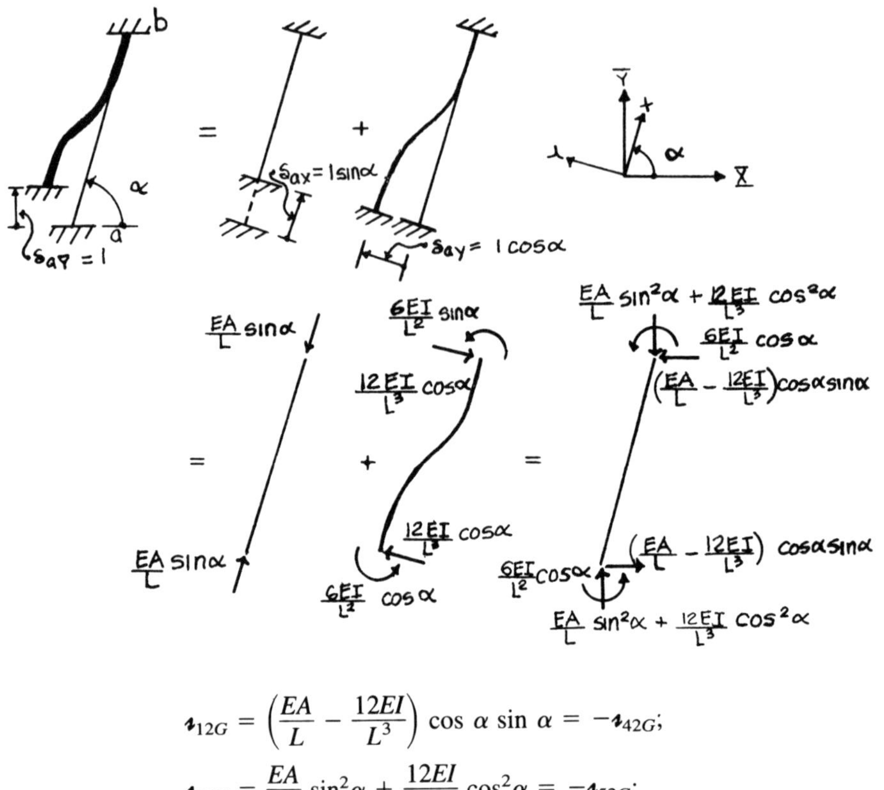

$$k_{12G} = \left(\frac{EA}{L} - \frac{12EI}{L^3}\right)\cos\alpha\,\sin\alpha = -k_{42G};$$

$$k_{22G} = \frac{EA}{L}\sin^2\alpha + \frac{12EI}{L^3}\cos^2\alpha = -k_{52G};$$

$$k_{32G} = k_{62G} = -\frac{6EI}{L^2}\cos\alpha$$

Figure 9.15

$$\theta_a = 1; \quad \delta_{a\underline{X}} = \delta_{a\overline{Y}} = \delta_{b\underline{X}} = \delta_{b\overline{Y}} = \theta_b = 0$$

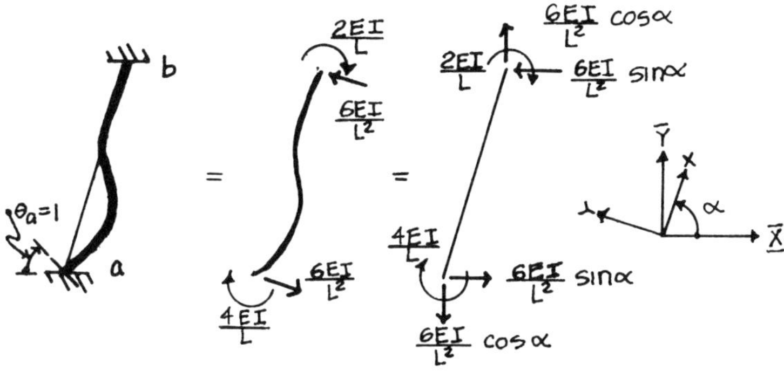

$$k_{13G} = \frac{6EI}{L^2}\sin\alpha = -k_{43G}; \quad k_{23G} = -\frac{6EI}{L^2}\cos\alpha = -k_{53G};$$

$$k_{33G} = \frac{4EI}{L}; \quad k_{63G} = \frac{2EI}{L}$$

Figure 9.16

$$\delta_{b\underline{X}} = 1; \quad \delta_{a\underline{X}} = \delta_{a\overline{Y}} = \delta_{b\overline{Y}} = \theta_a = \theta_b = 0$$

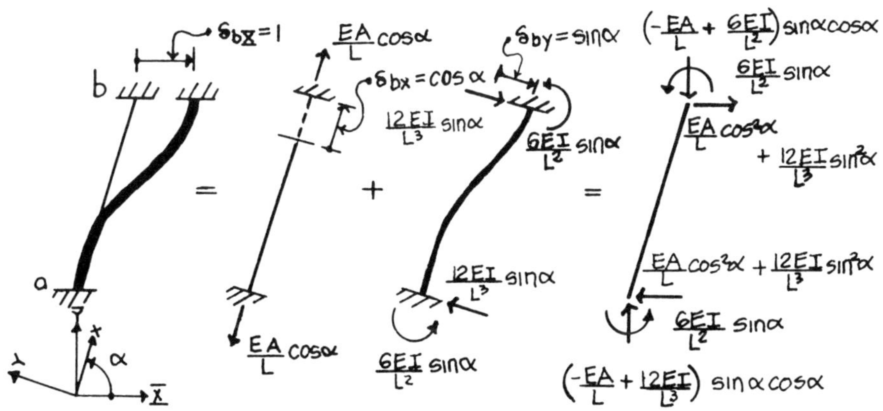

$$k_{14G} = -\frac{EA}{L}\cos^2\alpha - \frac{12EI}{L^3}\sin^2\alpha = -k_{44G};$$

$$k_{24G} = \left(-\frac{EA}{L} + \frac{12EI}{L^3}\right)\cos\alpha\,\sin\alpha = -k_{54G}; \quad k_{34G} = -\frac{6EI}{L^2}\sin\alpha = k_{64G}$$

Figure 9.17

$$\delta_{b\overline{Y}} = 1; \quad \delta_{a\underline{X}} = \delta_{a\overline{Y}} = \delta_{b\underline{X}} = \theta_a = \theta_b = 0$$

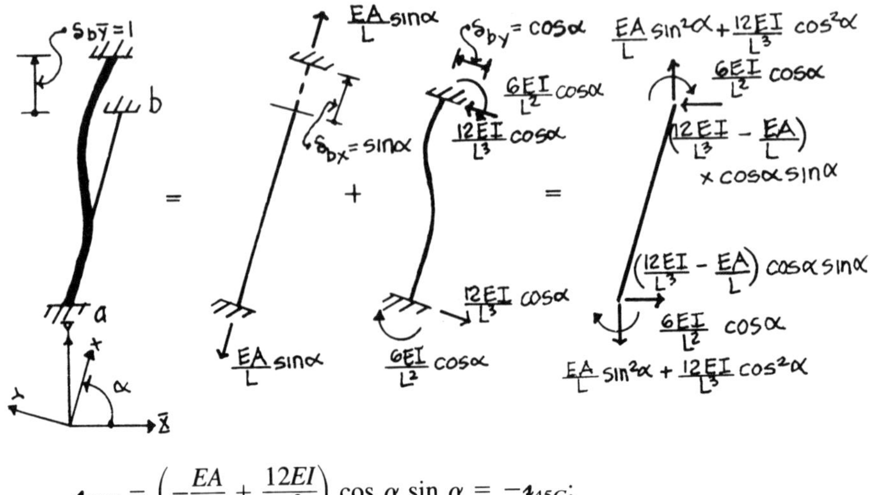

$$k_{15G} = \left(-\frac{EA}{L} + \frac{12EI}{L^3}\right)\cos\alpha\,\sin\alpha = -k_{45G};$$

$$k_{25G} = -\frac{EA}{L}\sin^2\alpha - \frac{12EI}{L^3}\cos^2\alpha = -k_{53G}; \quad k_{35G} = \frac{6EI}{L^2}\cos\alpha = k_{65G}$$

Figure 9.18

$$\theta_b = 1; \quad \delta_{a\underline{X}} = \delta_{a\overline{Y}} = \delta_{b\underline{X}} = \delta_{b\overline{Y}} = \theta_a = 0$$

$$k_{16G} = \frac{6EI}{L^2}\sin\alpha = -k_{46G}; \quad k_{26G} = -\frac{6EI}{L^2}\cos\alpha = -k_{56G};$$

$$k_{36G} = \frac{2EI}{L}; \quad k_{66G} = \frac{4EI}{L}$$

Figure 9.19

Shown in Figure 9.20 is the frame member stiffness matrix oriented with respect to the global axes (Note: $C = \cos \alpha$; $S = \sin \alpha$).

$$
\bar{\bar{k}}_{G(i)} = \begin{bmatrix}
(EA/L)C^2 + (12EI/L^3)S^2 & (EA/L - 12EI/L^3)CS & (6EI/L^2)S & -(EA/L)C^2 - (12EI/L^3)S^2 & (-EA/L + 12EI/L^3)CS & (6EI/L^2)S \\
 & (EA/L)S^2 + (12EI/L^3)C^2 & -(6EI/L^2)C & (-EA/L + 12EI/L^3)CS & -(EA/L)S^2 - (12EI/L^3)C^2 & -(6EI/L^2)C \\
 & & 4EI/L & -(6EI/L^2)S & (6EI/L^2)C & 2EI/L \\
 & \text{symmetric} & & (EA/L)C^2 + (12EI/L^3)S^2 & (EA/L - 12EI/L^3)CS & -(6EI/L^2)S \\
 & & & & (EA/L)S^2 + (12EI/L^3)C^2 & (6EI/L^2)C \\
 & & & & & 4EI/L
\end{bmatrix}
\begin{matrix} r_{ab\underline{X}} \\ r_{ab\bar{Y}} \\ M_{ab} \\ r_{ba\underline{X}} \\ r_{ba\bar{Y}} \\ M_{ba} \end{matrix}
$$

$$
\begin{matrix} \delta_{a\underline{X}} & \delta_{a\bar{Y}} & \theta_a & \delta_{b\underline{X}} & \delta_{b\bar{Y}} & \theta_b \end{matrix}
$$

Figure 9.20 Member stiffness matrix for frames (in global coordinates).

It can be seen by carrying out the matrix multiplication that the two member stiffness matrices, $\bar{\bar{k}}$ in local coordinates and $\bar{\bar{k}}_G$ in global coordinates, are related by

$$\bar{\bar{k}}_{(i)} = \bar{\bar{Q}}_{(i)}\bar{\bar{k}}_{G(i)}\bar{\bar{Q}}^T_{(i)} \quad \text{and} \quad \bar{\bar{k}}_{G(i)} = \bar{\bar{Q}}^T_{(i)}\bar{\bar{k}}_{(i)}\bar{\bar{Q}}_{(i)}$$

where $\bar{\bar{Q}}_{(i)}$, the complete angular rotation matrix, is

$$\bar{\bar{Q}}_{(i)} = \left[\begin{array}{ccc|ccc} \cos\alpha_i & \sin\alpha_i & 0 & 0 & 0 & 0 \\ -\sin\alpha_i & \cos\alpha_i & 0 & 0 & 0 & 0 \\ 0 & 0 & 1 & 0 & 0 & 0 \\ \hline 0 & 0 & 0 & \cos\alpha_i & \sin\alpha_i & 0 \\ 0 & 0 & 0 & -\sin\alpha_i & \cos\alpha_i & 0 \\ 0 & 0 & 0 & 0 & 0 & 1 \end{array}\right] = \begin{bmatrix} \bar{\bar{C}} & 0 \\ 0 & \bar{\bar{C}} \end{bmatrix}.$$

The 3×3 matrix, $\bar{\bar{C}}$, relates the two systems of axes at only one end of the element.

Now we can write the governing equations for frames:

$$0 = \bar{R}^{FM}_G + \bar{\bar{S}}_G\bar{D}_G, \tag{16}$$

$$\bar{r}_{(i)} = \bar{r}^{FM}_{(i)} + \bar{\bar{k}}_{(i)}\bar{d}_{(i)}. \tag{17}$$

In Equation (16) all the matrices are written with respect to the global $\overline{XY}$ axes. This means that $\bar{\bar{S}}_G$ can be formed by selective addition of terms from member stiffness matrices in $\overline{XY}$ coordinates (Figure 9.20). The fixed-end terms, $\bar{R}^{FM}_G$ must also be based on global axes, thus the resulting displacements, $\bar{D}_G$, will be referred to the global axes. The terms in Equation 17 can all be written in local (xy) coordinates or all in global ($\overline{XY}$) coordinates. The local are probably more meaningful because the force reactions will directly turn out as shears and axial forces. If local coordinates are used in Equation 17, then the found global displacements, $\bar{D}_G$ and $\bar{d}_G$ must be decomposed to form displacements, $\bar{d}$, in the local (xy) system. We note that

$$\bar{d}_{(i)} = \bar{\bar{Q}}_{(i)}\bar{d}_{G(i)}.$$

Example 2 illustrates the matrix stiffness process for frames.

EXAMPLE #2: STIFFNESS METHOD: FRAME

Given: Frame as shown

Find: 1) Displacements at b and c
2) The member reactions
3) V, M and axial force diagrams.

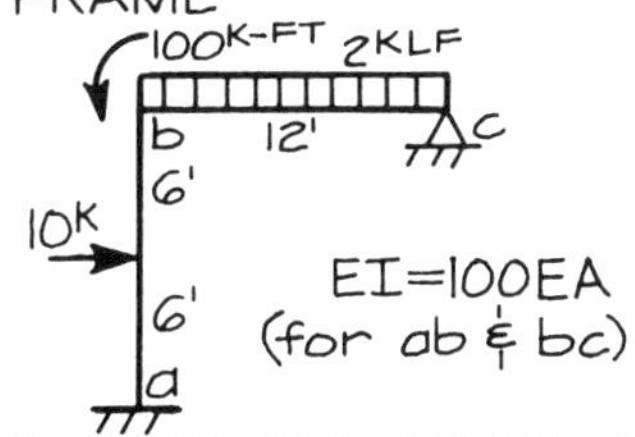

We arbitrarily take ab as member ①. With respect to global coordinate system we arbitrarily order the displacements as shown.

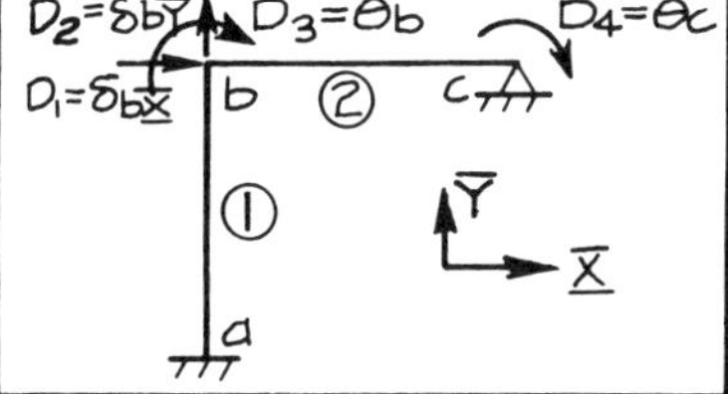

Set-up the member stiffness matrices in global coordinates (see fig. 20)

For member ①: $\alpha = 90°$; $L = 12'$; $A = I/100$

For member ②: $\alpha = 0°$; $L = 12'$; $A = I/100$

$$\bar{\bar{\mathcal{S}}}_G ① = \frac{EI}{(12)^2}\begin{bmatrix} 1 & 0 & 6 & -1 & 0 & 6 \\ 0 & .12 & 0 & 0 & -.12 & 0 \\ 6 & 0 & 48 & -6 & 0 & 24 \\ -1 & 0 & -6 & 1 & 0 & -6 \\ 0 & -.12 & 0 & 0 & .12 & 0 \\ 6 & 0 & 24 & -6 & 0 & 48 \end{bmatrix}\begin{matrix} r_{ab\bar{X}} \\ r_{ab\bar{Y}} \\ M_{ab} \\ r_{ba\bar{X}} \\ r_{ba\bar{Y}} \\ M_{ba} \end{matrix}$$

$$\begin{matrix} \delta_{a\bar{X}} & \delta_{a\bar{Y}} & \theta_a & \delta_{b\bar{X}} & \delta_{b\bar{Y}} & \theta_b \end{matrix}$$

$$\bar{\bar{\mathcal{S}}}_G ② = \frac{EI}{(12)^2}\begin{bmatrix} .12 & 0 & 0 & -.12 & 0 & 0 \\ 0 & 1 & -6 & 0 & -1 & -6 \\ 0 & -6 & 48 & 0 & 6 & 24 \\ -.12 & 0 & 0 & .12 & 0 & 0 \\ 0 & -1 & 6 & 0 & 1 & 6 \\ 0 & -6 & 24 & 0 & 6 & 48 \end{bmatrix}\begin{matrix} r_{bc\bar{X}} \\ r_{bc\bar{Y}} \\ M_{bc} \\ r_{cb\bar{X}} \\ r_{cb\bar{Y}} \\ M_{cb} \end{matrix}$$

$$\begin{matrix} \delta_{b\bar{X}} & \delta_{b\bar{Y}} & \theta_b & \delta_{c\bar{X}} & \delta_{c\bar{Y}} & \theta_c \end{matrix}$$

Form structure stiffness matrix in global coordinates from member stiffness matrices by choosing terms associated with $(\delta_{b\bar{X}}, \delta_{b\bar{Y}}, \theta_b, \theta_c)$ and $(r_{b\bar{X}}, r_{b\bar{Y}}, M_b, M_c)$.

$$\bar{\bar{S}}_G = \frac{EI}{(12)^2}\begin{bmatrix} (1+.12) & (0+0) & (-6+0) & 0 \\ (0+0) & (.12+1) & (0-6) & -6 \\ (-6+0) & (-6+0) & (48+48) & 24 \\ 0 & -6 & 24 & 48 \end{bmatrix} = \frac{EI}{(12)^2}\begin{bmatrix} 1.12 & 0 & -6 & 0 \\ 0 & 1.12 & -6 & -6 \\ -6 & -6 & 96 & 24 \\ 0 & -6 & 24 & 48 \end{bmatrix}\begin{matrix} r_{b\bar{X}} \\ r_{b\bar{Y}} \\ M_b \\ M_c \end{matrix}$$

$$\begin{matrix} \delta_{b\bar{X}} & \delta_{b\bar{Y}} & \theta_b & \theta_c \end{matrix}$$

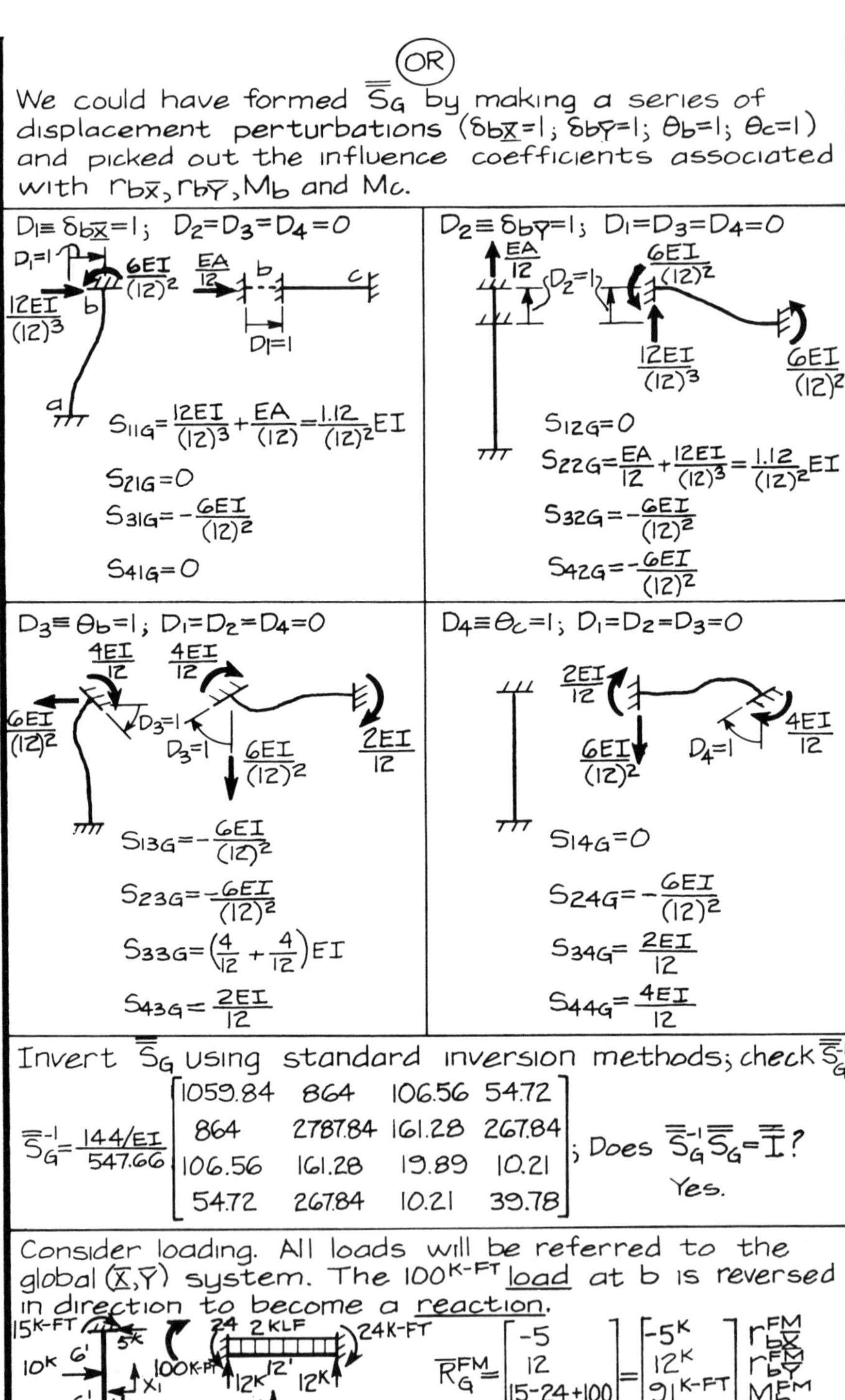

OR
We could have formed $\overline{\overline{S}}_G$ by making a series of displacement perturbations ($\delta b\overline{X}=1$; $\delta b\overline{Y}=1$; $\theta_b=1$; $\theta_c=1$) and picked out the influence coefficients associated with $r_{b\overline{X}}$, $r_{b\overline{Y}}$, M_b and M_c.
$D_1 \equiv \delta b\overline{X}=1$; $D_2=D_3=D_4=0$
$S_{11G}=\frac{12EI}{(12)^3}+\frac{EA}{(12)}=\frac{1.12}{(12)^2}EI$
$S_{21G}=0$
$S_{31G}=-\frac{6EI}{(12)^2}$
$S_{41G}=0$
$D_2 \equiv \delta b\overline{Y}=1$; $D_1=D_3=D_4=0$
$S_{12G}=0$
$S_{22G}=\frac{EA}{12}+\frac{12EI}{(12)^3}=\frac{1.12}{(12)^2}EI$
$S_{32G}=-\frac{6EI}{(12)^2}$
$S_{42G}=-\frac{6EI}{(12)^2}$
$D_3 \equiv \theta_b=1$; $D_1=D_2=D_4=0$
$S_{13G}=-\frac{6EI}{(12)^2}$
$S_{23G}=-\frac{6EI}{(12)^2}$
$S_{33G}=\left(\frac{4}{12}+\frac{4}{12}\right)EI$
$S_{43G}=\frac{2EI}{12}$
$D_4 \equiv \theta_c=1$; $D_1=D_2=D_3=0$
$S_{14G}=0$
$S_{24G}=-\frac{6EI}{(12)^2}$
$S_{34G}=\frac{2EI}{12}$
$S_{44G}=\frac{4EI}{12}$
Invert $\overline{\overline{S}}_G$ using standard inversion methods; check $\overline{\overline{S}}_G^{-1}$
$\overline{\overline{S}}_G^{-1}=\frac{144/EI}{547.66}\begin{bmatrix}1059.84 & 864 & 106.56 & 54.72\\ 864 & 2787.84 & 161.28 & 267.84\\ 106.56 & 161.28 & 19.89 & 10.21\\ 54.72 & 267.84 & 10.21 & 39.78\end{bmatrix}$; Does $\overline{\overline{S}}_G^{-1}\overline{\overline{S}}_G=\overline{\overline{I}}$?
Yes.
Consider loading. All loads will be referred to the global ($\overline{X},\overline{Y}$) system. The $100^{K\text{-}FT}$ load at b is reversed in direction to become a reaction.
$\overline{R}_G^{FM}=\begin{bmatrix}-5\\12\\15-24+100\\24\end{bmatrix}=\begin{bmatrix}-5^K\\12^K\\91^{K\text{-}FT}\\24^{K\text{-}FT}\end{bmatrix}\begin{matrix}r_{b\overline{X}}^{FM}\\r_{b\overline{Y}}^{FM}\\M_b^{FM}\\M_c^{FM}\end{matrix}$

Calculate displacements: $\overline{D}_G = -\overline{\overline{S}}_G^{-1}\overline{R}_G^{FM}$
(In global coordinates)

$$\overline{D}_G = -\frac{144/EI}{547.66}\begin{bmatrix} 1059.84 & 864 & 106.56 & 54.72 \\ 864 & 2787.84 & 161.28 & 267.84 \\ 106.56 & 161.28 & 19.89 & 10.21 \\ 54.72 & 267.84 & 10.21 & 39.78 \end{bmatrix}\begin{bmatrix} -5 \\ 12 \\ 91 \\ 24 \end{bmatrix} = \frac{1}{EI}\begin{bmatrix} -4227.77 \\ -13209.61 \\ -909.13 \\ -1268.49 \end{bmatrix}\begin{matrix} \delta_{b\overline{X}} \\ \delta_{b\overline{Y}} \\ \theta_b \\ \theta_c \end{matrix}$$

$\delta_{b\overline{X}}$, $\delta_{b\overline{Y}}$, θ_c, θ_b

Calculate the member reactions: $\overline{r}_{(i)} = \overline{r}_{(i)}^{FM} + \overline{\overline{\mathcal{S}}}_{(i)}\overline{d}_{(i)}$
For member (1) (in global coordinates):
The fixed-end terms are shown above and the displacements are taken from $\overline{D}_G$:

$$\overline{r}_{G(1)}^{FM} = \begin{bmatrix} -5 \\ 0 \\ -15 \\ -5 \\ 0 \\ 15 \end{bmatrix}\begin{matrix} r_{ab\overline{X}}^{FM} \\ r_{ab\overline{Y}}^{FM} \\ M_{ab}^{FM} \\ r_{ba\overline{X}}^{FM} \\ r_{ba\overline{Y}}^{FM} \\ M_{ba}^{FM} \end{matrix}; \quad \overline{d}_{G(1)} = \frac{1}{EI}\begin{bmatrix} 0 \\ 0 \\ 0 \\ -4227.77 \\ -13209.61 \\ -909.13 \end{bmatrix}$$

$\overline{Y}$, $\overline{X}$

$$\overline{r}_{G(1)} = \overline{r}_{G(1)}^{FM} + \overline{\overline{\mathcal{S}}}_{G(1)}\overline{d}_{G(1)}$$

$$\overline{r}_{G(1)} = \begin{bmatrix} -5 \\ 0 \\ -15 \\ -5 \\ 0 \\ 15 \end{bmatrix} + \frac{EI}{(12)^2}\begin{bmatrix} 1 & 0 & 6 & -1 & 0 & 6 \\ 0 & .12 & 0 & 0 & -.12 & 0 \\ 6 & 0 & 48 & -6 & 0 & 24 \\ -1 & 0 & -6 & 1 & 0 & -6 \\ 0 & -.12 & 0 & 0 & .12 & 0 \\ 6 & 0 & 24 & -6 & 0 & 48 \end{bmatrix}\frac{1}{EI}\begin{bmatrix} 0 \\ 0 \\ 0 \\ -4227.77 \\ -13209.61 \\ -909.13 \end{bmatrix} = \begin{bmatrix} -13.5^K \\ 11.0^K \\ 9.6^{K\text{-}FT} \\ 3.5^K \\ -11.0^K \\ -111.9^{K\text{-}FT} \end{bmatrix}$$

111.9K-FT, 11.0K, 3.5K, b, 6', 10K, 6', 13.5K, 9.6K-FT, 11.0K

$\overline{Y}$, $\overline{X}$

$\Sigma F_{\overline{X}} = 0$
$\Sigma F_{\overline{Y}} = 0$
$\Sigma M_a = 0$

(Note: reactions have been rounded-off to one decimal place.)

OR

We could calculate the member reactions for (1) in local coordinates: The fixed-end terms (referred to XY axes) are shown above; to find $\overline{d}_{(1)}$ use $\overline{d}_{(1)} = \overline{\overline{Q}}_{(1)}\overline{d}_{G(1)}$ (here $\alpha = 90°$). $\overline{r}_{(1)} = \overline{r}_{(1)}^{FM} + \overline{\overline{\mathcal{S}}}_{(1)}\overline{d}_{(1)}$

$$\bar{r}^{FM}_{①}=\begin{bmatrix}0\\5\\-15\\0\\5\\15\end{bmatrix}\begin{matrix}P^{FM}_{ab}\\V^{FM}_{ab}\\M^{FM}_{ab}\\P^{FM}_{ba}\\V^{FM}_{ba}\\M^{FM}_{ba}\end{matrix};\quad \bar{d}_{①}=\left[\begin{array}{ccc|ccc}0&1&0&0&0&0\\-1&0&0&0&0&0\\0&0&1&0&0&0\\\hline 0&0&0&0&1&0\\0&0&0&-1&0&0\\0&0&0&0&0&1\end{array}\right]\begin{bmatrix}0\\0\\0\\-4227.77\\-13209.61\\-909.13\end{bmatrix}\frac{1}{EI}=\frac{1}{EI}\begin{bmatrix}0\\0\\0\\-13209.61\\4227.77\\-909.13\end{bmatrix}\begin{matrix}\delta_{ax}\\\delta_{ay}\\\theta_a\\\delta_{bx}\\\delta_{by}\\\theta_b\end{matrix}$$

(We see that $\delta_{b\underline{X}}=-\delta_{by}$; $\delta_{b\overline{Y}}=\delta_{bx}$ and θ_b same for both systems.)

$$\bar{r}_{①}=\begin{bmatrix}0\\5\\-15\\0\\5\\15\end{bmatrix}+\frac{EI}{(12)^2}\begin{bmatrix}.12&0&0&-.12&0&0\\0&1&-6&0&-1&-6\\0&-6&48&0&6&24\\-.12&0&0&.12&0&0\\0&-1&6&0&1&6\\0&-6&24&0&6&48\end{bmatrix}\frac{1}{EI}\begin{bmatrix}0\\0\\0\\-13209.61\\4227.77\\-909.13\end{bmatrix}=\begin{bmatrix}11.0^K\\13.5^K\\9.6^{K\text{-}FT}\\-11.0^K\\-3.5^K\\-111.9^{K\text{-}FT}\end{bmatrix}\begin{matrix}P_{ab}\\V_{ab}\\M_{ab}\\P_{ba}\\V_{ba}\\M_{ba}\end{matrix}$$

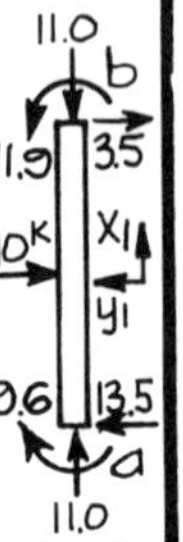

(here $\bar{\bar{\Delta}}_{①}$ is from Figure 12)

For member ② the local and global coincide so $\bar{r}_{G②}=\bar{r}_{②}$, etc.

$$\bar{r}^{FM}_{②}=\begin{bmatrix}0\\12\\-24\\0\\12\\24\end{bmatrix}\begin{matrix}P^{FM}_{bc}\\V^{FM}_{bc}\\M^{FM}_{bc}\\P^{FM}_{cb}\\V^{FM}_{cb}\\M^{FM}_{cb}\end{matrix};\quad \bar{d}_{②}=\frac{1}{EI}\begin{bmatrix}-4227.77\\-13209.61\\-909.13\\0\\0\\-1268.49\end{bmatrix}\begin{matrix}\delta_{bx}\\\delta_{by}\\\theta_b\\\delta_{cx}\\\delta_{cy}\\\theta_c\end{matrix}$$

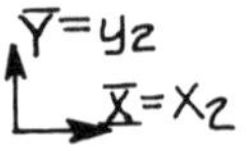

$\overline{Y}=y_2$, $\overline{X}=x_2$

$$\bar{r}_{②}=\bar{r}^{FM}_{②}+\bar{\bar{\Delta}}_{②}\bar{d}_{②}$$

$$\bar{r}_{②}=\begin{bmatrix}0\\12\\-24\\0\\12\\24\end{bmatrix}+\frac{EI}{(12)^2}\begin{bmatrix}.12&0&0&-.12&0&0\\0&1&-6&0&-1&-6\\0&-6&48&0&6&24\\-.12&0&0&.12&0&0\\0&-1&6&0&1&6\\0&-6&24&0&6&48\end{bmatrix}\frac{1}{EI}\begin{bmatrix}-4227.77\\-13209.61\\-909.13\\0\\0\\-1268.49\end{bmatrix}=\begin{bmatrix}-3.5^K\\11.0^K\\11.9^{K\text{-}FT}\\3.5^K\\13.0^K\\0\end{bmatrix}\begin{matrix}P_{bc}\\V_{bc}\\M_{bc}\\P_{cb}\\V_{cb}\\M_{cb}\end{matrix}$$

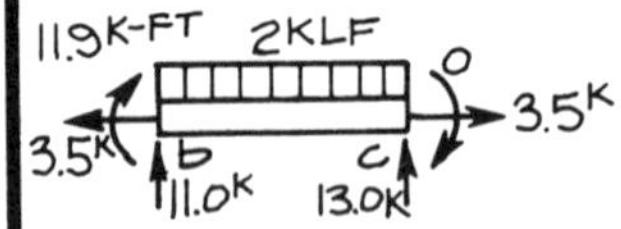

11.9 K-FT, 2 KLF, 0, 3.5K, 3.5K, b, c, 11.0K, 13.0K

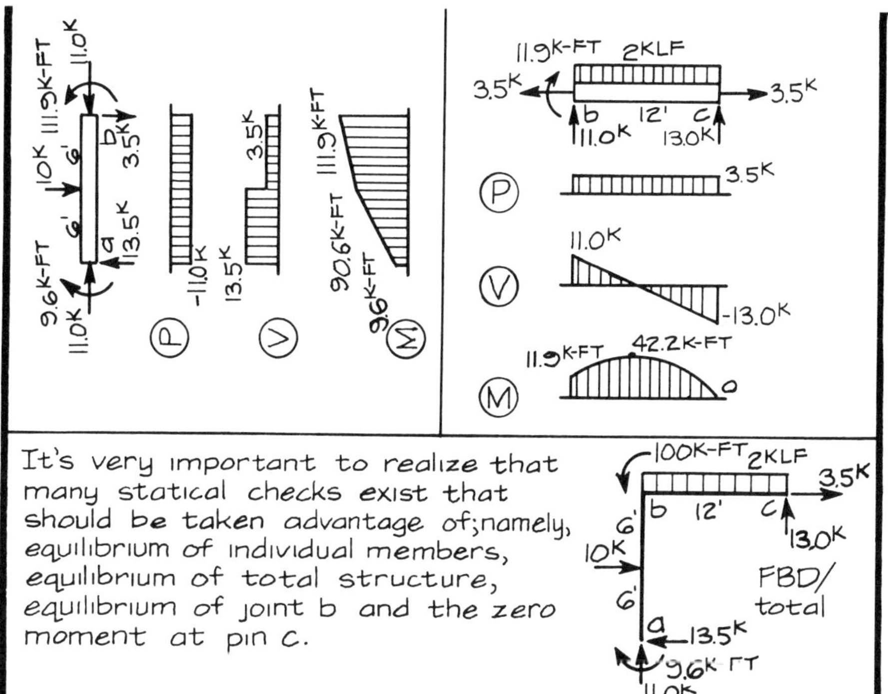

It's very important to realize that many statical checks exist that should be taken advantage of; namely, equilibrium of individual members, equilibrium of total structure, equilibrium of joint b and the zero moment at pin c.

TRUSSES

Since truss members by definition cannot bend, the stiffness coefficients are very simple. As with frames, the geometry of trusses implies the use of local (xy) and global ($\overline{X}\overline{Y}$) coordinate systems.

First the member stiffness influence coefficients in local coordinates: we have four displacements (all translations) to consider: $d_1 = \delta_{ax}$; $d_2 = \delta_{ay}$; $d_3 = \delta_{bx}$; $d_4 = \delta_{by}$ (Figures 9.21, 9.22).

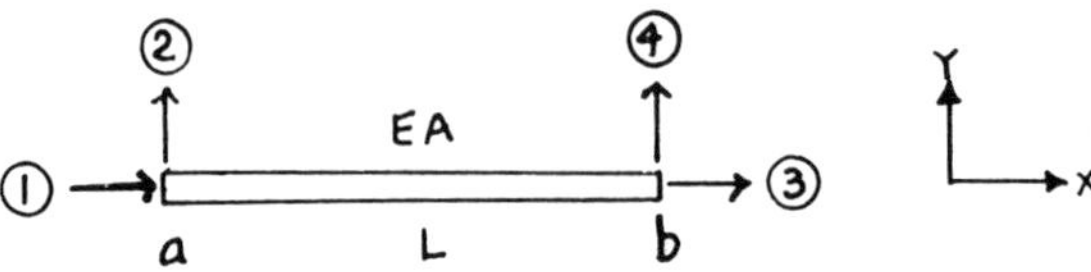

Figure 9.21

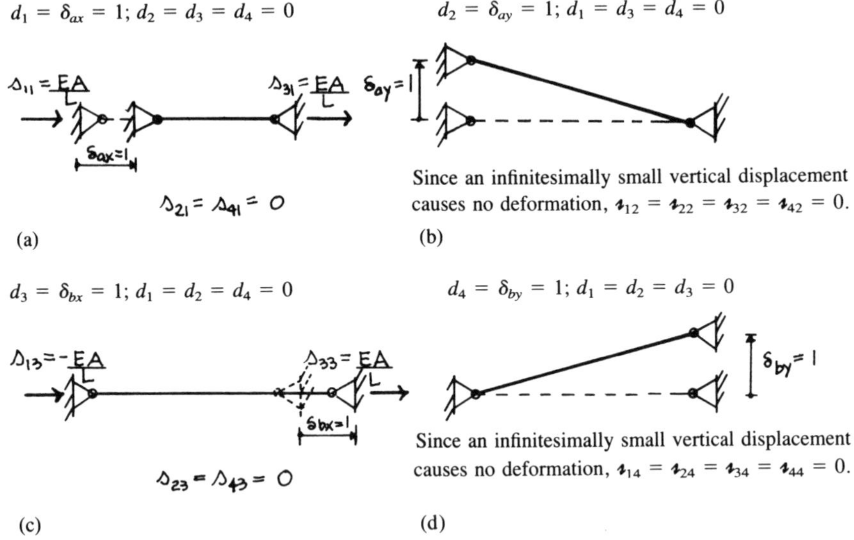

Figure 9.22

Thus the member stiffness matrix in local coordinates for trusses is shown in Figure 9.23.

$$\bar{\bar{s}}_{(i)} = \begin{bmatrix} EA/L & 0 & -EA/L & 0 \\ 0 & 0 & 0 & 0 \\ -EA/L & 0 & EA/L & 0 \\ 0 & 0 & 0 & 0 \end{bmatrix} \begin{matrix} r_{abx} \\ r_{aby} \\ r_{bax} \\ r_{bay} \end{matrix}$$

$$\delta_{ax} \qquad \delta_{ay} \qquad \delta_{bx} \qquad \delta_{by}$$

Figure 9.23 Member stiffness matrix for trusses (in local coordinates).

To form the member stiffness matrix in global coordinates, we look at an inclined member and cause horizontal and vertical unit displacements (Figures 9.24–9.27).

$$\delta_{a\bar{X}} = 1; \quad \delta_{a\bar{Y}} = \delta_{b\bar{X}} = \delta_{b\bar{Y}} = 0$$

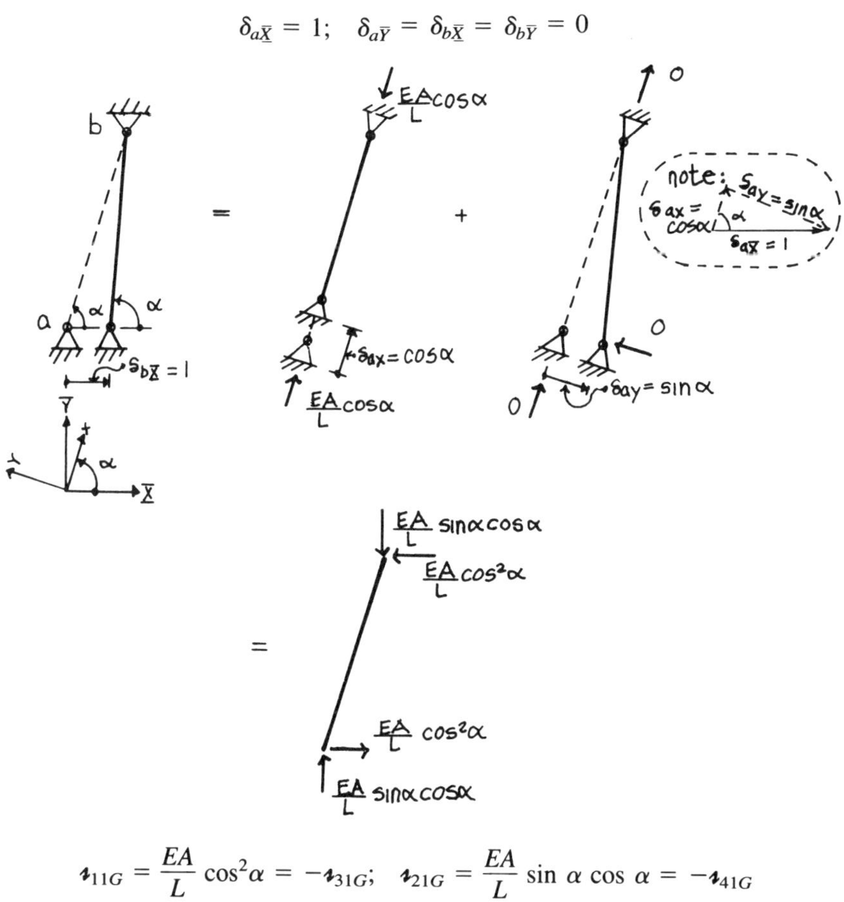

$$s_{11G} = \frac{EA}{L}\cos^2\alpha = -s_{31G}; \quad s_{21G} = \frac{EA}{L}\sin\alpha\cos\alpha = -s_{41G}$$

Figure 9.24

$$\delta_{a\bar{Y}} = 1; \quad \delta_{a\underline{X}} = \delta_{b\underline{X}} = \delta_{b\bar{Y}} = 0$$

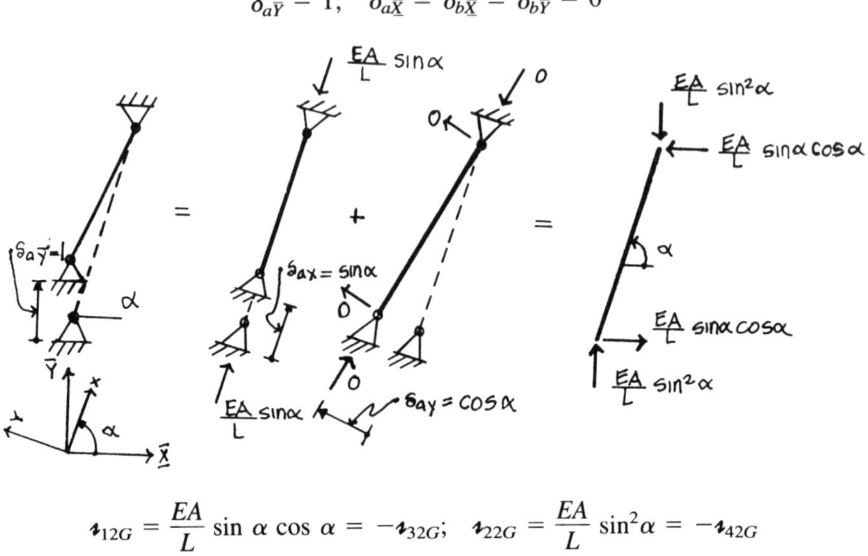

$$k_{12G} = \frac{EA}{L} \sin\alpha \cos\alpha = -k_{32G}; \quad k_{22G} = \frac{EA}{L} \sin^2\alpha = -k_{42G}$$

Figure 9.25

$$\delta_{b\underline{X}} = 1; \quad \delta_{a\underline{X}} = \delta_{a\bar{Y}} = \delta_{b\bar{Y}} = 0$$

$$k_{13G} = -\frac{EA}{L} \cos^2\alpha = -k_{33G}; \quad k_{23G} = -\frac{EA}{L} \sin\alpha \cos\alpha = -k_{43G}$$

Figure 9.26

$$\delta_{b\overline{Y}} = 1; \quad \delta_{a\underline{X}} = \delta_{a\overline{Y}} = \delta_{b\underline{X}} = 0$$

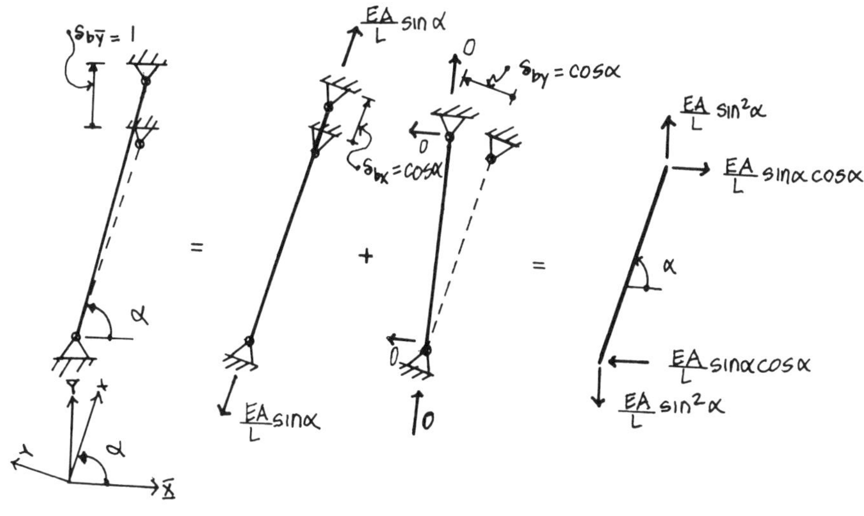

$$ɹ_{14G} = -\frac{EA}{L} \sin\alpha \cos\alpha = -ɹ_{34G}; \quad ɹ_{24G} = -\frac{EA}{L} \sin^2\alpha = -ɹ_{44G}$$

Figure 9.27

The truss member stiffness matrix in global coordinates is as depicted in Figure 9.28.

$$\overline{\overline{ɹ}}_{G\textcircled{i}} = \frac{E_iA_i}{L_i}\begin{bmatrix} \cos^2\alpha & \sin\alpha\cos\alpha & -\cos^2\alpha & -\sin\alpha\cos\alpha \\ & \sin^2\alpha & -\sin\alpha\cos\alpha & -\sin^2\alpha \\ & \text{symmetric} & \cos^2\alpha & \sin\alpha\cos\alpha \\ & & & \sin^2\alpha \end{bmatrix} \begin{matrix} r_{ab\underline{X}} \\ r_{ab\overline{Y}} \\ r_{ba\underline{X}} \\ r_{ba\overline{Y}} \end{matrix}$$

$$\delta_{a\underline{X}} \qquad \delta_{a\overline{Y}} \qquad \delta_{b\underline{X}} \qquad \delta_{b\overline{Y}}$$

Figure 9.28 Member stiffness matrix for trusses (in global coordinates).

The governing equations for trusses are:

$$0 = \overline{R}_G^{FM} + \overline{\overline{S}}_G\overline{D}_G, \tag{18}$$

$$\overline{r}_{\textcircled{i}} = \overline{\overline{ɹ}}_{\textcircled{i}}d_{\textcircled{i}}. \tag{19}$$

Since for trusses all loads, by definition, are placed at the joints, the fixed-end terms in Equation 18 are just negatives of the joint loads. These joint loads do not directly come into the member reaction Equation 19, thus 19 is completely devoid of fixed-end terms. Care must be taken in forming the structure stiffness matrix, $\overline{\overline{S}}_G$, by selective addition of member stiffness terms due to the fact that for trusses it is not uncommon for three or more members to meet at a joint. If local coordinates are used in Eq. 19, the resulting member reactions will be strictly axial (x) forces. Using local coordinates in Eq. 19 means decomposing the global displacements ($\overline{d}_{ⓘ} = \overline{\overline{Q}}_{ⓘ}\overline{d}_{Gⓘ}$). If global coordinates are used in Eq. 19, no displacement decomposition is necessary but the member forces, $\overline{r}_{ⓘ}$, will, in general, have both $\underline{\overline{X}}$ and $\overline{Y}$ components. Example 3 treats both of these approaches.

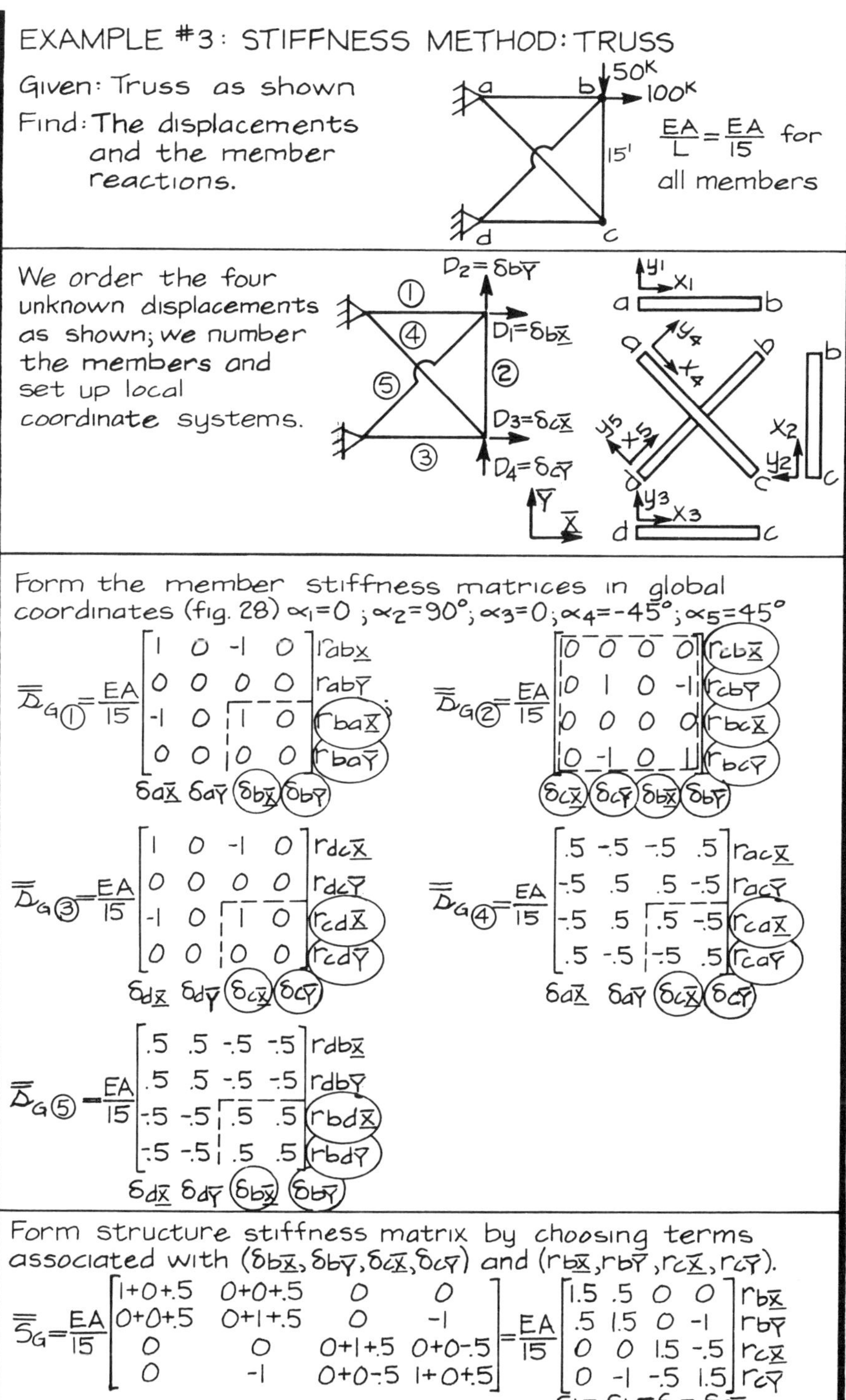
EXAMPLE #3: STIFFNESS METHOD: TRUSS
Given: Truss as shown
Find: The displacements and the member reactions.
50K
100K
a
b
15'
c
d
EA/L = EA/15 for all members
We order the four unknown displacements as shown; we number the members and set up local coordinate systems.
D2=δbY
D1=δbX
D3=δcX
D4=δcY
Form the member stiffness matrices in global coordinates (fig. 28) α1=0 ; α2=90°; α3=0; α4=-45°; α5=45°
Form structure stiffness matrix by choosing terms associated with (δbX, δbY, δcX, δcY) and (rbX, rbY, rcX, rcY).
SG = EA/15 [1+0+.5 0+0+.5 0 0; 0+0+.5 0+1+.5 0 -1; 0 0 0+1+.5 0+0-.5; 0 -1 0+0-.5 1+0+.5] = EA/15 [1.5 .5 0 0; .5 1.5 0 -1; 0 0 1.5 -.5; 0 -1 -.5 1.5]

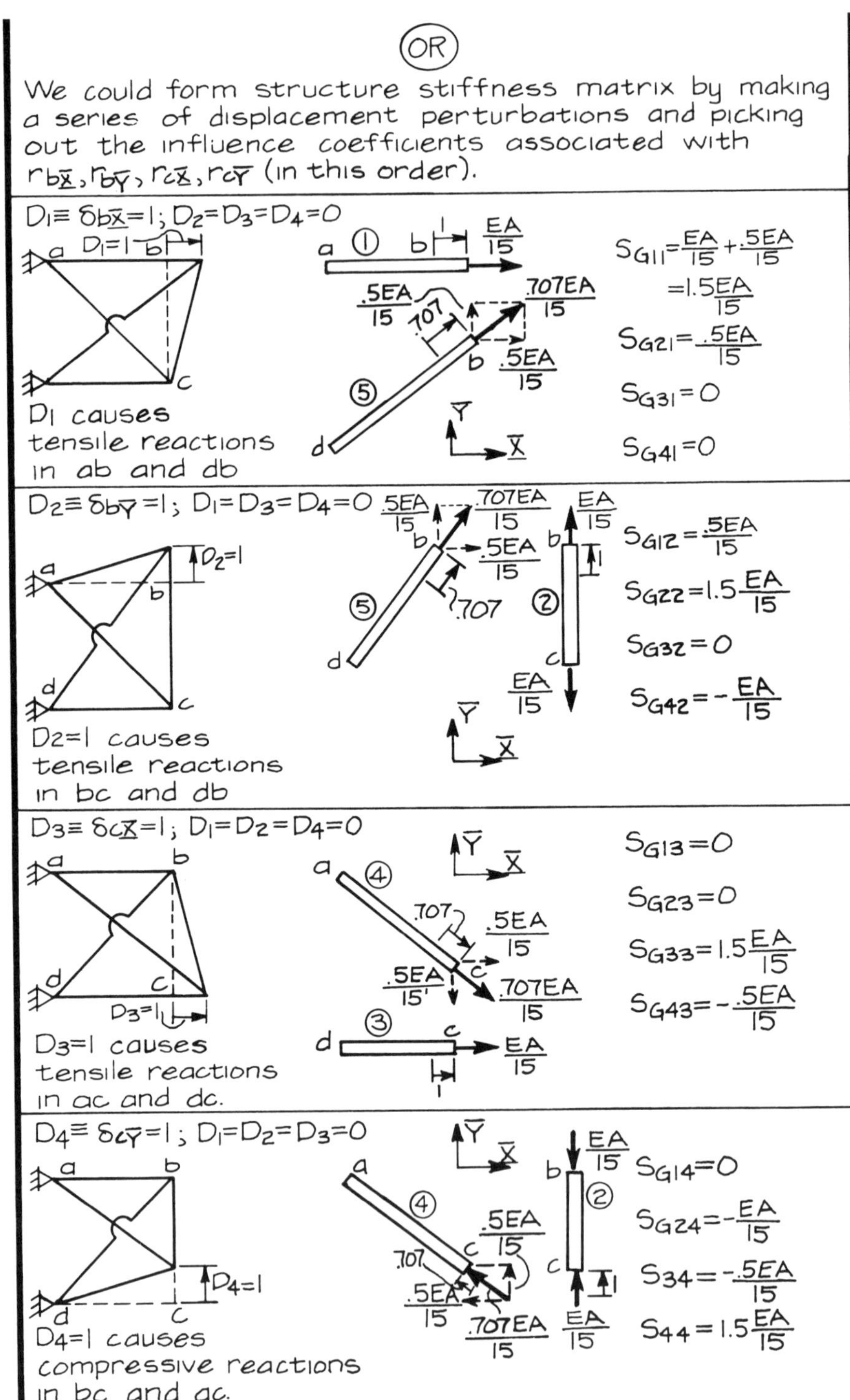
OR
We could form structure stiffness matrix by making a series of displacement perturbations and picking out the influence coefficients associated with $r_{b\bar{X}}, r_{b\bar{Y}}, r_{c\bar{X}}, r_{c\bar{Y}}$ (in this order).
$D_1 \equiv \delta_{b\bar{X}} = 1; D_2 = D_3 = D_4 = 0$
D1 causes tensile reactions in ab and db
$S_{G11} = \frac{EA}{15} + \frac{.5EA}{15} = 1.5\frac{EA}{15}$
$S_{G21} = \frac{.5EA}{15}$
$S_{G31} = 0$
$S_{G41} = 0$
$D_2 \equiv \delta_{b\bar{Y}} = 1; D_1 = D_3 = D_4 = 0$
D2=1 causes tensile reactions in bc and db
$S_{G12} = \frac{.5EA}{15}$
$S_{G22} = 1.5\frac{EA}{15}$
$S_{G32} = 0$
$S_{G42} = -\frac{EA}{15}$
$D_3 \equiv \delta_{c\bar{X}} = 1; D_1 = D_2 = D_4 = 0$
D3=1 causes tensile reactions in ac and dc.
$S_{G13} = 0$
$S_{G23} = 0$
$S_{G33} = 1.5\frac{EA}{15}$
$S_{G43} = -\frac{.5EA}{15}$
$D_4 \equiv \delta_{c\bar{Y}} = 1; D_1 = D_2 = D_3 = 0$
D4=1 causes compressive reactions in bc and ac.
$S_{G14} = 0$
$S_{G24} = -\frac{EA}{15}$
$S_{34} = -\frac{.5EA}{15}$
$S_{44} = 1.5\frac{EA}{15}$

Invert $\bar{\bar{S}}_G$ using standard inversion methods; check $\bar{\bar{S}}_G^{-1}$

$$\bar{\bar{S}}^{-1} = \frac{15/EA}{1.75}\begin{bmatrix} 1.5 & -1.0 & -.25 & -.75 \\ -1.0 & 3.0 & .75 & 2.25 \\ -.25 & .75 & 1.5 & 1.0 \\ -.75 & 2.25 & 1.0 & 3.0 \end{bmatrix}; \text{ Does } \bar{\bar{S}}_G \bar{\bar{S}}_G^{-1} = \bar{\bar{I}}\text{? Yes.}$$

Calculate displacements: $\bar{D}_G = -\bar{\bar{S}}_G^{-1}\bar{R}_G^{FM} = \bar{\bar{S}}_G^{-1}(-\bar{R}_G^{FM})$. The vector $(-\bar{R}_G^{FM})$ is made up of the <u>loads</u> at joints b & c.

$$\bar{D}_G = \frac{15/EA}{1.75}\begin{bmatrix} 1.5 & -1.0 & -.25 & -.75 \\ -1.0 & 3.0 & .75 & 2.25 \\ -.25 & .75 & 1.5 & 1.0 \\ -.75 & 2.25 & 1.0 & 3.0 \end{bmatrix}\begin{bmatrix} 100^K \\ -50^K \\ 0 \\ 0 \end{bmatrix} = \frac{1}{EA}\begin{bmatrix} 1714.29 \\ -2142.86 \\ -535.71 \\ -1607.14 \end{bmatrix}\begin{matrix} \delta_{b\bar{X}} \\ \delta_{b\bar{Y}} \\ \delta_{c\bar{X}} \\ \delta_{c\bar{Y}} \end{matrix}$$

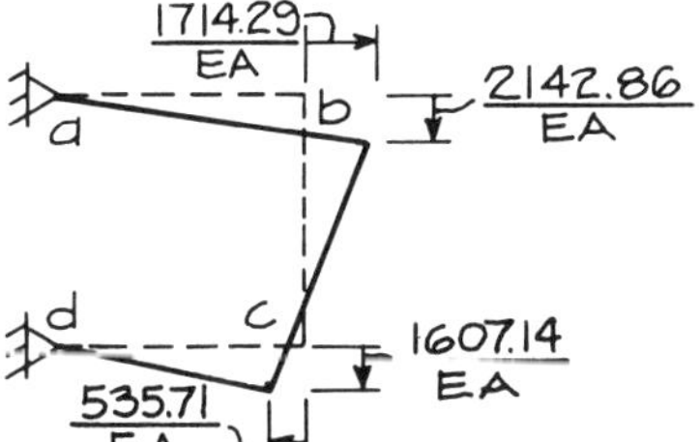

The member reactions in global coordinates are calculated: $\bar{r}_{G(i)} = \bar{\bar{s}}_{G(i)}\bar{d}_{G(i)}$. For member ① the global and local coordinates coincide.

$$\bar{d}_{G(1)} = \frac{1}{EA}\begin{bmatrix} 0 \\ 0 \\ 1714.29 \\ -2142.86 \end{bmatrix}\begin{matrix} \delta_{a\bar{X}} \\ \delta_{a\bar{Y}} \\ \delta_{b\bar{X}} \\ \delta_{b\bar{Y}} \end{matrix}$$

114.29K ← a ① b → 114.29K ; $\bar{Y} = y_1$, $\bar{X} = x_1$

$$\bar{r}_{G(1)} = \frac{EA}{15}\begin{bmatrix} 1 & 0 & -1 & 0 \\ 0 & 0 & 0 & 0 \\ -1 & 0 & 1 & 0 \\ 0 & 0 & 0 & 0 \end{bmatrix}\frac{1}{EA}\begin{bmatrix} 0 \\ 0 \\ 1714.29 \\ -2142.86 \end{bmatrix} = \begin{bmatrix} -114.29^K \\ 0 \\ 114.29^K \\ 0 \end{bmatrix}\begin{matrix} r_{ab\bar{X}} \equiv P_{ab} \\ r_{ab\bar{Y}} \equiv V_{ab} \\ r_{ba\bar{X}} \equiv P_{ba} \\ r_{ba\bar{Y}} \equiv V_{ba} \end{matrix}$$

For member ②

$$\bar{d}_{G②} = \frac{1}{EA}\begin{bmatrix} -535.71 \\ -1607.14 \\ 1714.29 \\ -2142.86 \end{bmatrix} \begin{matrix} \delta_{c\bar{X}} \\ \delta_{c\bar{Y}} \\ \delta_{b\bar{X}} \\ \delta_{b\bar{Y}} \end{matrix}$$

$$\bar{r}_{G②} = \frac{EA}{15}\begin{bmatrix} 0 & 0 & 0 & 0 \\ 0 & 1 & 0 & -1 \\ 0 & 0 & 0 & 0 \\ 0 & -1 & 0 & 1 \end{bmatrix} \frac{1}{EA}\begin{bmatrix} -535.71 \\ -1607.14 \\ 1714.29 \\ -2142.86 \end{bmatrix} = \begin{bmatrix} 0 \\ 35.71^K \\ 0 \\ -35.71^K \end{bmatrix} \begin{matrix} r_{cb\bar{X}} \\ r_{cb\bar{Y}} \\ r_{bc\bar{X}} \\ r_{bc\bar{Y}} \end{matrix}$$

For member ③ the global and local coordinates coincide.

$$\bar{d}_{G③} = \frac{1}{EA}\begin{bmatrix} 0 \\ 0 \\ -535.71 \\ -1607.14 \end{bmatrix} \begin{matrix} \delta_{d\bar{X}} \\ \delta_{d\bar{Y}} \\ \delta_{c\bar{X}} \\ \delta_{c\bar{Y}} \end{matrix}$$

$$\bar{r}_{G③} = \frac{EA}{15}\begin{bmatrix} 1 & 0 & -1 & 0 \\ 0 & 0 & 0 & 0 \\ -1 & 0 & 1 & 0 \\ 0 & 0 & 0 & 0 \end{bmatrix} \frac{1}{EA}\begin{bmatrix} 0 \\ 0 \\ -535.71 \\ -1607.14 \end{bmatrix} = \begin{bmatrix} 35.71^K \\ 0 \\ -35.71^K \\ 0 \end{bmatrix} \begin{matrix} r_{dc\bar{X}} \\ r_{dc\bar{Y}} \\ r_{cd\bar{X}} \\ r_{cd\bar{Y}} \end{matrix}$$

For member ④

$$\bar{d}_{G④} = \frac{1}{EA}\begin{bmatrix} 0 \\ 0 \\ -535.71 \\ -1607.14 \end{bmatrix} \begin{matrix} \delta_{a\bar{X}} \\ \delta_{a\bar{Y}} \\ \delta_{c\bar{X}} \\ \delta_{c\bar{Y}} \end{matrix}$$

$$\bar{r}_{G④} = \frac{.5EA}{15}\begin{bmatrix} 1 & -1 & -1 & 1 \\ -1 & 1 & 1 & -1 \\ -1 & 1 & 1 & -1 \\ 1 & -1 & -1 & 1 \end{bmatrix} \frac{1}{EA}\begin{bmatrix} 0 \\ 0 \\ -535.71 \\ -1607.14 \end{bmatrix} = \begin{bmatrix} -35.71 \\ 35.71 \\ 35.71 \\ -35.71 \end{bmatrix} \begin{matrix} r_{ac\bar{X}} \\ r_{ac\bar{Y}} \\ r_{ca\bar{X}} \\ r_{ca\bar{Y}} \end{matrix}$$

For member ⑤

$$\bar{d}_{G⑤} = \frac{1}{EA}\begin{bmatrix} 0 \\ 0 \\ 1714.29 \\ -2142.86 \end{bmatrix} \begin{matrix} \delta_{d\bar{X}} \\ \delta_{d\bar{Y}} \\ \delta_{b\bar{X}} \\ \delta_{b\bar{Y}} \end{matrix}$$

$$\bar{r}_{G⑤} = \frac{.5EA}{15}\begin{bmatrix} 1 & 1 & -1 & -1 \\ 1 & 1 & -1 & -1 \\ -1 & -1 & 1 & 1 \\ -1 & -1 & 1 & 1 \end{bmatrix} \frac{1}{EA}\begin{bmatrix} 0 \\ 0 \\ 1714.29 \\ -2142.86 \end{bmatrix} = \begin{bmatrix} 14.29^K \\ 14.29^K \\ -14.29^K \\ -14.29^K \end{bmatrix} \begin{matrix} r_{db\bar{X}} \\ r_{db\bar{Y}} \\ r_{bd\bar{X}} \\ r_{bd\bar{Y}} \end{matrix}$$

(OR)

The member reactions could be found in local coordinates: $\bar{r}_{(i)} = \bar{\bar{\Delta}}_{(i)} \bar{d}_{(i)}$ where $\bar{d}_{(i)} = \bar{\bar{Q}}_{(i)} d_{G(i)}$ and $\bar{\bar{\Delta}}_{(i)}$ is found from Fig. 23 and $\bar{\bar{Q}}_{(i)}$ is two-dimensional. Since for members ① and ③ the local and global coordinates coincide we only have to consider members ②, ④ and ⑤.

For member ② $\alpha = 90°$

$$\bar{d}_{(2)} = \bar{\bar{Q}}_{(2)} d_{G(2)} = \begin{bmatrix} 0 & 1 & 0 & 0 \\ -1 & 0 & 0 & 0 \\ 0 & 0 & 0 & 1 \\ 0 & 0 & -1 & 0 \end{bmatrix} \frac{1}{EA} \begin{bmatrix} -535.71 \\ -1607.14 \\ 1714.29 \\ -2142.86 \end{bmatrix} = \frac{1}{EI} \begin{bmatrix} -1607.14 \\ 535.71 \\ -2142.86 \\ -1714.29 \end{bmatrix} \begin{matrix} \delta_{cx} \\ \delta_{cy} \\ \delta_{bx} \\ \delta_{by} \end{matrix}$$

$$\bar{r}_{(2)} = \bar{\bar{\Delta}}_{(2)} \bar{d}_{(2)} = \frac{EA}{15} \begin{bmatrix} 1 & 0 & -1 & 0 \\ 0 & 0 & 0 & 0 \\ -1 & 0 & 1 & 0 \\ 0 & 0 & 0 & 0 \end{bmatrix} \frac{1}{EA} \begin{bmatrix} -1607.14 \\ 535.71 \\ -2142.86 \\ -1714.29 \end{bmatrix} = \begin{bmatrix} 35.71^K \\ 0 \\ -35.71^K \\ 0 \end{bmatrix} \begin{matrix} P_{cb} \\ V_{cb} \\ P_{bc} \\ V_{bc} \end{matrix}$$

35.71^K, b, ②, x_2, $\bar{X}$, y_2, $\bar{Y}$, c, 35.71^K

For member ④ $\alpha = -45°$

$$\bar{d}_{(4)} = \bar{\bar{Q}}_{(4)} d_{G(4)} = \begin{bmatrix} .707 & -.707 & 0 & 0 \\ .707 & .707 & 0 & 0 \\ 0 & 0 & .707 & -.707 \\ 0 & 0 & .707 & .707 \end{bmatrix} \frac{1}{EA} \begin{bmatrix} 0 \\ 0 \\ -535.71 \\ -1607.14 \end{bmatrix} = \frac{1}{EA} \begin{bmatrix} 0 \\ 0 \\ 757.50 \\ -1514.99 \end{bmatrix} \begin{matrix} \delta_{ax} \\ \delta_{ay} \\ \delta_{cx} \\ \delta_{cy} \end{matrix}$$

$$\bar{r}_{(4)} = \bar{\bar{\Delta}}_{(4)} \bar{d}_{(4)} = \frac{EA}{15} \begin{bmatrix} 1 & 0 & -1 & 0 \\ 0 & 0 & 0 & 0 \\ -1 & 0 & 1 & 0 \\ 0 & 0 & 0 & 0 \end{bmatrix} \frac{1}{EA} \begin{bmatrix} 0 \\ 0 \\ 757.50 \\ -1514.99 \end{bmatrix} = \begin{bmatrix} -50.5^K \\ 0 \\ 50.5^K \\ 0 \end{bmatrix} \begin{matrix} P_{ac} \\ V_{ac} \\ P_{ca} \\ V_{ca} \end{matrix}$$

50.5^K, $\bar{Y}$, y_4, a, $\bar{X}$, x_4, ④, c, 50.5^K

For member ⑤ $\alpha = 45°$

$$\bar{d}_{(5)} = \bar{\bar{Q}}_{(5)} d_{G(5)} = \begin{bmatrix} .707 & .707 & 0 & 0 \\ -.707 & .707 & 0 & 0 \\ 0 & 0 & .707 & .707 \\ 0 & 0 & -.707 & .707 \end{bmatrix} \frac{1}{EA} \begin{bmatrix} 0 \\ 0 \\ 1714.29 \\ -2142.86 \end{bmatrix} = \frac{1}{EA} \begin{bmatrix} 0 \\ 0 \\ -303.00 \\ -2727.00 \end{bmatrix} \begin{matrix} \delta_{dx} \\ \delta_{dy} \\ \delta_{bx} \\ \delta_{by} \end{matrix}$$

$$\bar{r}_{(5)} = \bar{\bar{\Delta}}_{(5)} \bar{d}_{(5)} = \frac{EA}{15} \begin{bmatrix} 1 & 0 & -1 & 0 \\ 0 & 0 & 0 & 0 \\ -1 & 0 & 1 & 0 \\ 0 & 0 & 0 & 0 \end{bmatrix} \frac{1}{EA} \begin{bmatrix} 0 \\ 0 \\ -303.00 \\ -2727.00 \end{bmatrix} = \begin{bmatrix} 20.20^K \\ 0 \\ -20.20^K \\ 0 \end{bmatrix} \begin{matrix} P_{dc} \\ V_{dc} \\ P_{cd} \\ V_{cd} \end{matrix}$$

20.20^K, c, ⑤, d, $\bar{Y}$, x_5, 20.20^K, y_5, $\bar{X}$

Note: All joints are in equilibrium as is the total structure. These are valuable statical checks.

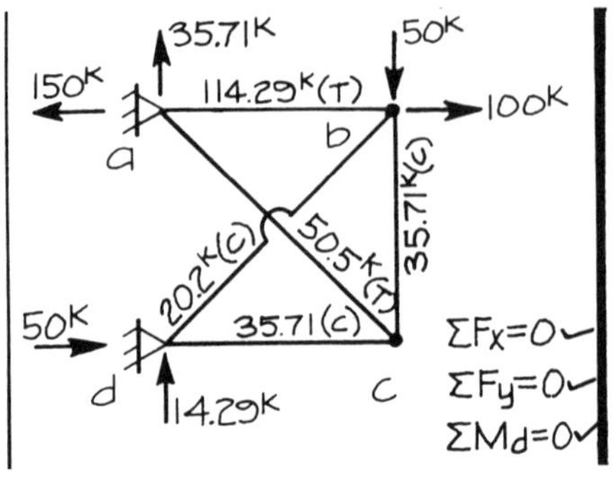

PROBLEMS

Problems 1–15

For each structure shown below:

a) Identify and order the unknown joint displacements; and
b) form the structure stiffness matrix.

Problems 16–30

For each structure 1–15, using stiffness methods:

a) Find the unknown joint displacements,
b) find the member reactions; and
c) for the beams and frames draw the shear force and bending moment diagrams.

Note: ◯ = member number

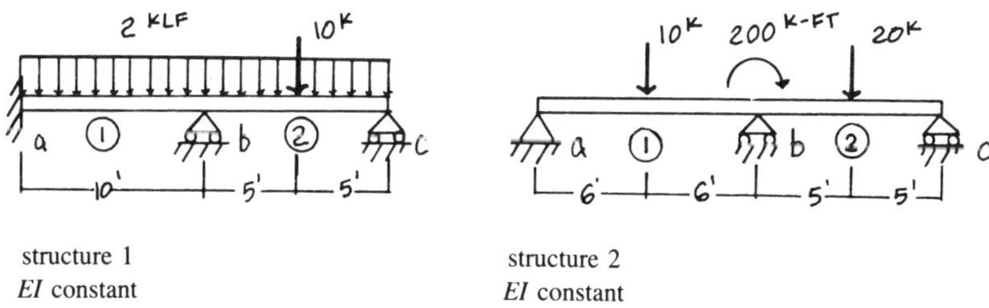

structure 1
EI constant

structure 2
EI constant

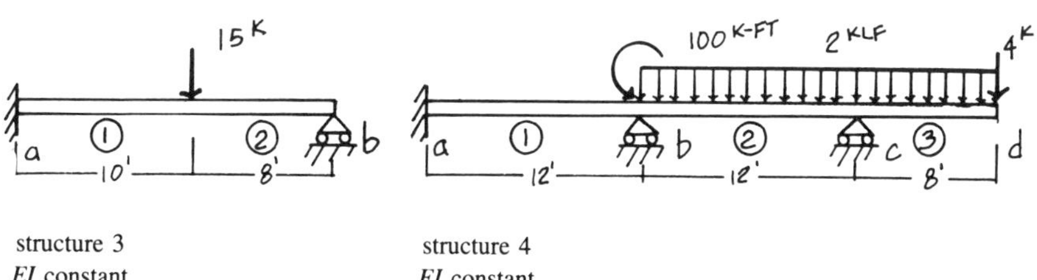

structure 3
EI constant

structure 4
EI constant

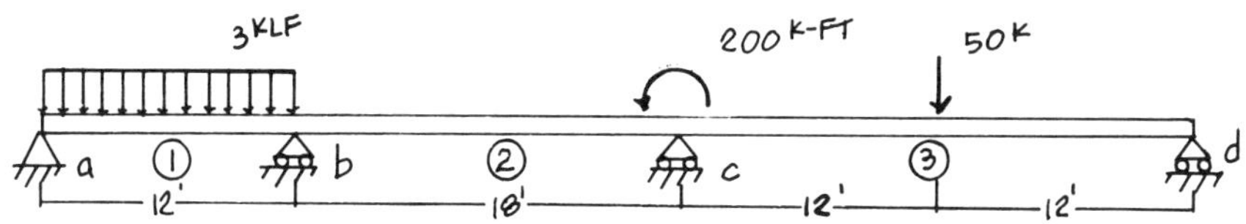

structure 5
EI constant

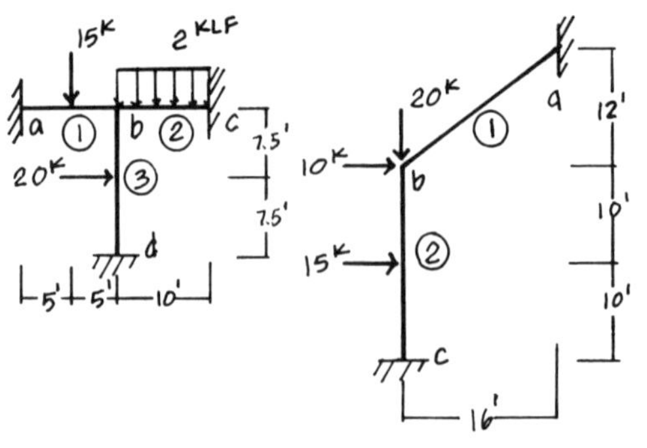

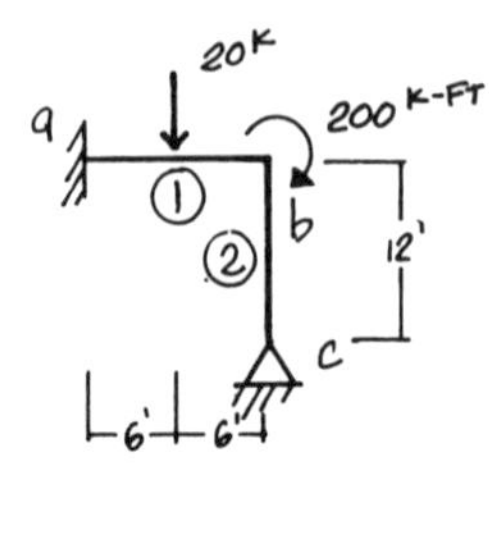

structure 6
$EI = 500\ EA$
EI constant

structure 7
$EI = 500\ EA$
EI constant

structure 8
$EI = 100\ EA$
EI constant

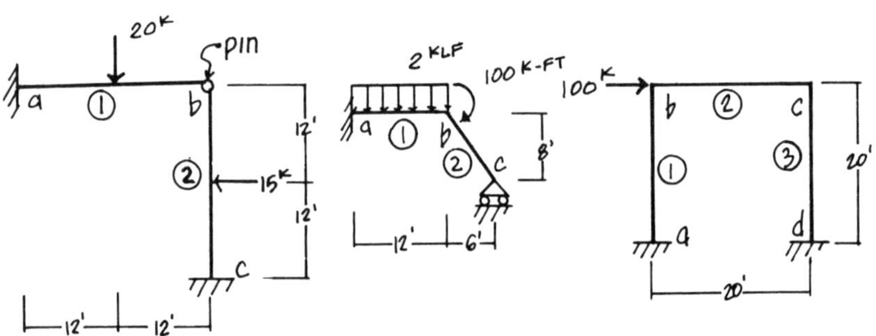

structure 9
$EI = 50\ EA$
EI constant

structure 10
$EI = 500\ EA$
EI constant

structure 11
$EI = 50\ EA$
EI constant

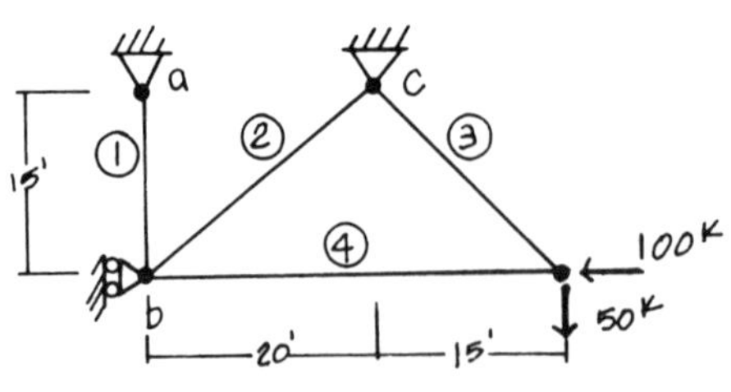

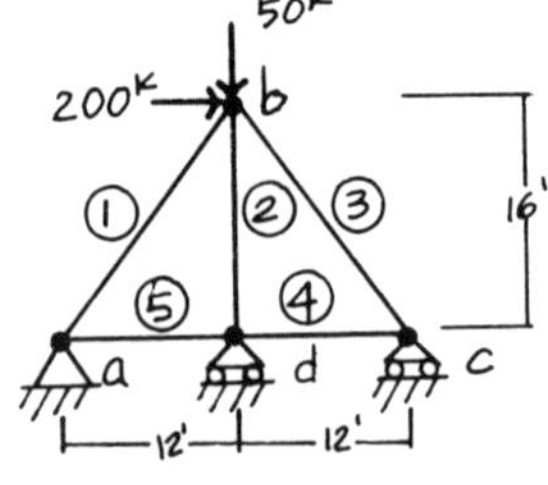

structure 12
AE/L constant

structure 13
AE/L constant

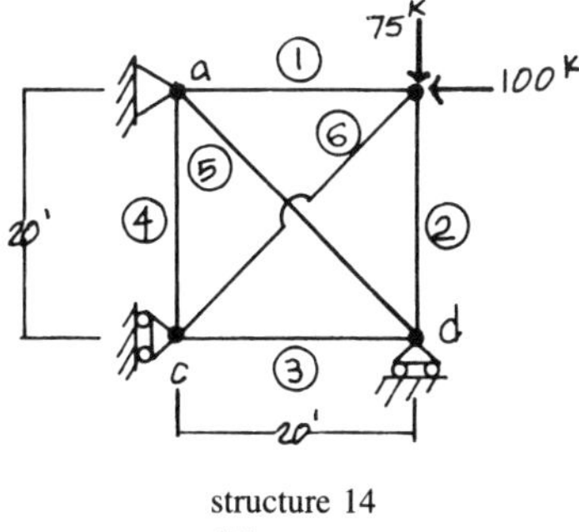

structure 14
AE constant

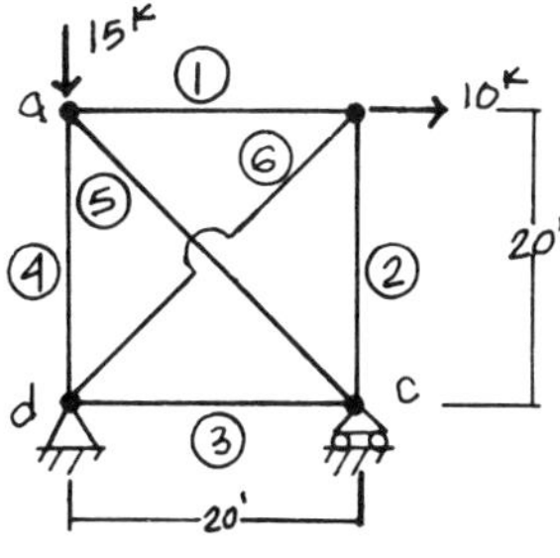

structure 15
AE constant

10

APPROXIMATE SOLUTIONS

INTRODUCTION

First, a clarification of the word "approximate." All solutions are, in reality, approximations if one considers the uncertainties of loadings, the inexactness of material response, as well as the problems inherent in fabrication and joint fixity. "Exact" solutions require ideal conditions. By "approximate" solutions, we mean solutions that reduce, due to simplifying assumptions, indeterminate structures to determinate ones. Most approximate solutions are based on conventional analysis.

In the analysis of determinate structures, contrary to that of indeterminate structures, the elastic properties (areas, moment of inertia, etc.) do not play a factor. Thus, in approximate solutions the elastic properties are of no (or little) concern, making approximate solutions useful in preliminary planning. They are also convenient for quick but rough checks of "exact" solutions. While there are many approximations available, we shall restrict ourselves to three: two dealing with wind loads on multistory structures and one dealing with gravity loads on multistory structures. References for various other approximate solutions will be made at chapter's end.

WIND LOADS ON MULTISTORY FRAMES

In both methods investigated here, as well as many other approximate solutions, a basic assumption is made that locates the point of inflection. The point of inflection (or point of contraflecture) is where the member changes bending curvature from positive to negative. That is, the point of inflection is where the bending moment is zero. These points can be treated as internal pins.

Cantilever Method

This approximation is best used for tall, slender, rigid-jointed, rectangular, multi-story frames; those frames that act like a vertical cantilever beam (Figure 10.1).

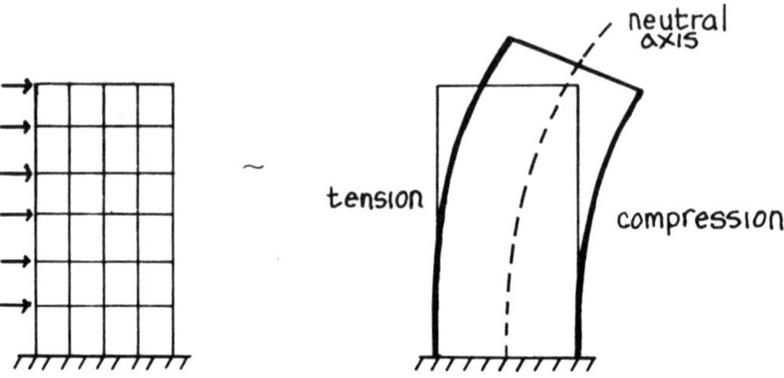

Figure 10.1

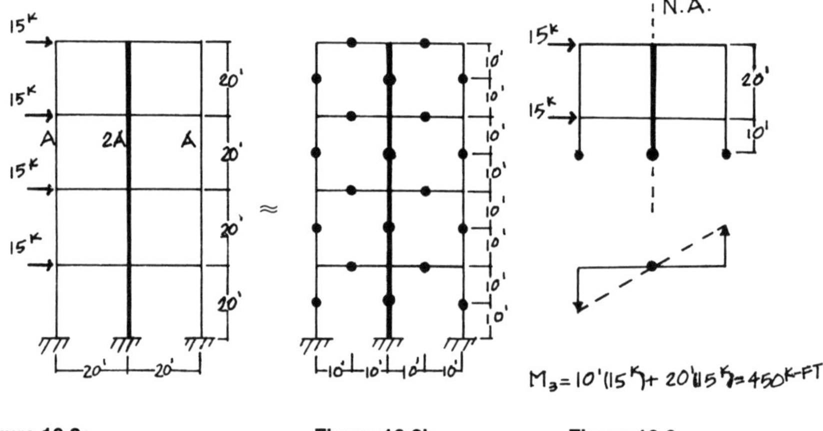

Figure 10.2a

Figure 10.2b

Figure 10.2c

Two sets of assumptions are made to reduce the frame to a determinate one:

1) Inflection points occur at mid-span of all girders and at mid-height of all rigidly connected columns (of course, inflection points for pinned-end columns occur at the pins). See Figure 10.2b.
2) The axial column forces, P_{ij}, at the inflection points are found by analogy with beam theory, i.e.,

$$\sigma_{ij} = \frac{P_{ij}}{A_j} = \frac{M_i y_j}{I_i},$$

where i refers to the story and j to the column. Here M_i is the moment at the "pin" level of story i and y_j is the distance from the centroid (neutral axis) of the floor to the j column. I_i is the moment of inertia of the floor (see Figure 10.2c).

Example 1 shows how the process is carried out. Note that the two sets of assumptions, along with statics, supply excess equations, thus statical checks result. Note also that some statement about relative column areas must be made.

Portal Method

This very useful method (actually there are several variations of this approach, all of which are referred to as the portal method) is mostly applied to squat, multistory frames, that is, frames whose horizontal dimensions are larger than their vertical dimensions. The two assumptions are:

1) Points of inflection occur at beam mid-spans and column mid-heights (or at the pin for pinned-end columns). This is the same assumption as for the cantilever method.
2) An assumption is made on the horizontal shear at the column inflection points. There are several assumptions in use. We shall illustrate (Example 2) the one that assigns column shear in the ratio of one part for exterior columns and two parts for interior columns. Other horizontal shear assumptions proportion the shear according to the floor area supported by the column, or according to the column cross-sectional area, if known.

Example 2 illustrates the portal method for a two-story frame with the same loading and geometry as the top two stories of the frame in Example 1. Again, note statical checks exist at all story levels.

GRAVITY LOADS ON FRAMES

The approximation presented here is easily applied to rigid-jointed frames subjected to gravity loads. Concentrated, as well as distributed loads, can be handled. Two assumptions are made:

1) Points of inflection for girders occur at 1/10 of the span from both beam ends.
2) Axial forces in beam are neglected.

These two assumptions, in effect, insert internal rollers at the 1/10 span points for all beams, thereby making the frame determinate. Example 3 illustrates the procedure.

Other Approximations

The entire book by Benjamin (Reference 6) is devoted to approximate methods. For internally statically indeterminate trusses with two diagonals per panel, see McCormac (Reference 3).

An approximation for the Vierendeel "truss" (not a true truss) is also illustrated in McCormac (Reference 3).

A simplified structure approach for frames where only local effects are considered is discussed in Mantell and Marion (Reference 2).

A two-cycle moment distribution truncation for frames is also presented by Mantell and Marion (Reference 2).

Moment and shear coefficients for uniform but critically placed loads on frames have been developed by the American Concrete Institute.

One-story mill buildings using a variation of the portal method are treated by McCormac (Reference 3).

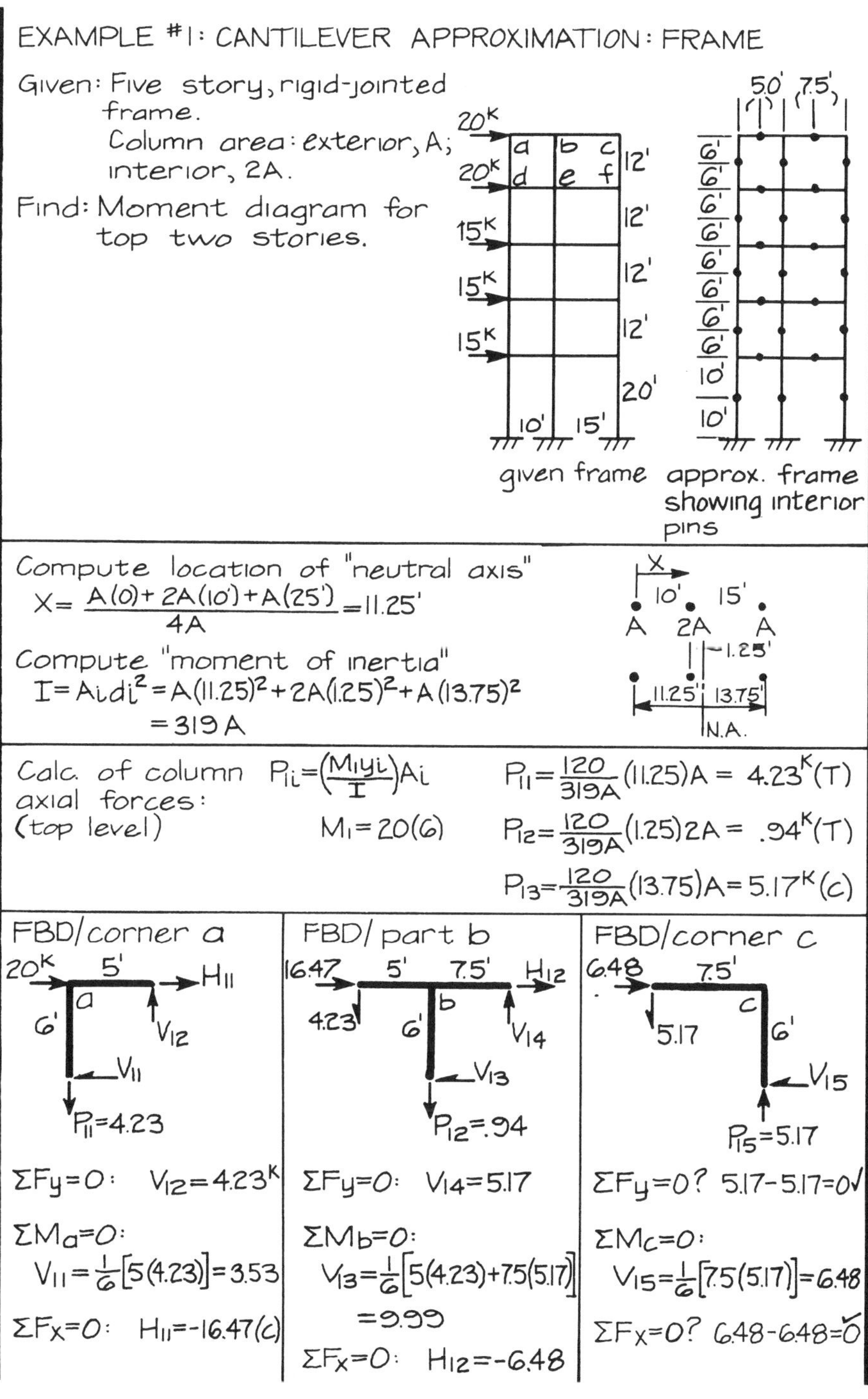

EXAMPLE #1: CANTILEVER APPROXIMATION: FRAME

Given: Five story, rigid-jointed frame.
Column area: exterior, A; interior, 2A.

Find: Moment diagram for top two stories.

Compute location of "neutral axis"

$$X = \frac{A(0) + 2A(10') + A(25')}{4A} = 11.25'$$

Compute "moment of inertia"

$$I = A_i d_i^2 = A(11.25)^2 + 2A(1.25)^2 + A(13.75)^2 = 319A$$

Calc. of column axial forces: (top level)

$$P_{1i} = \left(\frac{M_1 y_i}{I}\right) A_i \qquad M_1 = 20(6)$$

$$P_{11} = \frac{120}{319A}(11.25)A = 4.23^K (T)$$

$$P_{12} = \frac{120}{319A}(1.25)2A = .94^K (T)$$

$$P_{13} = \frac{120}{319A}(13.75)A = 5.17^K (C)$$

FBD/corner a

$\Sigma F_y = 0$: $V_{12} = 4.23^K$

$\Sigma M_a = 0$:
$V_{11} = \frac{1}{6}[5(4.23)] = 3.53$

$\Sigma F_x = 0$: $H_{11} = -16.47 (C)$

FBD/part b

$\Sigma F_y = 0$: $V_{14} = 5.17$

$\Sigma M_b = 0$:
$V_{13} = \frac{1}{6}[5(4.23) + 7.5(5.17)]$
$= 9.99$

$\Sigma F_x = 0$: $H_{12} = -6.48$

FBD/corner c

$\Sigma F_y = 0$? $5.17 - 5.17 = 0$ ✓

$\Sigma M_c = 0$:
$V_{15} = \frac{1}{6}[7.5(5.17)] = 6.48$

$\Sigma F_x = 0$? $6.48 - 6.48 = 0$ ✓

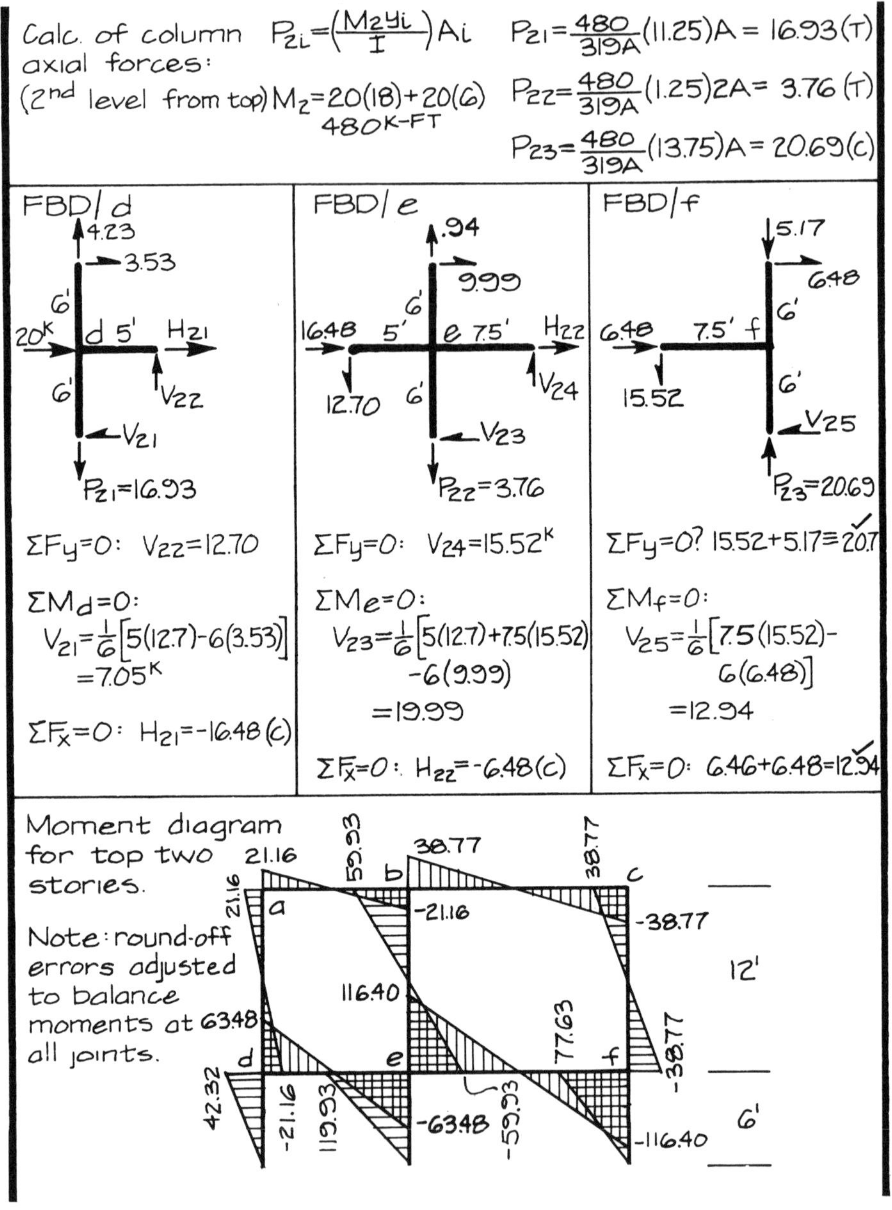
Calc. of column axial forces:
P2i=(M2yi/I)Ai
(2nd level from top) M2=20(18)+20(6)
480K-FT
P21=480/319A(11.25)A = 16.93(T)
P22=480/319A(1.25)2A= 3.76 (T)
P23=480/319A(13.75)A= 20.69(C)
FBD/d
4.23
3.53
6'
20K
d 5'
H21
6'
V22
V21
P21=16.93
ΣFy=0: V22=12.70
ΣMd=0:
V21=1/6[5(12.7)-6(3.53)]
=7.05K
ΣFx=0: H21=-16.48 (C)
FBD/e
.94
9.99
6'
16.48
5'
e 7.5'
H22
12.70
6'
V24
V23
P22=3.76
ΣFy=0: V24=15.52K
ΣMe=0:
V23=1/6[5(12.7)+7.5(15.52)
-6(9.99)
=19.99
ΣFx=0: H22=-6.48(C)
FBD/f
5.17
6.48
6'
6.48
7.5' f
15.52
6'
V25
P23=20.69
ΣFy=0? 15.52+5.17≅20.7
ΣMf=0:
V25=1/6[7.5(15.52)-
6(6.48)]
=12.94
ΣFx=0: 6.46+6.48=12.94
Moment diagram for top two stories.
Note: round-off errors adjusted to balance moments at all joints.
21.16
21.16
a
59.93
b
38.77
-21.16
38.77
c
-38.77
12'
116.40
63.48
77.63
-38.77
d
e
f
42.32
-21.16
119.93
-63.48
-59.93
-116.40
6'

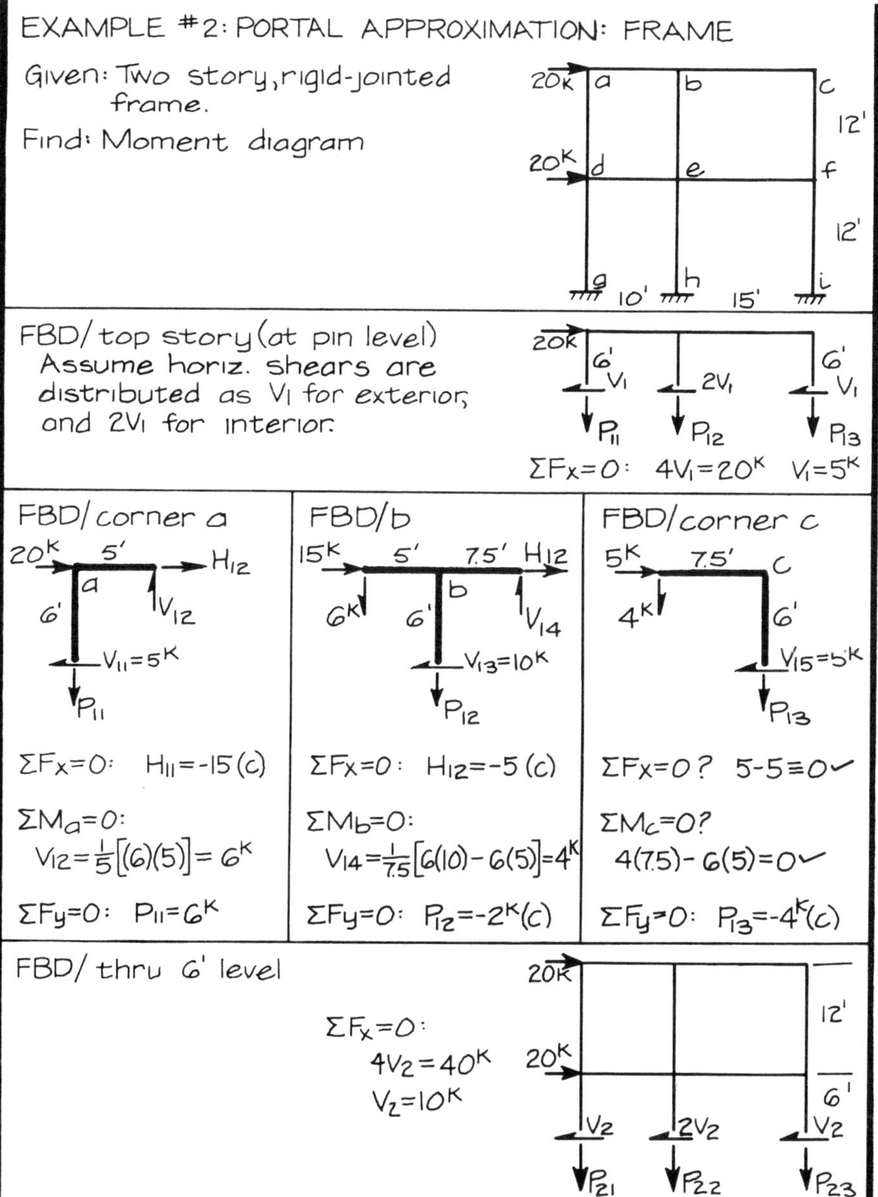
EXAMPLE #2: PORTAL APPROXIMATION: FRAME
Given: Two story, rigid-jointed frame.
Find: Moment diagram
20K
a
b
c
12'
20K
d
e
f
12'
g
h
i
10'
15'
FBD/ top story (at pin level)
Assume horiz. shears are distributed as V1 for exterior, and 2V1 for interior.
20K
6'
V1
2V1
6'
V1
P11
P12
P13
ΣFx=0: 4V1=20K V1=5K
FBD/ corner a
20K
5'
H12
a
6'
V12
V11=5K
P11
ΣFx=0: H11=-15 (c)
ΣMa=0:
V12=1/5[(6)(5)]=6K
ΣFy=0: P11=6K
FBD/b
15K
5'
7.5'
H12
b
6K
6'
V14
V13=10K
P12
ΣFx=0: H12=-5 (c)
ΣMb=0:
V14=1/7.5[6(10)-6(5)]=4K
ΣFy=0: P12=-2K(c)
FBD/corner c
5K
7.5'
c
4K
6'
V15=5K
P13
ΣFx=0? 5-5≡0✓
ΣMc=0?
4(7.5)-6(5)=0✓
ΣFy=0: P13=-4K(c)
FBD/ thru 6' level
ΣFx=0:
4V2=40K
V2=10K
20K
12'
20K
6'
V2
2V2
V2
P21
P22
P23

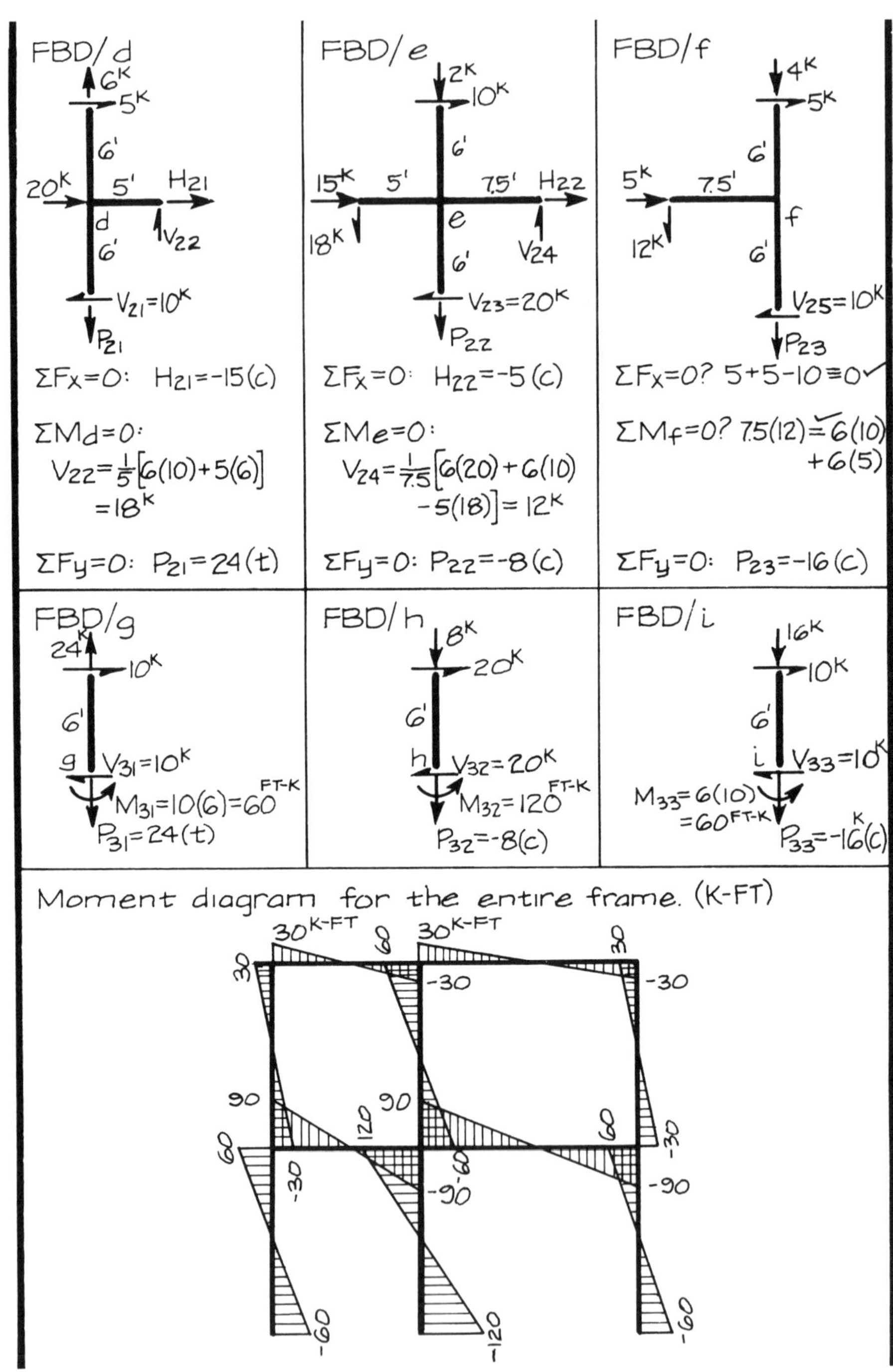

FBD/d
6K
5K
6'
20K
5'
H21
d
V22
6'
V21=10K
P21
ΣFx=0: H21=-15(c)
ΣMd=0:
V22=1/5[6(10)+5(6)]
=18K
ΣFy=0: P21=24(t)
FBD/e
2K
10K
6'
15K
5'
7.5'
H22
18K
e
V24
6'
V23=20K
P22
ΣFx=0: H22=-5(c)
ΣMe=0:
V24=1/7.5[6(20)+6(10)
-5(18)]=12K
ΣFy=0: P22=-8(c)
FBD/f
4K
5K
6'
5K
7.5'
f
12K
6'
V25=10K
P23
ΣFx=0? 5+5-10≡0
ΣMf=0? 7.5(12)=6(10)
+6(5)
ΣFy=0: P23=-16(c)
FBD/g
24K
10K
6'
g
V31=10K
M31=10(6)=60 FT-K
P31=24(t)
FBD/h
8K
20K
6'
h
V32=20K
M32=120 FT-K
P32=-8(c)
FBD/i
16K
10K
6'
i
V33=10K
M33=6(10)
=60 FT-K
P33=-16K(c)
Moment diagram for the entire frame. (K-FT)
30K-FT
60
30K-FT
30
30
-30
-30
90
90
120
60
60
-30
-60
-90
-90
-30
-60
-120
-60

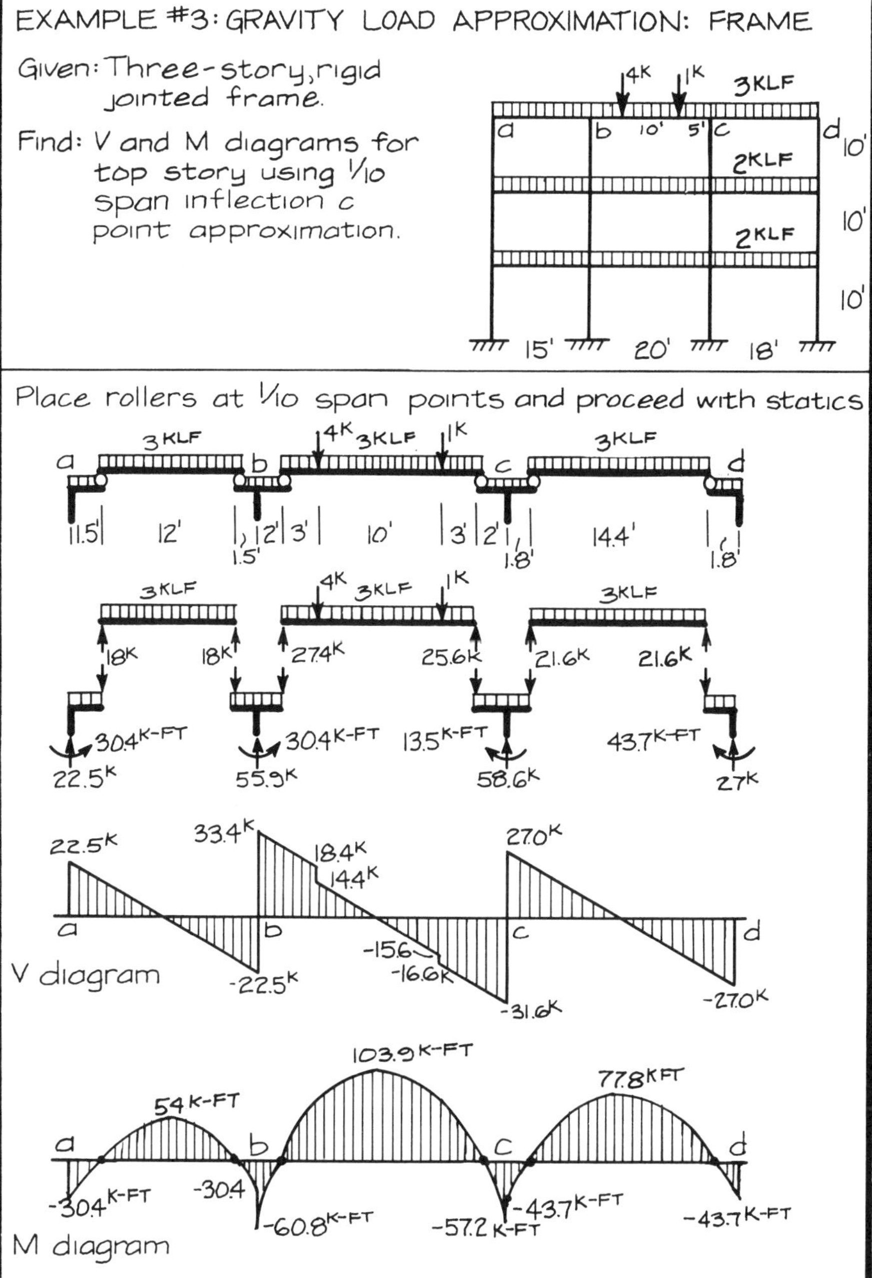
EXAMPLE #3: GRAVITY LOAD APPROXIMATION: FRAME
Given: Three-story, rigid jointed frame.
Find: V and M diagrams for top story using 1/10 span inflection point approximation.
4K 1K 3KLF
a b 10' 5' c d
2KLF
2KLF
10' 10' 10'
15' 20' 18'
Place rollers at 1/10 span points and proceed with statics
3KLF 4K 3KLF 1K 3KLF
1.5' 12' 1.5' 2' 3' 10' 3' 2' 1.8' 14.4' 1.8'
18K 18K 27.4K 25.6K 21.6K 21.6K
30.4K-FT 30.4K-FT 13.5K-FT 43.7K-FT
22.5K 55.9K 58.6K 27K
22.5K 33.4K 18.4K 14.4K 27.0K
-15.6 -16.6K
-22.5K -31.6K -27.0K
V diagram
103.9K-FT 54K-FT 77.8KFT
-30.4K-FT -30.4 -60.8K-FT -57.2K-FT -43.7K-FT -43.7K-FT
M diagram

PROBLEMS

Problems 1–3

Compute moment, shears, and axial forces for all members in frames 1–3 using the cantilever method. Draw moment diagrams also.

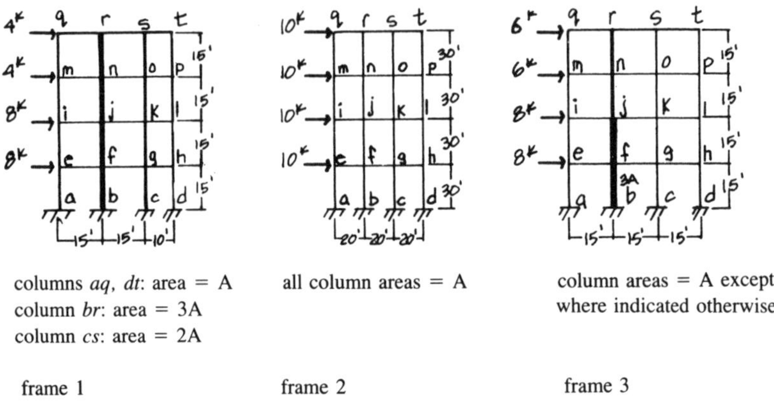

columns *aq, dt*: area = A
column *br*: area = 3A
column *cs*: area = 2A

frame 1

all column areas = A

frame 2

column areas = A except where indicated otherwise

frame 3

Problems 4–6

Compute moments, shears, and axial forces for all members in frames 4–6 using the portal method. Draw moment diagrams also (all column areas = *A*).

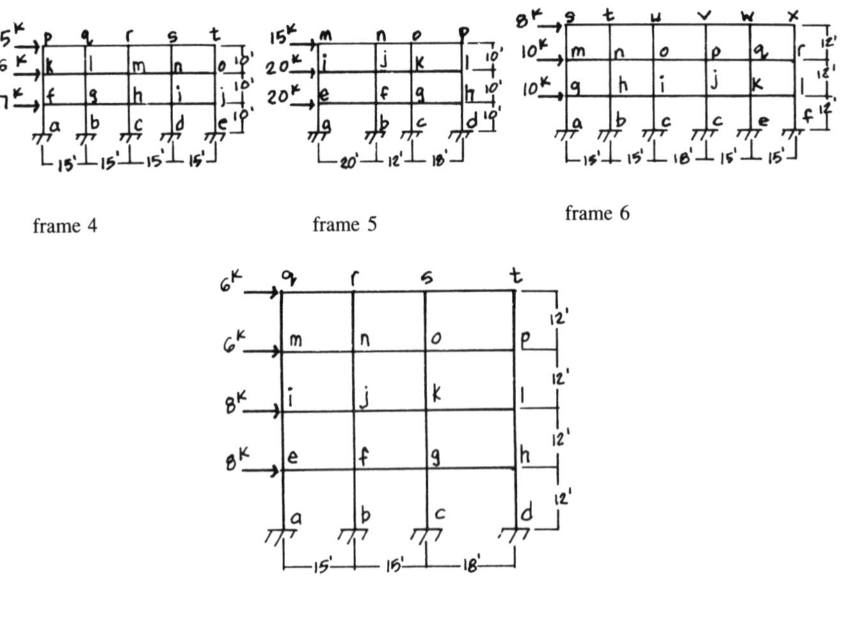

frame 4

frame 5

frame 6

frame 7

Problems 7 and 8

For frame 7, compute same information as for Problems 4–6 but use cantilever (7) and portal (8) methods (all column areas = A).

Problems 9–11

Find moment diagrams for all floors of frames 9–11 using the 1/10 span inflection point approximation.

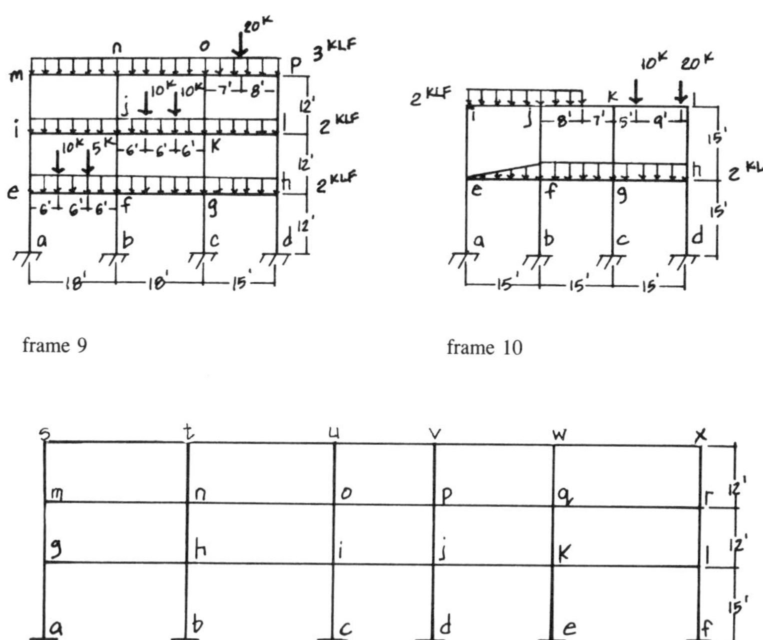

frame 9

frame 10

all floors loaded uniformly with 2*klj*

frame 11

11

INFLUENCE LINES

INTRODUCTION

An influence line is a diagram that shows the effect on one specific reaction (internal or external) of a unit load as it travels across the structure. This "specific" reaction may be a support reaction, an internal shear force at a particular point of a beam, an internal bending moment at a particular point of a beam, an axial force in a particular truss member, and so forth.

Influence lines are used in analysis and design of structures that are subjected to moving live loads—this includes most structures. Clearly, as an 18-wheel truck traverses a bridge the magnitude of, say, the left bridge support reaction depends upon the position of the truck. The question "What is the maximum value of the support reaction?" is of concern to the designer. An influence line for the left bridge support reaction will aid in answering this question.

While vehicle-involved problems are the obvious examples for the use of influence lines, other situations are just as common. Two such cases are industrial shop buildings with movable overhead pulleys and classrooms with constantly changing occupancy.

For determinate structures, influence lines involve nothing more than equilibrium equations. There is nothing difficult or new about influence lines for determinate structures. A major problem for students, however, is confusing influence lines with shear and moment diagrams for beams. Even though these two sets of diagrams may look somewhat alike they are very different in concept. An *influence line* for a beam shows what is happening to, say, the shear at *one* point as the *load moves* while a *shear diagram* for a beam shows values of the shear at *all* points of the beam for a *fixed loading* situation.

In this chapter we shall investigate influence lines for two determinate beams and a truss and discuss critical loading situations for a series of live loads as well as influence lines for indeterminate structures.

EXAMPLE #1: INFLUENCE LINES FOR A SIMPLE SPAN

Given: A simply supported beam

Find: Influence lines for the support reactions (R_a, R_b); the shear at the left quarter point ($V_{10'}$); the moment at mid-span ($M_{20'}$).

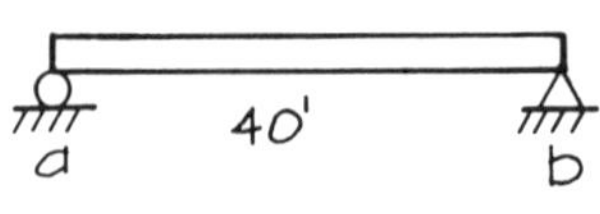

We can find the influence functions for R_a and R_b together with a single FBD. Let the variable ρ locate the unit load.

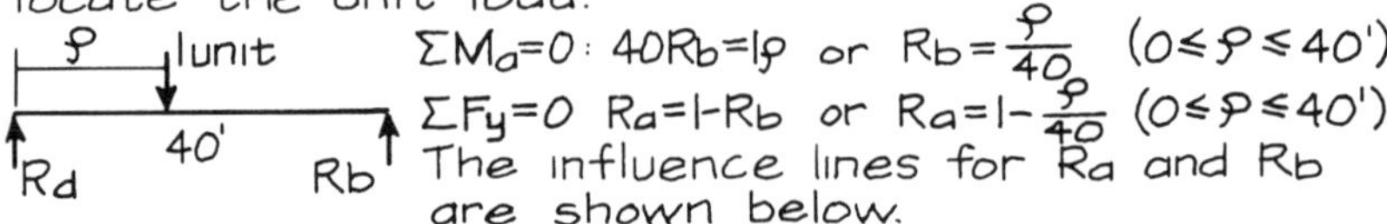

$\Sigma M_a = 0$: $40R_b = 1\rho$ or $R_b = \frac{\rho}{40}$ $(0 \le \rho \le 40')$

$\Sigma F_y = 0$ $R_a = 1 - R_b$ or $R_a = 1 - \frac{\rho}{40}$ $(0 \le \rho \le 40')$

The influence lines for R_a and R_b are shown below.

The influence function for $V_{10'}$ requires two FBDs since the unit load can be to the right or to the left of the 10' point. We shall use the R_a and R_b found above; again ρ locates the load.

FBD/ cut at 10' point; load to left of 10' point

$\Sigma F_y = 0$: $V_{10'} = R_a - 1 = -R_b = \frac{\rho}{40}$ $(0 \le \rho \le 10')$

FBD/ cut at 10' point; load to right of 10' point

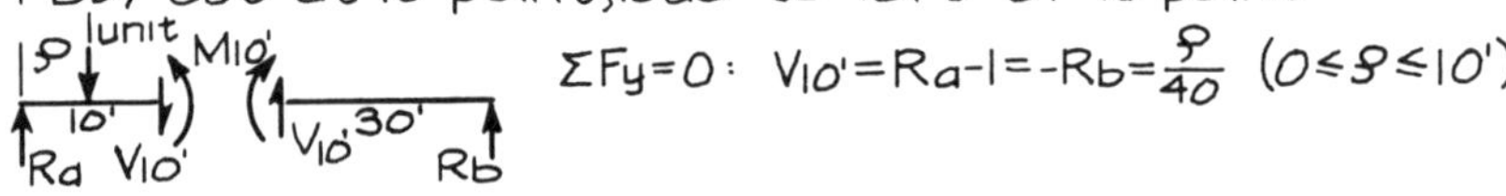

$\Sigma F_y = 0$: $V_{10'} = R_a = 1 - R_b = \frac{40-\rho}{40}$ $(10' \le \rho \le 40')$

These results are graphed below

The influence function of $M_{20'}$ requires two FBDs.

FBD/ cut at 20' point; load to left of 20' point.

$\Sigma M_{mid\text{-}span} = 0$ $M_{20'} = 20R_b = \frac{\rho}{2}$ $(0 \le \rho \le 20')$

FBD/ cut at 20' point; load to right of 20' point.

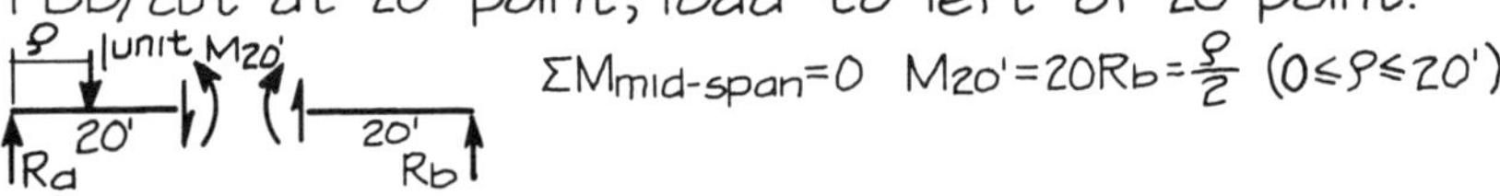

$\Sigma M_{mid\text{-}span} = 0$ $M_{20'} = 20R_a = \frac{40-\rho}{2}$ $(20' \le \rho \le 40')$

These results are graphed below

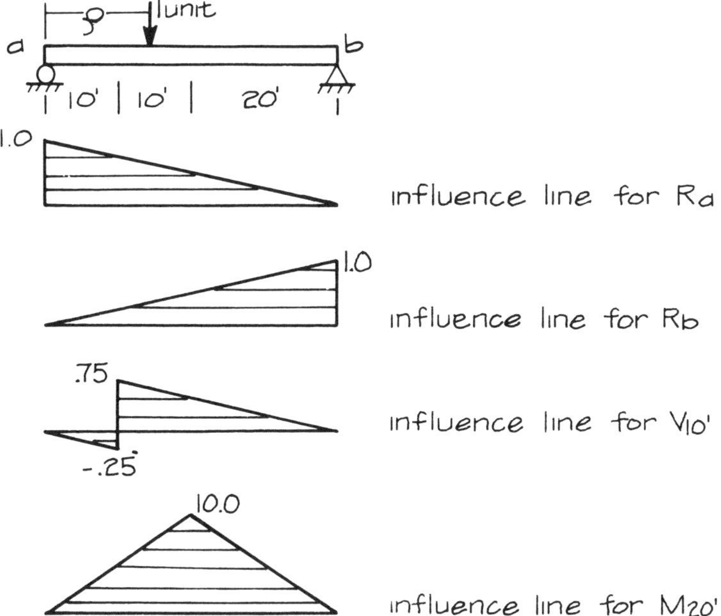

A few basic conclusions from the influence lines: The left support reaction is maximum when the load is placed over the left support; likewise the right support reaction is maximum when the load is placed over the right support. The shear force at the 10' point is maximum when the load is placed slightly to the right of the 10' point; the value of the maximum shear at the 10' point is equal to 3/4 of the concentrated load. The bending moment at mid-span is maximum when the load is placed at mid-span.

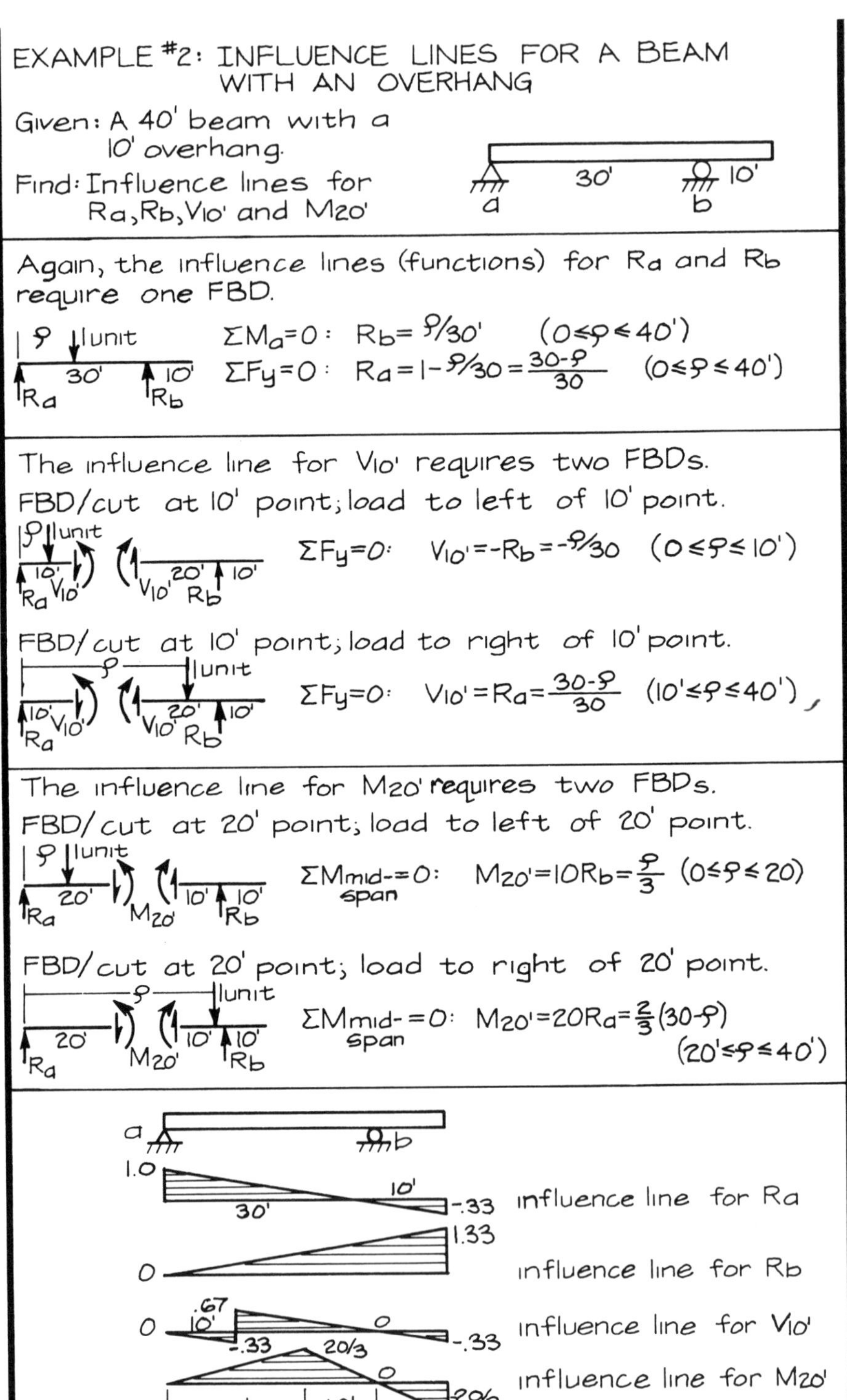

EXAMPLE #2: INFLUENCE LINES FOR A BEAM WITH AN OVERHANG

Given: A 40' beam with a 10' overhang.

Find: Influence lines for R_a, R_b, $V_{10'}$ and $M_{20'}$

Again, the influence lines (functions) for R_a and R_b require one FBD.

$\Sigma M_a = 0: \quad R_b = \rho/30' \qquad (0 \le \rho \le 40')$

$\Sigma F_y = 0: \quad R_a = 1 - \rho/30 = \frac{30-\rho}{30} \qquad (0 \le \rho \le 40')$

The influence line for $V_{10'}$ requires two FBDs.

FBD/cut at 10' point; load to left of 10' point.

$\Sigma F_y = 0: \quad V_{10'} = -R_b = -\rho/30 \qquad (0 \le \rho \le 10')$

FBD/cut at 10' point; load to right of 10' point.

$\Sigma F_y = 0: \quad V_{10'} = R_a = \frac{30-\rho}{30} \qquad (10' \le \rho \le 40')$

The influence line for $M_{20'}$ requires two FBDs.

FBD/cut at 20' point; load to left of 20' point.

$\Sigma M_{\text{mid-span}} = 0: \quad M_{20'} = 10R_b = \frac{\rho}{3} \qquad (0 \le \rho \le 20)$

FBD/cut at 20' point; load to right of 20' point.

$\Sigma M_{\text{mid-span}} = 0: \quad M_{20'} = 20R_a = \frac{2}{3}(30-\rho) \qquad (20' \le \rho \le 40')$

EXAMPLE #3: INFLUENCE LINES FOR TRUSS

Given: Truss as shown

Find: Influence lines for members cd, cf and fg.

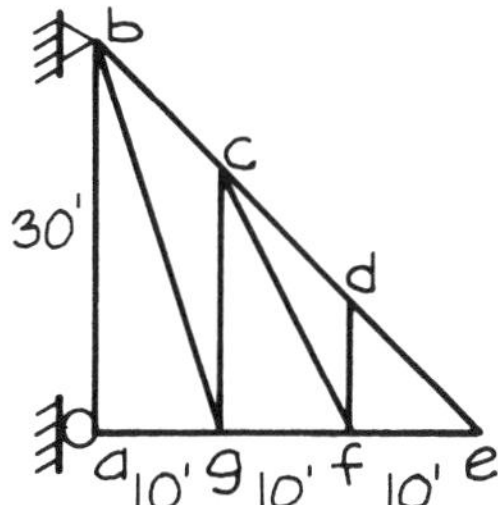

Here the unit load is not continuously moved; the load can be applied at the joints only — it is usual to take only the bottom chord as load points, ie, points a, g, f and e.

A section cut will allow the finding of all three member values.

FBD/ load at a or g

$\Sigma M_f = 0$: $F_{cd} = 0$

$\Sigma F_y = 0$: $F_{cf} = 0$

$\Sigma F_x = 0$: $F_{gf} = 0$

FBD/ load at point f

$\Sigma M_c = 0$: $20F_{gf} + 10(1) = 0$: $F_{gf} = -.5$ (c)

$\Sigma M_f = 0$: $F_{cd} = 0$

$\Sigma F_x = 0$: $\frac{1}{\sqrt{5}} F_{cf} - .5 = 0$: $F_{cf} = 1.12$ (t)

FBD/ load at point e

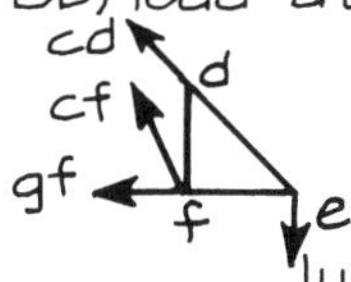

$\Sigma M_c = 0$: $20F_{gf} + 20(1) = 0$: $F_{gf} = -1.0$ (c)

$\Sigma M_f = 0$: $\frac{1}{\sqrt{2}} F_{cd}(10) - 10(1) = 0$: $F_{cd} = 1.414$ (t)

$\Sigma M_e = 0$: $F_{cf} = 0$

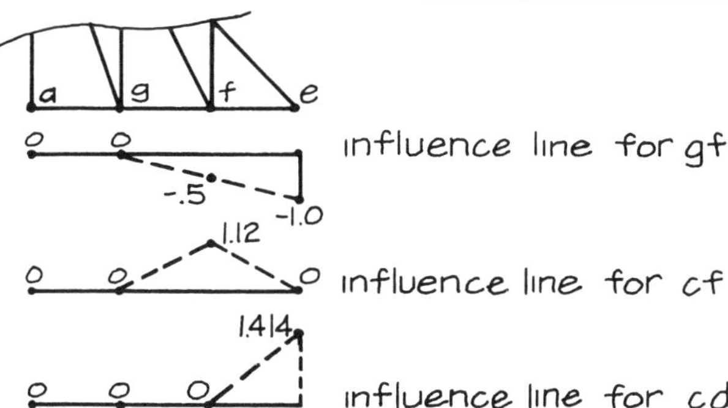

USE OF INFLUENCE LINES

There are five basic points to be made about the use of influence lines corresponding to structures subjected to either a single concentrated load or distributed loads (the more complex problem of a series of concentrated loads will be dealt with later and in Example 5). Example 1 illustrates our point; we restrict the loading to be vertical.

1) A single concentrated load produces a reaction equal to the magnitude of the load times the value of the influence line coordinate at the point of application of the load. Example: By placing a 20 kip load 12′ from the left of the beam in Example 1, we'd find the shear force at the left quarter point to be (Figure 11.1):

$$V_{10'}\Big|_{12'}^{20^k @} = 20^k\left[\left(\frac{28}{30}\right)(0.75)\right] = 14^k$$

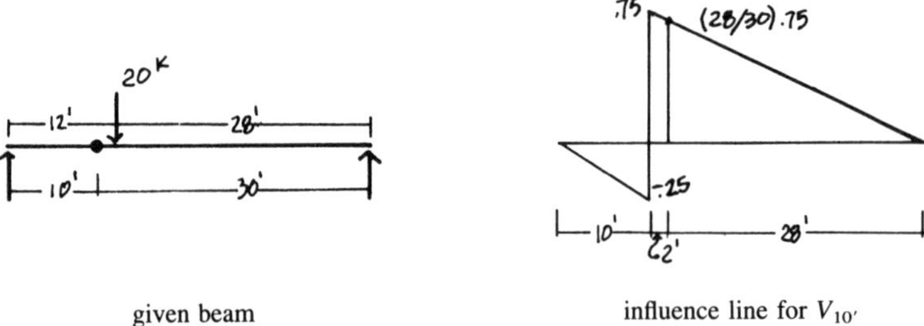

given beam

influence line for $V_{10'}$

Figure 11.1

2) A single concentrated load produces an extreme reaction when the load is placed at the point corresponding to the peak value of the influence line.

3) A uniformly distributed load produces a reaction equal to the product of the load intensity and the area under the influence line corresponding to the load. Example: By placing a uniform load of 2klf over the right 20′ of the beam in Example 1 we'd find the left support reaction to be (Figure 11.2):

$$R_a = (2\text{klf})[(20)(1/2)(0.50)] = 10.0 \text{ kips}$$

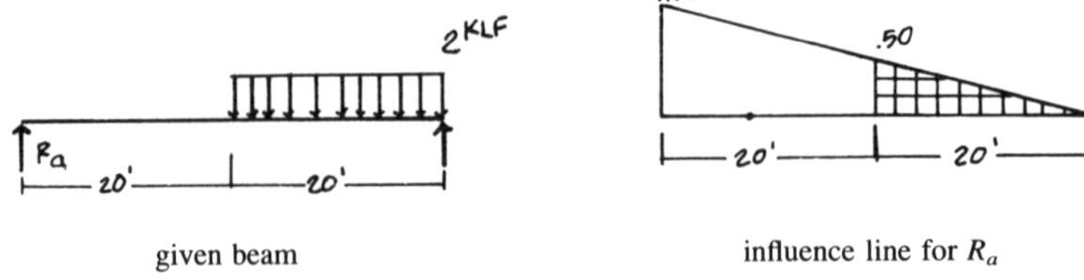

given beam

influence line for R_a

Figure 11.2

4) If the distributed load isn't uniform then the reaction is found by integrating, over the loading distance, the product of the load intensity function and the influence function. Example: By placing a ramp load over the right 30′ of the beam in Example 1 we'd find the left support reaction to be (Figure 11.3):

$$R_a = \int_0^{30} (0.1\zeta)\left(0.75 - \frac{0.75}{30}\zeta\right) d\zeta = 11.25 \text{ kips}$$

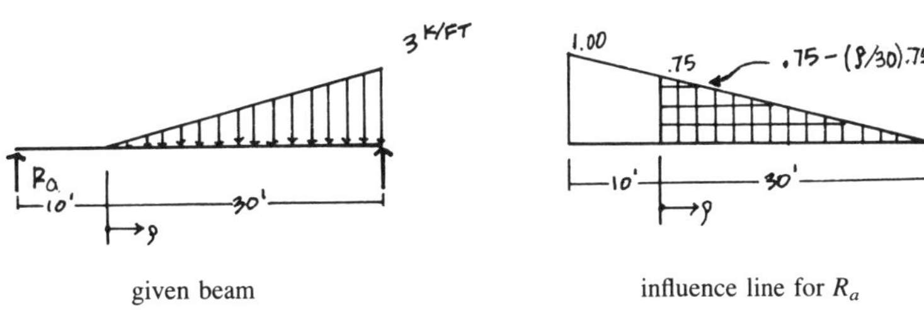

given beam

influence line for R_a

Figure 11.3

5) A uniformly distributed load produces a maximum positive reaction when the load is placed over all those portions of the structure for which the influence line is positive. Example: Placing a uniform load of, say, 2klf, over only the right 30′ of the beam in Example 1 causes $V_{10'}$ to be a maximum (Figure 11.4):

$$V_{10'}|_{max} = (2\text{klf})[(1/2)(30)(0.75)] = 22.5 \text{ kips}$$

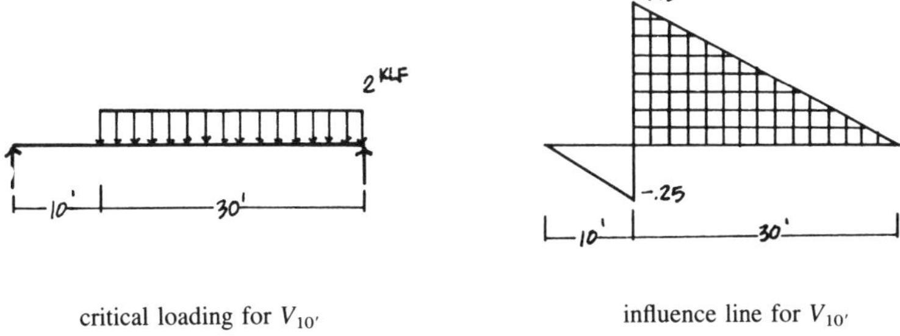

critical loading for $V_{10'}$

influence line for $V_{10'}$

Figure 11.4

Example 4 deals with uniform loads and critical loading patterns. Due to its nature, the dead load must be applied over the entire beam; however, the live load is applied only over those portions of the beam that will lead to a maximum value of the desired reaction.

Example 5 investigates maximum values of specific reactions that occur due to a series of concentrated loads. Such concentrated loads and influence lines are much more complicated than if the loadings were uniform. There are specific methods available for such loads (Reference 5) but trial-and-error for many patterns is easily applied. In all three parts of Example 5, the critical loading pattern is one in which a concentrated load is placed at the apex of the influence line, that is, a concentrated load is placed at the point where the reaction is sought. In Example 5 it is obvious, considering the shapes of the influence lines, that a load must be placed at support *a* to maximize R_a and that a load be placed slightly to the right of the 10′ point in order for $V_{10'}$ to be maximum. However it may not be obvious that $M_{20'}$ will be maximum when a load is placed at the 20′ point (the apex of the influence line).

Consider a general case with, for simplicity, three concentrated loads (Figure 11.5). Say, $P_3 > P_1 + P_2$. For the situation shown, M_q is calculated. If the loads are moved *x* feet to the left, P_1 and P_2 will have smaller contributions to M_q while P_3's contribution will increase. Thus M_q decreases $x(b/\ell)(P_1 + P_2)$ while M_q increases $x(a/\ell)P_3$. Since we chose $P_3 > P_1 + P_2$, by moving the loads to the left M_q increases by $x[(a/\ell)P_3 - (b/\ell)(P_1 + P_2)]$. Thus M_q, for these three loads, is maximum when the loads are moved so that P_3 is at point *q*—if "*x*" is larger, then all three loads will detract from M_q; if "*x*" is smaller, then the largest increase in M_q has not been found.

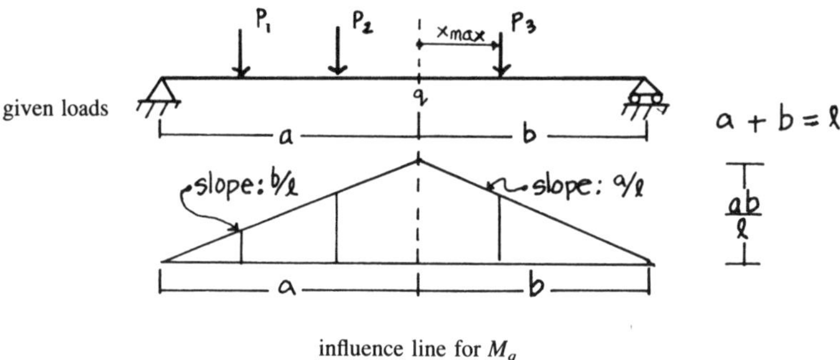

Figure 11.5

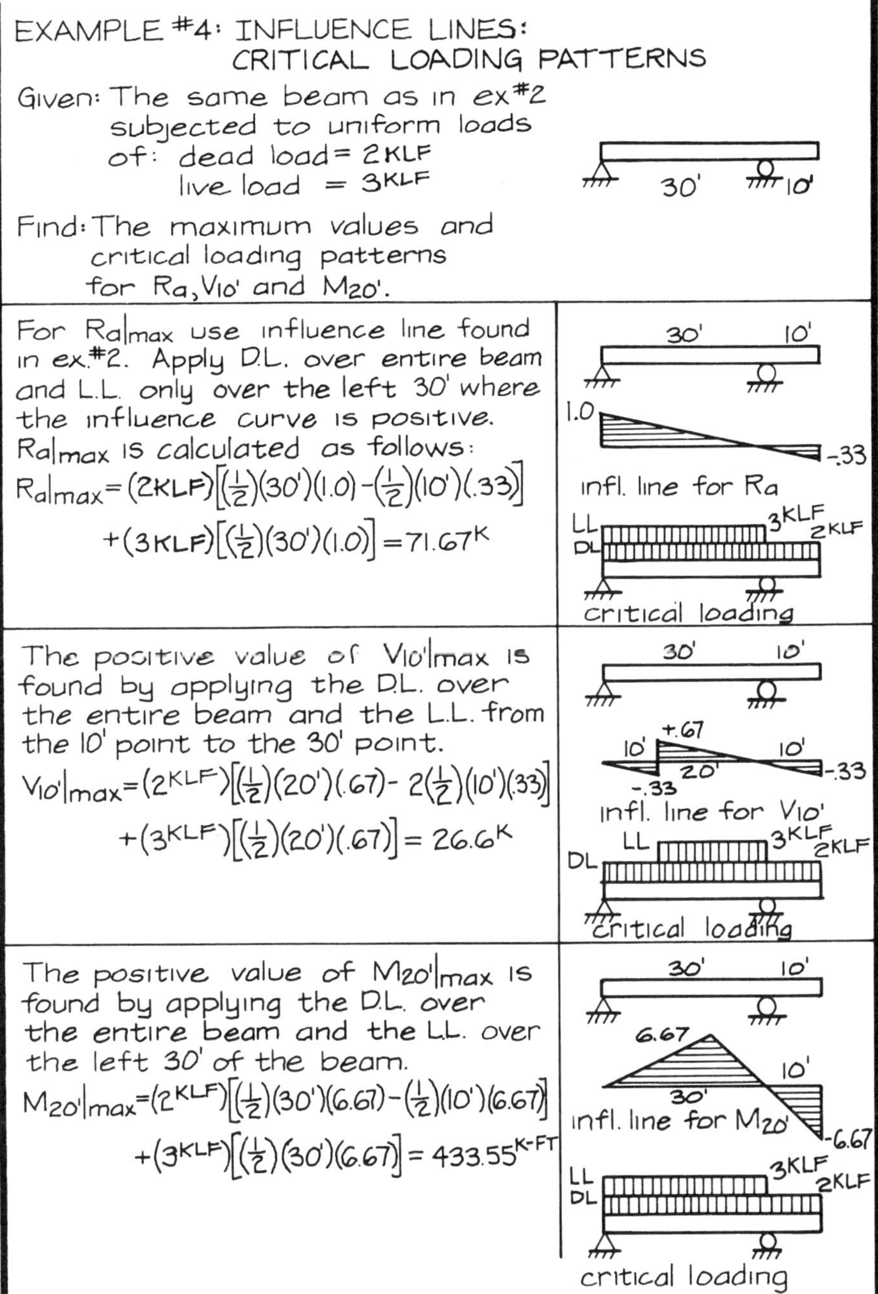
EXAMPLE #4: INFLUENCE LINES: CRITICAL LOADING PATTERNS
Given: The same beam as in ex#2 subjected to uniform loads of: dead load = 2KLF live load = 3KLF
30'
10'
Find: The maximum values and critical loading patterns for Ra, V10' and M20'.
For Ra|max use influence line found in ex.#2. Apply D.L. over entire beam and L.L. only over the left 30' where the influence curve is positive. Ra|max is calculated as follows:
Ra|max = (2KLF)[(1/2)(30')(1.0) - (1/2)(10')(.33)] + (3KLF)[(1/2)(30')(1.0)] = 71.67K
1.0
-.33
infl. line for Ra
LL
DL
3KLF
2KLF
critical loading
The positive value of V10'|max is found by applying the D.L. over the entire beam and the L.L. from the 10' point to the 30' point.
V10'|max = (2KLF)[(1/2)(20')(.67) - 2(1/2)(10')(.33)] + (3KLF)[(1/2)(20')(.67)] = 26.6K
+.67
20'
-.33
infl. line for V10'
The positive value of M20'|max is found by applying the D.L. over the entire beam and the L.L. over the left 30' of the beam.
M20'|max = (2KLF)[(1/2)(30')(6.67) - (1/2)(10')(6.67)] + (3KLF)[(1/2)(30')(6.67)] = 433.55K-FT
6.67
-6.67
infl. line for M20'

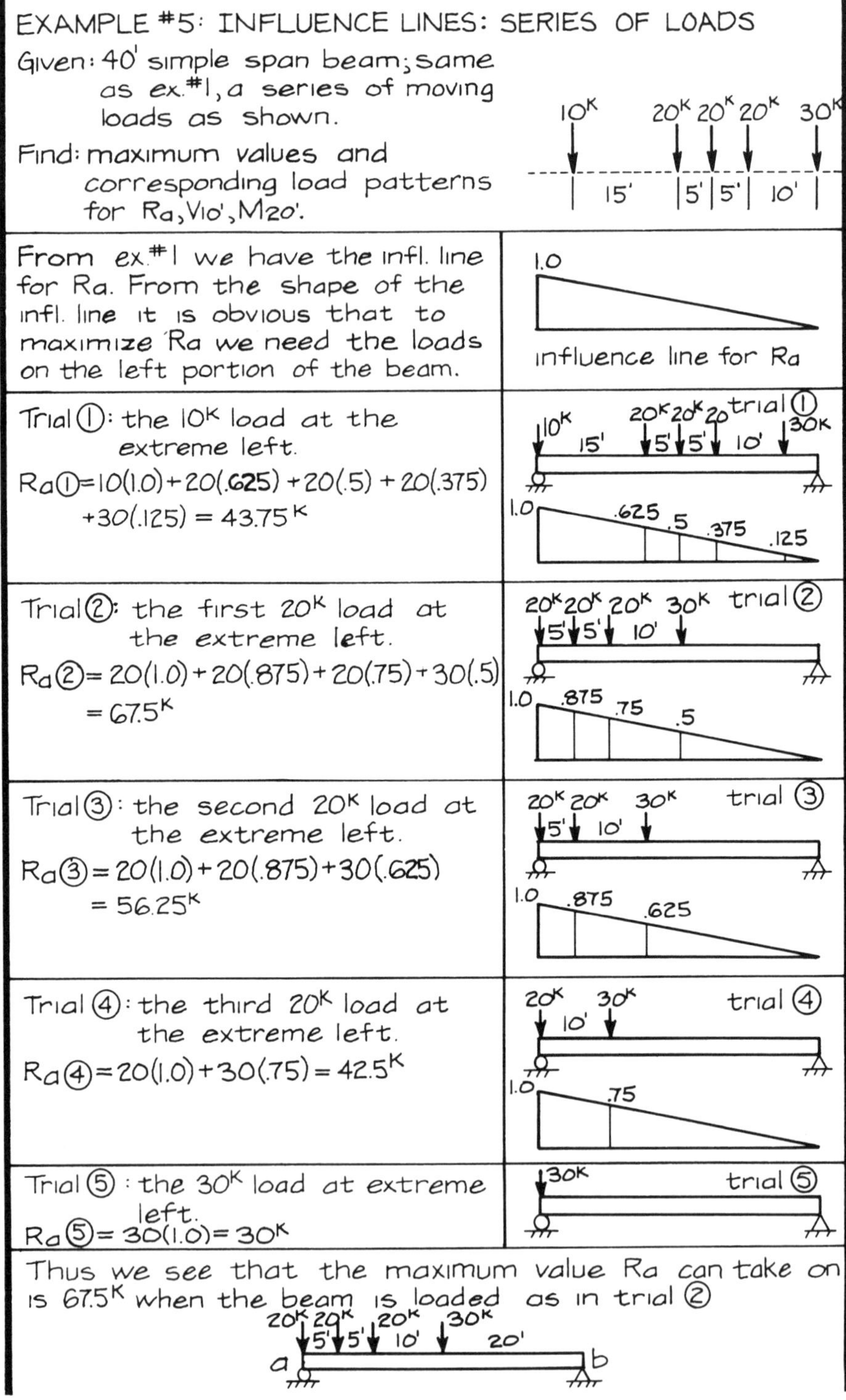

EXAMPLE #5: INFLUENCE LINES: SERIES OF LOADS

Given: 40' simple span beam; same as ex. #1, a series of moving loads as shown.

Find: maximum values and corresponding load patterns for R_a, $V_{10'}$, $M_{20'}$.

From ex. #1 we have the infl. line for R_a. From the shape of the infl. line it is obvious that to maximize R_a we need the loads on the left portion of the beam.

influence line for R_a

Trial ①: the 10^K load at the extreme left.

$R_a① = 10(1.0) + 20(.625) + 20(.5) + 20(.375) + 30(.125) = 43.75^K$

Trial ②: the first 20^K load at the extreme left.

$R_a② = 20(1.0) + 20(.875) + 20(.75) + 30(.5) = 67.5^K$

Trial ③: the second 20^K load at the extreme left.

$R_a③ = 20(1.0) + 20(.875) + 30(.625) = 56.25^K$

Trial ④: the third 20^K load at the extreme left.

$R_a④ = 20(1.0) + 30(.75) = 42.5^K$

Trial ⑤: the 30^K load at extreme left.

$R_a⑤ = 30(1.0) = 30^K$

Thus we see that the maximum value R_a can take on is 67.5^K when the beam is loaded as in trial ②

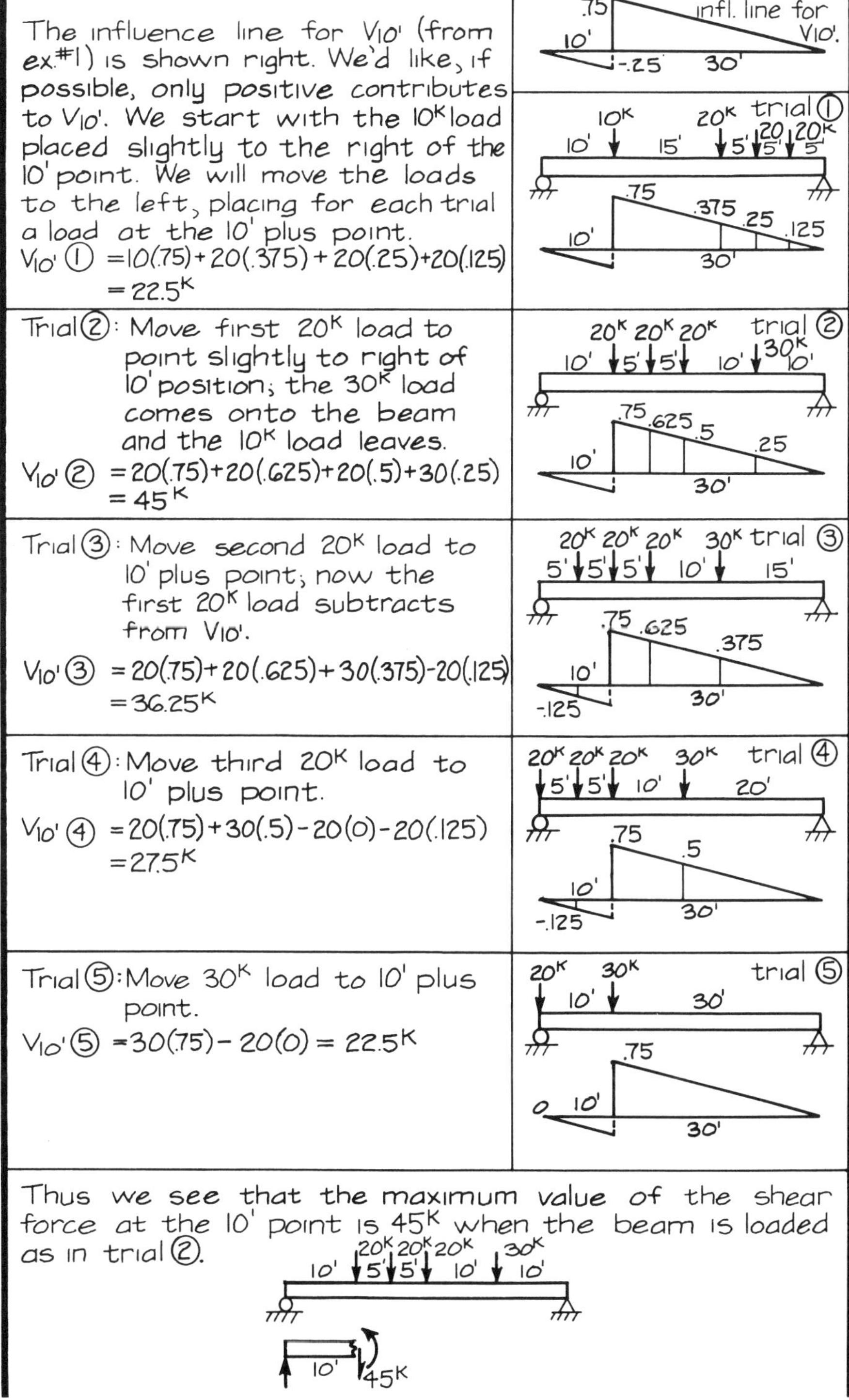

The influence line for $V_{10'}$ (from ex.#1) is shown right. We'd like, if possible, only positive contributes to $V_{10'}$. We start with the 10^K load placed slightly to the right of the 10' point. We will move the loads to the left, placing for each trial a load at the 10' plus point.

$V_{10'}$ ① $= 10(.75) + 20(.375) + 20(.25) + 20(.125) = 22.5^K$

Trial ②: Move first 20^K load to point slightly to right of 10' position; the 30^K load comes onto the beam and the 10^K load leaves.

$V_{10'}$ ② $= 20(.75) + 20(.625) + 20(.5) + 30(.25) = 45^K$

Trial ③: Move second 20^K load to 10' plus point; now the first 20^K load subtracts from $V_{10'}$.

$V_{10'}$ ③ $= 20(.75) + 20(.625) + 30(.375) - 20(.125) = 36.25^K$

Trial ④: Move third 20^K load to 10' plus point.

$V_{10'}$ ④ $= 20(.75) + 30(.5) - 20(0) - 20(.125) = 27.5^K$

Trial ⑤: Move 30^K load to 10' plus point.

$V_{10'}$ ⑤ $= 30(.75) - 20(0) = 22.5^K$

Thus we see that the maximum value of the shear force at the 10' point is 45^K when the beam is loaded as in trial ②.

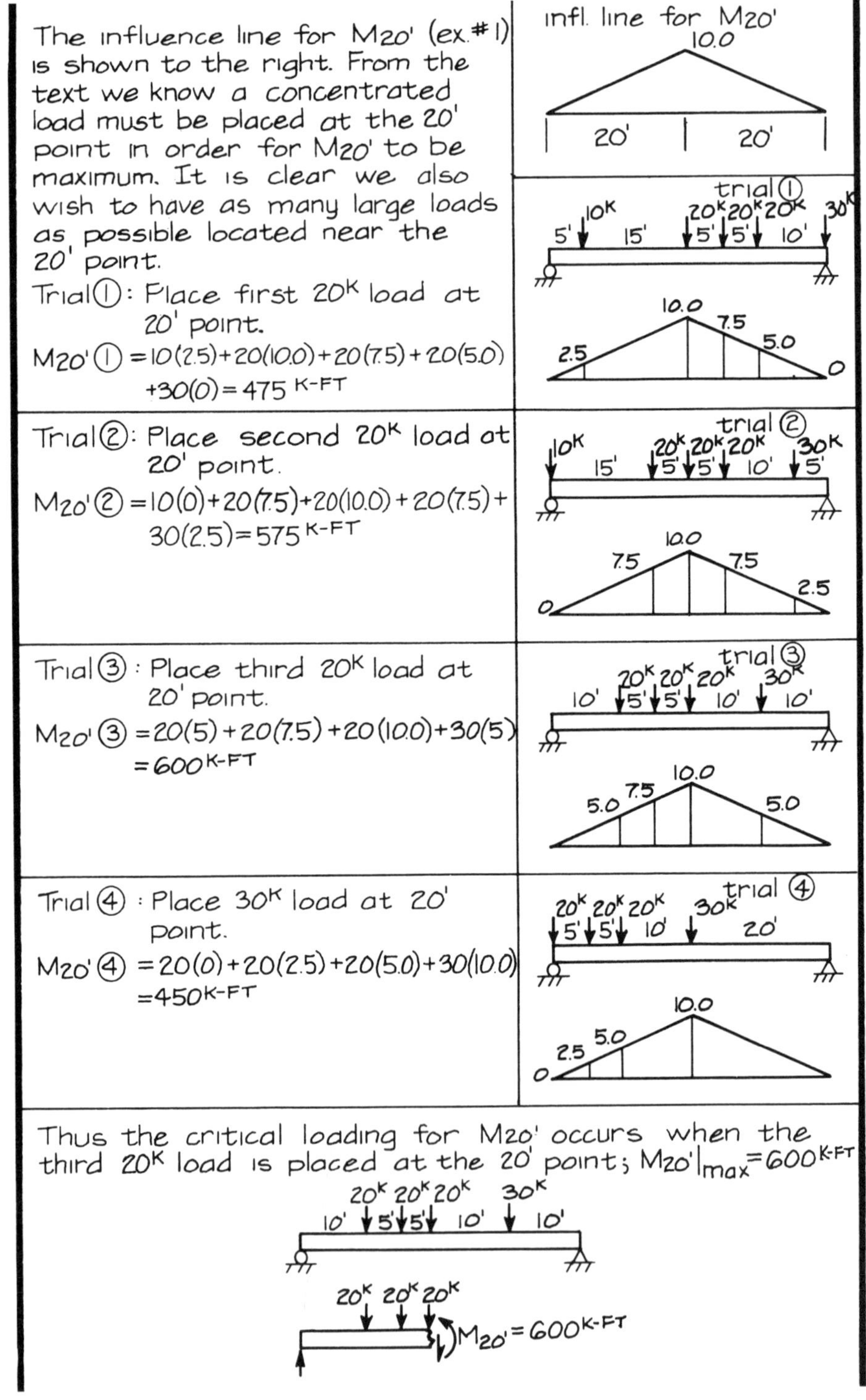

The influence line for $M_{20'}$ (ex. #1) is shown to the right. From the text we know a concentrated load must be placed at the 20' point in order for $M_{20'}$ to be maximum. It is clear we also wish to have as many large loads as possible located near the 20' point.

Trial ①: Place first 20^K load at 20' point.

$M_{20'}$ ① $= 10(2.5) + 20(10.0) + 20(7.5) + 20(5.0) + 30(0) = 475^{K\text{-}FT}$

Trial ②: Place second 20^K load at 20' point.

$M_{20'}$ ② $= 10(0) + 20(7.5) + 20(10.0) + 20(7.5) + 30(2.5) = 575^{K\text{-}FT}$

Trial ③: Place third 20^K load at 20' point.

$M_{20'}$ ③ $= 20(5) + 20(7.5) + 20(10.0) + 30(5) = 600^{K\text{-}FT}$

Trial ④: Place 30^K load at 20' point.

$M_{20'}$ ④ $= 20(0) + 20(2.5) + 20(5.0) + 30(10.0) = 450^{K\text{-}FT}$

Thus the critical loading for $M_{20'}$ occurs when the third 20^K load is placed at the 20' point; $M_{20'}|_{max} = 600^{K\text{-}FT}$

Influence Lines for Indeterminate Beams

Influence lines for determinate beams involve nothing more than equilibrium and the resulting curves are made up of straight lines. As one would expect, indeterminate beam influence lines require deformation analysis as well as equilibrium (the resulting curves are curvilinear).

There are two approaches toward influence lines for indeterminate beams: 1) the direct approach—superposing the "unit load" beam and the "redundant" beam and using compatibility of deformations; 2) an indirect approach called the Müller-Breslau Principle, which is based on the Maxwell theorem of reciprocal deformation. The direct approach is laborious and time-consuming. The Müller-Breslau Principle is clever and easily applied once it is understood. We shall develop here only the indirect approach, starting with the statement of *Maxwell's Theorem of Reciprocal Deformation*. Simply, the theorem says that for any given beam loaded by a single force, P, the deflection at point b due to force P being situated at point a is exactly the same as the deflection at point a when the load P is situated at point b (Figure 11.6).

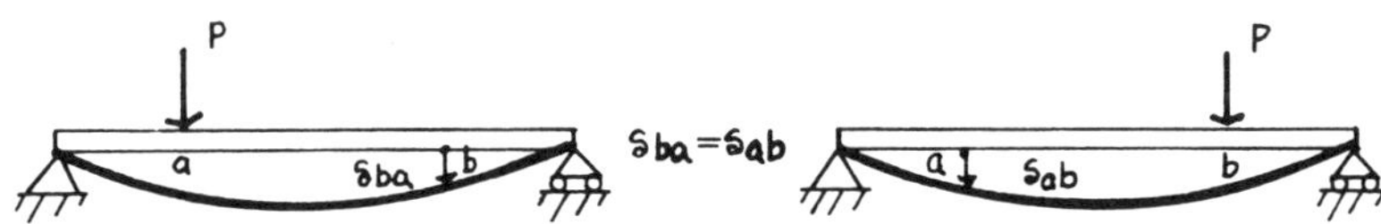

Figure 11.6

Although we stated the theorem for beams, it can be applied to trusses and frames as well. Further, the theorem can be worded in terms of rotations (instead of deflections) if the word "couple" replaces "load." We will not prove the theorem here; for a proof see Reference 1.

Back to the problem at hand: indeterminate beam influence lines. Say we wish to find the influence line of support reaction b for the beam shown in Figure 11.7a. We can superpose two beams, as shown in Figures 11.7b and 11.7c, using R_b as the redundant. Compatibility implies

$$\sum \delta_b = 0 \Rightarrow \delta_{b\zeta} + [-(\delta_{bb})R_b] = 0 \qquad \text{or} \qquad R_b = \frac{\delta_{b\zeta}}{\delta_{bb}}.$$

Here $\delta_{b\zeta}$ is the deflection at point b due to a unit (downward) load at point ζ (load moves), δ_{bb} is the deflection at point b due to a unit (downward) load at point b (one loading). But Maxwell's Theorem states $\delta_{b\zeta} = \delta_{\zeta b}$, where $\delta_{\zeta b}$ is the deflection at point ζ due to a unit load at point b (one loading). Thus:

$$R_b = \frac{\delta_{\zeta b}}{\delta_{bb}}.$$

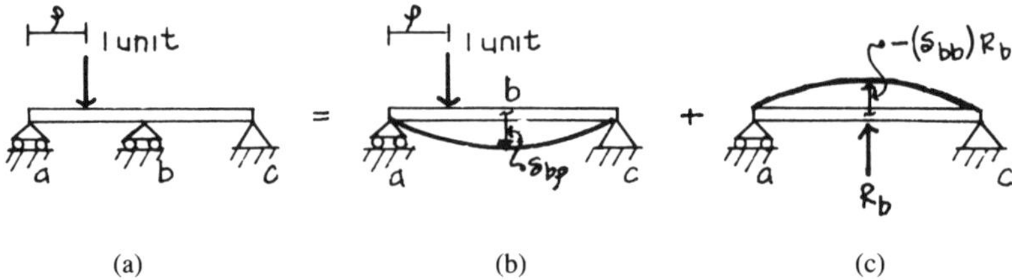

(a) (b) (c)

Figure 11.7

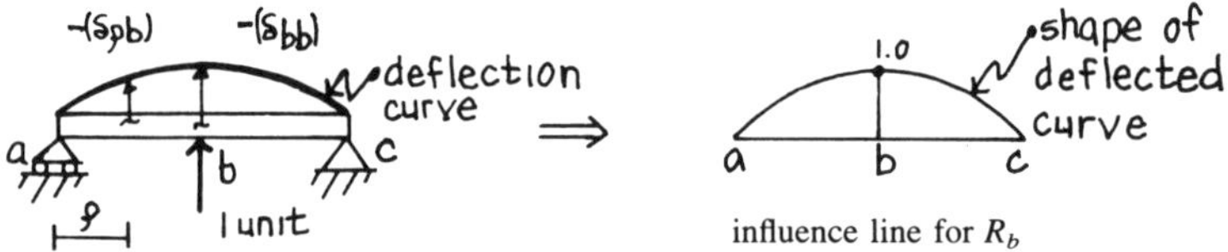

influence line for R_b

Figure 11.8

This result is quite startling for we see that the influence line for R_b is of the same shape as the deflection curve for the *determinate* beam subjected to a unit upward load at point b. Note that when the unit load is at point b the reaction R_b is unity. Thus the Müller-Breslau Principle: When a unit deflection or rotation is inserted in an unloaded beam, the resulting deflection curve is the influence line for the reaction corresponding to the unit deflection or rotation.

The Müller-Breslau Principle makes it easy to sketch influence lines. To calculate values for them, deflection analysis must be done. The illustrations show how influence lines are sketched (Figures 11.9, 11.10). Example 6 will show how values are found for the curves.

Sketch influence lines for reaction a and the shear at point q for the beam shown. To find the influence line for R_a, remove support at a and allow a to go through a unit deflection. The resulting deflection curve which is consistent with the supports will be the influence line for R_a. The influence line for V_q is a bit tricky. First cut the beam at q into two parts and place the shear force V_q on both parts. Allow point q to go through a total of one unit deflection which is consistent with the shears. The resulting deflection curves will also be the influence line for V_q.

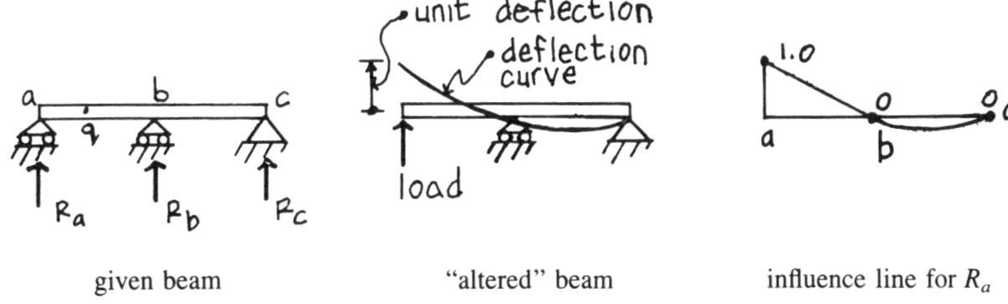

given beam

"altered" beam

influence line for R_a

Figure 11.9

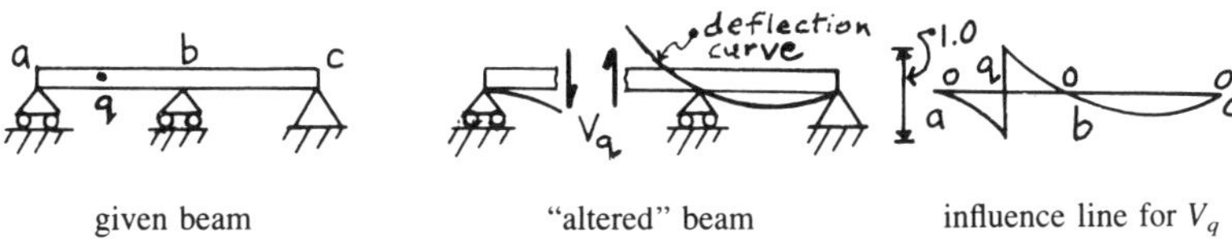

given beam

"altered" beam

influence line for V_q

Figure 11.10

EXAMPLE #6: INFLUENCE LINES: INDETERMINATE BEAM

Given: Indeterminate beam as shown.

Find: Influence lines for support reaction at b (R_b) and bending moment at 10'.

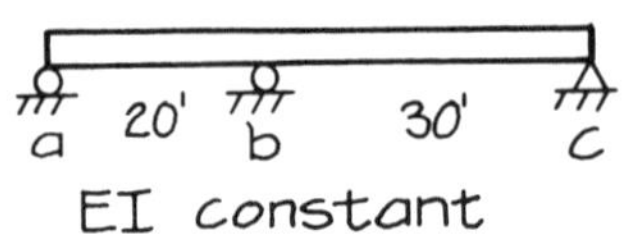

The shape of the infl. line for R_b is easily drawn: remove support b and impose a unit displacement at b.

1.0
o
o
a
b
c

We shall find ordinates at the 10' and 35' points. Remove support b; place a unit load at b only. Using moment-area we calculate deflections at 10', 35' and at point b (due to load at b).

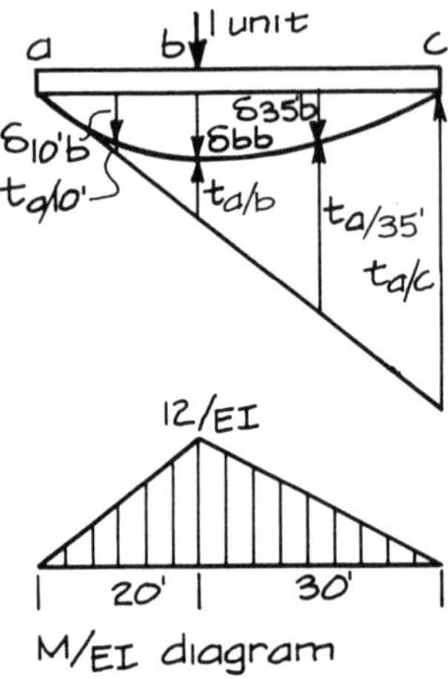

$$R_b = \frac{\delta_{gb}}{\delta_{bb}}$$

$$\delta_{10'b} = \left[\tfrac{1}{5} t_{a/c} - t_{a/10'}\right]$$

$$t_{a/c} = \frac{1}{EI}\left[\left(\frac{30}{2}\right)(12)(30)\left(\frac{2}{3}\right) + \left(\frac{20}{2}\right)(12)\left(30 + \frac{20}{3}\right)\right]$$

$$= \frac{8000}{EI}$$

$$t_{a/10'} = \frac{1}{EI}\left[\frac{1}{2}(10)(6)\left(\frac{10}{3}\right)\right] = \frac{100}{EI}\text{, thus,}$$

$$\delta_{10'b} = \frac{1500}{EI} \quad \text{similarly}$$

$$\delta_{35'b} = \left[\frac{35}{50} t_{a/c} - t_{a/35'}\right] = \frac{1875}{EI} \quad \text{and}$$

$$\delta_{bb} = \left[\frac{2}{5} t_{a/c} - t_{a/b}\right] = \frac{2400}{EI}$$

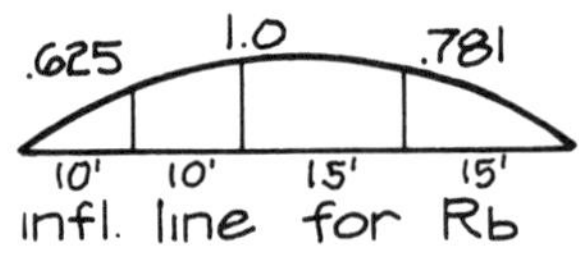

$$R_b\Big|_{10'} = \frac{\delta_{10'b}}{\delta_{bb}} = .625 \quad (R_b \uparrow)$$

$$R_b\Big|_{35'} = \frac{\delta_{35b}}{\delta_{bb}} = .781 \quad (R_b \uparrow)$$

The shape of the influence line for $M_{10'}$ is easily sketched: cut beam at 10'; allow a total relative rotation of one unit.

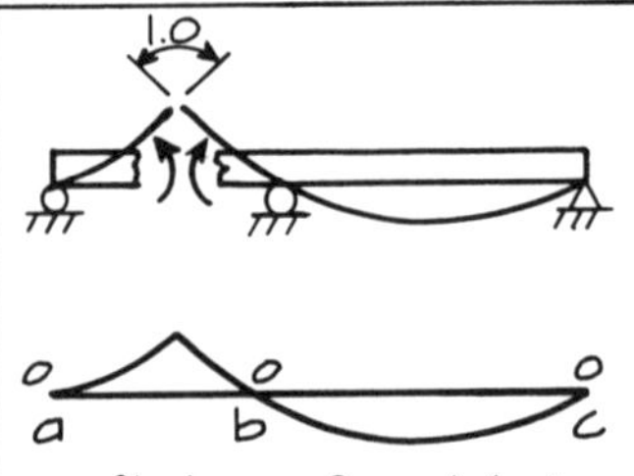

We shall find ordinates for the influence line of $M_{10'}$ at 10' and 35' points: we use the results found above for R_b and statics. When the unit load is at 10' we know $R_b = .625$; therefore by statics we find $R_a = .425 (\uparrow)$; $R_c = -.05 (\downarrow)$. A cut at 10' exposes $M_{10'}$ (load at 10'). By statics $M_{10'}$ (load at 10') $= 4.25$ unit-ft

When the load is at 35' we know $R_b = .781$; statics assesses $R_a = -.169 (\downarrow)$ and $R_c = .388 (\uparrow)$. A cut at 10' exposes M_{10} (load at 35'). By statics $M_{10'}$ (load at 35') $= -1.69$

The influence line for $M_{10'}$ is shown

PROBLEMS

Problems 1, 2, and 3

For the beams shown find the influence lines for:

a) The support reactions,
b) the shear forces at points 10′ and 15′ from the left end, and
c) the bending moments at points 10′ and 15′ from the left end.

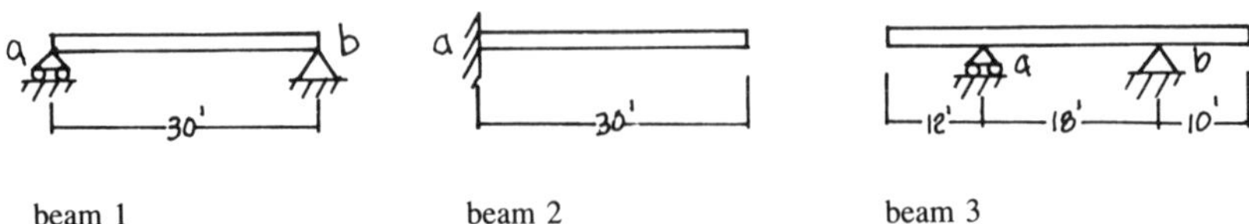

beam 1 beam 2 beam 3

Problems 4, 5, and 6

Referring to Problems 1, 2, and 3, find the maximum values (and corresponding loading patterns) for the sought reactions if the beams are subjected to a dead load of 2 klf; and live loads of 3 klf.

Problems 7, 8, and 9

Referring to Problems 1, 2, and 3, find the maximum values (and corresponding loading patterns) for the sought reactions if the beams are subjected to a series of concentrated loads as shown.

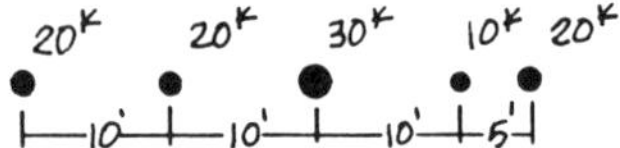

Problems 10, 11, and 12

For the beams shown find the influence lines for:

a) The support reactions,
b) the shear force at 10′ from the left end, and
c) the bending moment at 10′ from the left end.

Establish values at supports and at least two interior points.

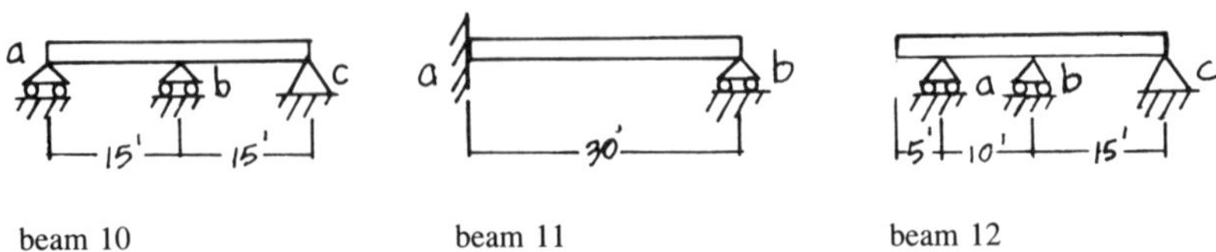

beam 10 beam 11 beam 12

Problems 13, 14, and 15

Referring to the beams in Problems 10, 11, and 12, find the maximum values (approximate), (and corresponding loading patterns) for the support reactions if the beams are subjected to dead loads of 2 klf and live loads of 3 klf.

APPENDICES

A BRIEF REVIEW OF MATRICES

Definitions

A *matrix* is a rectangular array of discrete elements arranged in rows and columns. If there are m rows and n columns, the matrix is said to be of *order* $m \times n$. The element in the ith row and jth column of matrix $\bar{\bar{A}}$ is called a_{ij}. We shall use two bars above a symbol to indicate a general matrix. For example, the matrix $\bar{\bar{A}}$ below is of order 3×4; the element $a_{23} = 7$; the element $a_{33} = 2$, etc.

$$\bar{\bar{A}} = \begin{bmatrix} 1 & 2 & -3 & 1 \\ 4 & 6 & 7 & 6 \\ -8 & 4 & 2 & 2 \end{bmatrix}$$

A matrix with only one column is called a *vector* or a *column matrix*. We shall use one bar over a symbol to indicate a column matrix. $\bar{B}$ below is a column matrix of order 5×1.

$$\bar{B} = \begin{bmatrix} 6 \\ -3 \\ 2 \\ -5 \\ 4 \end{bmatrix}$$

A *square matrix* has the same number of rows as columns. The elements a_{11}, a_{22}, a_{33}, a_{44}, etc., which are positioned along the diagonal of a square matrix, are called the *diagonal elements*; all the other terms are called *off-diagonal elements*.

A special type of square matrix is one in which all the off-diagonal terms are zero; such a matrix is called a *diagonal matrix*. If each of the diagonal terms of a diagonal matrix has the value of 1, we call the matrix a *unit* or *identity matrix*, $\bar{\bar{I}}$. This identity matrix has order 3×3:

$$\bar{\bar{I}} = \begin{bmatrix} 1 & 0 & 0 \\ 0 & 1 & 0 \\ 0 & 0 & 1 \end{bmatrix}$$

The *transpose of the matrix* $\bar{\bar{C}}$ is found by interchanging the rows and columns of $\bar{\bar{C}}$. We write the transpose of $\bar{\bar{C}}$ as $\bar{\bar{C}}^T$. In the example below $\bar{\bar{C}}$ is 4×2, thus $\bar{\bar{C}}^T$ is 2×4.

$$\text{If} \quad \bar{\bar{C}} = \begin{bmatrix} 2 & -4 \\ 6 & -1 \\ 3 & 7 \\ 4 & 5 \end{bmatrix} \quad \text{then} \quad \bar{\bar{C}}^T = \begin{bmatrix} 2 & 6 & 3 & 4 \\ -4 & -1 & 7 & 5 \end{bmatrix}.$$

A *symmetric matrix* is a square matrix that is equal to its own transpose. The matrix $\bar{\bar{A}}$ below is a symmetric 3×3 matrix; note $a_{12} = a_{21} = -2$; $a_{13} = a_{31} = 7$; $a_{23} = a_{32} = 2$.

$$\bar{\bar{A}} = \begin{bmatrix} 6 & -2 & 7 \\ -2 & -1 & 2 \\ 7 & 2 & 4 \end{bmatrix}$$

Matrix Algebra

Addition and Subtraction

Only matrices of the same order can be added or subtracted. The sum (or difference) of two matrices ($\bar{\bar{A}}$ and $\bar{\bar{B}}$) of order $m \times n$ is a third matrix ($\bar{\bar{C}}$) of order $m \times n$. Each term c_{ij} in $\bar{\bar{C}}$ is formed by adding (or subtracting) the corresponding terms in $\bar{\bar{A}}$ and $\bar{\bar{B}}$, i.e., $c_{ij} = a_{ij} \pm b_{ij}$. For example, if

$$\bar{\bar{A}} = \begin{bmatrix} 4 & 5 & -1 \\ 6 & 1 & 4 \end{bmatrix}; \qquad \bar{\bar{B}} = \begin{bmatrix} 5 & -6 & 8 \\ 2 & 3 & 7 \end{bmatrix} \quad \text{then}$$

$$\bar{\bar{A}} + \bar{\bar{B}} = \begin{bmatrix} 9 & -1 & 7 \\ 8 & 4 & 11 \end{bmatrix} \quad \text{and} \quad \bar{\bar{A}} - \bar{\bar{B}} = \begin{bmatrix} -1 & 11 & -9 \\ 4 & -2 & -3 \end{bmatrix}.$$

Note that addition is commutative, i.e., $\bar{\bar{A}} + \bar{\bar{B}} = \bar{\bar{B}} + \bar{\bar{A}}$, but subtraction is not commutative, i.e., $\bar{\bar{A}} - \bar{\bar{B}} \neq \bar{\bar{B}} - \bar{\bar{A}}$—instead, $\bar{\bar{A}} - \bar{\bar{B}} = -(\bar{\bar{B}} - \bar{\bar{A}})$. Both addition and subtraction are associative, i.e., $(\bar{\bar{A}} \pm \bar{\bar{B}}) \pm \bar{\bar{C}} = \bar{\bar{A}} \pm (\bar{\bar{B}} \pm \bar{\bar{C}})$.

Multiplication

There are two forms of matrix multiplication we shall treat. Multiplication of a matrix by a *scalar* is performed by multiplying each element of the matrix by the scalar. For example, if

$$\bar{\bar{A}} = \begin{bmatrix} 6 & 4 \\ -1 & 2 \end{bmatrix} \quad \text{then} \quad 8\bar{\bar{A}} = \begin{bmatrix} 48 & 32 \\ -8 & 16 \end{bmatrix}.$$

The *product* of two matrices, $\bar{\bar{A}}$ times $\bar{\bar{B}}$, is only defined if the number of columns in $\bar{\bar{A}}$ equals the number of rows in $\bar{\bar{B}}$. If $\bar{\bar{A}}$ is of order $m \times n$ and $\bar{\bar{B}}$ is of order $n \times p$ then their product $\bar{\bar{C}}$ will be of order $m \times p$ and each element

$$c_{ij} = \sum_{k=1}^{k=q} a_{ik} b_{kj}.$$

For example, if

$$\bar{\bar{A}}_{(2\times3)} = \begin{bmatrix} 4 & -5 & 1 \\ 6 & 7 & 2 \end{bmatrix}; \qquad \bar{\bar{B}}_{(3\times3)} = \begin{bmatrix} 6 & 3 & -2 \\ 1 & 6 & 5 \\ -2 & 7 & 4 \end{bmatrix} \quad \text{then}$$

$$\bar{\bar{A}}_{(2\times3)}\bar{\bar{B}}_{(3\times3)} = \bar{\bar{C}}_{(2\times3)} = \begin{bmatrix} 4 & -5 & 1 \\ 6 & 7 & 2 \end{bmatrix} \begin{bmatrix} 6 & 3 & -2 \\ 1 & 6 & 5 \\ -2 & 7 & 4 \end{bmatrix}$$

$$\bar{\bar{C}} = \begin{bmatrix} (4)(6) + (-5)(1) + (1)(-2) & (4)(3) + (-5)(6) + (1)(7) & (4)(-2) + (-5)(5) + (1)(4) \\ (6)(6) + (7)(1) + (2)(-2) & (6)(3) + (7)(6) + (2)(7) & (6)(-2) + (7)(5) + (2)(4) \end{bmatrix}$$

$$\bar{\bar{C}} = \begin{bmatrix} 17 & -11 & -29 \\ 39 & 74 & 31 \end{bmatrix}.$$

We see for this example that $\bar{\bar{B}}\bar{\bar{A}}$ makes no sense because the number of columns in $\bar{\bar{B}}$, which is 3, does not equal the number of rows in $\bar{\bar{A}}$, 2. So in general $\bar{\bar{A}}\bar{\bar{B}} \neq \bar{\bar{B}}\bar{\bar{A}}$, i.e., matrix multiplication is not commutative.

However, matrix multiplication is both associative and distributive:

$$(\bar{\bar{A}}\bar{\bar{B}})\bar{\bar{C}} = \bar{\bar{A}}(\bar{\bar{B}}\bar{\bar{C}}) \qquad \text{and} \qquad \bar{\bar{A}}(\bar{\bar{B}} + \bar{\bar{C}}) = \bar{\bar{A}}\bar{\bar{B}} + \bar{\bar{A}}\bar{\bar{C}}.$$

The transpose of the product of two or more matrices has the following identities

$$(\bar{\bar{A}}\bar{\bar{B}})^T = \bar{\bar{B}}^T\bar{\bar{A}}^T; \qquad (\bar{\bar{A}}\bar{\bar{B}}\bar{\bar{C}})^T = \bar{\bar{C}}^T\bar{\bar{B}}^T\bar{\bar{A}}^T.$$

Matrix division is not defined in matrix algebra but a similar concept, matrix inversion, is. As a prelude to matrix inversion we will first investigate determinants of matrices.

Determinants of Matrices

Actually, determinants are not restricted to matrices but our interest here is confined to matrices. Only square matrices have determinants. While a matrix is composed of many numbers, the determinant of a matrix is *one* number (called the value of the determinant). Finding this value involves a process called expansion. Let's start with a 2×2 matrix, $\bar{\bar{A}}$. If

$$\bar{\bar{A}} = \begin{bmatrix} 2 & 4 \\ 6 & -5 \end{bmatrix}$$ then the determinant of $\bar{\bar{A}}$, written $|\bar{\bar{A}}|$, is $$|\bar{\bar{A}}| = \begin{vmatrix} 2 & 4 \\ 6 & -5 \end{vmatrix}.$$

The expansion of $|\bar{\bar{A}}|$ is accomplished by the difference of two products:

$$|\bar{\bar{A}}| = \begin{vmatrix} 2 & 4 \\ 6 & -5 \end{vmatrix} = [(2)(-5)] - [(4)(6)] = -34.$$

$\ominus \qquad \oplus$

Note when multiplying upper left and lower right elements we associate a positive sign with this product; upper right times lower left is associated with a negative sign.

The value of a 3×3 or larger determinant is a bit more involved; there are many ways to expand such a determinant. We shall expand about the first row:

If $$\bar{\bar{A}} = \begin{bmatrix} -6 & 1 & 2 \\ 4 & 3 & 3 \\ 2 & 7 & -1 \end{bmatrix}$$ then

$$|\bar{\bar{A}}| = (-6)\begin{vmatrix} 3 & 3 \\ 7 & -1 \end{vmatrix} - (1)\begin{vmatrix} 4 & 3 \\ 2 & -1 \end{vmatrix} + (2)\begin{vmatrix} 4 & 3 \\ 2 & 7 \end{vmatrix}$$

$$= (-6)[(-3) - (21)] - (1)[(-4) - (6)] + (2)[(28) - (6)] = 198.$$

We call each sub-determinant (here 2 × 2) in the expansion a *minor*, M_{ij}. In particular,

$$\text{the minor of } a_{11} = M_{11} = \begin{vmatrix} 3 & 3 \\ 7 & -1 \end{vmatrix} \equiv \begin{vmatrix} a_{22} & a_{23} \\ a_{32} & a_{33} \end{vmatrix};$$

$$\text{the minor of } a_{12} = M_{12} = \begin{vmatrix} 4 & 3 \\ 2 & -1 \end{vmatrix} \equiv \begin{vmatrix} a_{21} & a_{23} \\ a_{31} & a_{33} \end{vmatrix};$$

$$M_{13} = \begin{vmatrix} 4 & 3 \\ 2 & 7 \end{vmatrix} \equiv \begin{vmatrix} a_{21} & a_{22} \\ a_{31} & a_{32} \end{vmatrix}.$$

Thus the above expansion can be written $|\bar{\bar{A}}| = a_{11}M_{11} - a_{12}M_{12} + a_{13}M_{13}$. We see that second product is subtracted; looking at its subscripts we see the sum of the subscripts is odd. Consequently, we define the *cofactor*, co_{ij}, of element a_{ij} as $co_{ij} = (-1)^{i+j}M_{ij}$. Thus, we write the expansion of determinant $\bar{\bar{A}}$ as

$$|\bar{\bar{A}}| = a_{11}co_{11} + a_{12}co_{12} + a_{13}co_{13}.$$

Here, of course, $co_{11} = M_{11}$ since $1 + 1 = 2$ is even; $co_{12} = -M_{12}$ since $1 + 2 = 3$ is odd and $co_{13} = M_{13}$ since $1 + 3 = 4$ is even.

To generalize, the expansion of a determinant about the first row can be written

$$|\bar{\bar{A}}| = \sum_{j=1}^{n} a_{ij}co_{ij}.$$

We note that if $\bar{\bar{A}}$ is a 4 × 4 then each co_{ij} will be formed from a 3 × 3 determinant.

Inversion

The inverse of a matrix $\bar{\bar{B}}$ is defined as the matrix, written $\bar{\bar{B}}^{-1}$, that when multiplied by $\bar{\bar{B}}$ yields the identity matrix, i.e., $\bar{\bar{B}}\bar{\bar{B}}^{-1} = \bar{\bar{I}} = \bar{\bar{B}}^{-1}\bar{\bar{B}}$. Only square matrices possess inverses. The inverse of $\bar{\bar{B}}$ can be found by

$$\bar{\bar{B}}^{-1} = \frac{(co\bar{\bar{B}})^T}{|\bar{\bar{B}}|}$$

where $|\bar{\bar{B}}|$ is the determinant of $\bar{\bar{B}}$ and $(co\bar{\bar{B}})^T$ is the transpose of the cofactor matrix of $\bar{\bar{B}}$. The *cofactor matrix* of $\bar{\bar{B}}$ is composed of the cofactors terms co_{11}, co_{12}, etc.

We note that if $|\bar{\bar{B}}| = 0$ then no inverse of $\bar{\bar{B}}$ exists; $\bar{\bar{B}}$ is said to be *singular*. Let's look at an example. If

$$\bar{\bar{B}} = \begin{bmatrix} 4 & 2 & -3 \\ 0 & 6 & 0 \\ 5 & 1 & -1 \end{bmatrix}$$

then
$$|\bar{\bar{B}}| = (4)\begin{vmatrix} 6 & 0 \\ 1 & -1 \end{vmatrix} - (2)\begin{vmatrix} 0 & 0 \\ 5 & -1 \end{vmatrix} + (-3)\begin{vmatrix} 0 & 6 \\ 5 & 1 \end{vmatrix}$$

$$= (4)(-6) - (2)(0) + (-3)(-30) = 66$$

and

$$co\bar{\bar{B}} = \begin{bmatrix} \begin{vmatrix} 6 & 0 \\ 1 & -1 \end{vmatrix} & -\begin{vmatrix} 0 & 0 \\ 5 & -1 \end{vmatrix} & \begin{vmatrix} 0 & 6 \\ 5 & 1 \end{vmatrix} \\ -\begin{vmatrix} 2 & -3 \\ 1 & -1 \end{vmatrix} & \begin{vmatrix} 4 & -3 \\ 5 & -1 \end{vmatrix} & -\begin{vmatrix} 4 & 2 \\ 5 & 1 \end{vmatrix} \\ \begin{vmatrix} 2 & -3 \\ 6 & 0 \end{vmatrix} & -\begin{vmatrix} 4 & -3 \\ 0 & 0 \end{vmatrix} & \begin{vmatrix} 4 & 2 \\ 0 & 6 \end{vmatrix} \end{bmatrix} = \begin{bmatrix} -6 & 0 & -30 \\ -1 & 11 & 6 \\ 18 & 0 & 24 \end{bmatrix}$$

thus,

$$\bar{\bar{B}}^{-1} = \frac{(C_0\bar{\bar{B}})^T}{|\bar{\bar{B}}|} = \begin{bmatrix} -6 & -1 & 18 \\ 0 & 11 & 0 \\ -30 & 6 & 24 \end{bmatrix}\left(\frac{1}{66}\right).$$

Check: does $\bar{\bar{B}}\bar{\bar{B}}^{-1} = \bar{\bar{I}}$?

$$\begin{bmatrix} 4 & 2 & -3 \\ 0 & 6 & 0 \\ 5 & 1 & -1 \end{bmatrix}\begin{bmatrix} -6 & -1 & 18 \\ 0 & 11 & 0 \\ -30 & 6 & 24 \end{bmatrix}\left(\frac{1}{66}\right) \equiv \begin{bmatrix} 1 & 0 & 0 \\ 0 & 1 & 0 \\ 0 & 0 & 1 \end{bmatrix}. \qquad \text{Yes.}$$

REFERENCES

1. Kinney, J. S. *Indeterminate Structural Analysis*. Reading, Massachusetts: Addison-Wesley Publishers, 1957.
2. Mantell, M. I., and John F. Marron. *Structural Analysis*. New York: Ronald Press Co., 1962.
3. McCormac, J. C. *Structural Analysis*, 4th ed. New York: Intex Educational Publishers, 1983.
4. Willems, N., and William M. Lucas, Jr. *Structural Analysis for Engineers*. New York: McGraw-Hill Book Co., 1978.
5. Hsieh, Y. Y. *Elementary Theory of Structures*, 2nd ed. Englewood Cliffs, New Jersey: Prentice-Hall, 1982.
6. Benjamin, J. R. *Statically Indeterminate Structures*. New York: McGraw-Hill Book Co., 1959.

ANSWERS TO SELECTED PROBLEMS

(Forces in kips, moments in kip feet, stress in ksi.)

CHAPTER 2

2.1 $P_{ab} = 10(t)$
$P_{bc} = 20(t)$
$P_{cd} = 20(t)$
$P_{ad} = -11.2(c)$
$P_{bd} = -21.2(c)$

2.2 $P_{ab} = 20(t)$
$P_{bc} = 50(t)$
$P_{cd} = 20(t)$
$P_{ad} = -20.6(c)$
$P_{bd} = -33.5(c)$

2.3 $P_{ab} = P_{bc} = 5(t)$
$P_{cd} = P_{de} = 15(t)$
$P_{ef} = 10.6(t)$
$P_{ah} = -5.6(c)$
$P_{hg} = -11.2(c)$
$P_{gf} = -23.8(c)$
$P_{bh} = -5(c)$
$P_{ch} = 5.6(t)$
$P_{cd} = -9(c)$
$P_{dg} = -7.5(c)$
$P_{eg} = 11.3(t)$

2.4 $P_{ab} = P_{bc} = 10(t)$
$P_{cd} = P_{de} = 30(t)$
$P_{ef} = 10.6(t)$
$P_{ah} = -10.3(c)$
$P_{hg} = -20.6(c)$
$P_{gf} = -43.8(c)$
$P_{bh} = -5(c)$
$P_{ch} = 10.3(t)$
$P_{cd} = -12.5(c)$
$P_{dg} = -7.5(c)$
$P_{eg} = 15.6(t)$

2.5 $P_{ab} = -42.4(c)$
$P_{bc} = -40(c)$
$P_{cd} = 0$
$P_{ae} = P_{ed} = 30(t)$
$P_{be} = 20(t)$
$P_{bd} = 14.14(t)$

2.6 $P_{ab} = -60(c)$
$P_{bc} = -69.3(c)$
$P_{cd} = 0$

$P_{ae} = P_{ed} = 52(t)$
$P_{be} = 20(t)$
$P_{bd} = 20(t)$

2.7 $P_{ab} = -39.1(c)$
$P_{bc} = P_{cd} = -30(c)$
$P_{de} = -40(c)$
$P_{ch} = P_{ef} = 0$
$P_{bi} = P_{dg} = 10(t)$
$P_{hg} = P_{gf} = 37.5(t)$
$P_{bh} = 28(t)$
$P_{dh} = -16.8(c)$
$P_{df} = 5.6(t)$
$P_{ji} = P_{ih} = 17.5(t)$

2.8 $P_{ab} = -46.3(c)$
$P_{bc} = P_{cd} = -52(c)$
$P_{de} = -69.3(c)$
$P_{ch} = P_{ef} = 0$
$P_{bi} = P_{dg} = 10(t)$
$P_{hg} = P_{gf} = 65(t)$
$P_{bh} = 33.1(t)$
$P_{dh} = -19.8(c)$
$P_{df} = 6.6(t)$
$P_{ji} = P_{ih} = 30.3(t)$

2.13 597.5 lbs.

2.14 607.8 lbs.

2.15 651.5 lbs.

2.16 525.1 lbs.

CHAPTER 3

3.1 $V_1 = 4.73;\ M_1 = 4.73x\ (0 \le x \le 4')$
$V_2 = 1.73;\ M_2 = 1.73x + 12\ (4' \le x \le 8')$
$V_3 = -2.27;\ M_3 = -2.27x + 44\ (8' \le x \le 13')$
$V_4 = -7.27;\ M_4 = -7.27x + 109\ (13' \le x \le 15')$

3.5 $V_1 = 4.1;\ M_1 = 4.1x\ (0 \le x \le 3')$
$V_2 = 5.3 - 0.4x;\ M_2 = -0.2x^2 + 5.3x - 1.8\ (3' \le x \le 6')$
$V_3 = 2.3 - 4x;\ M_3 = -0.2x^2 + 2.3x + 16.2\ (6' \le x \le 12')$
$V_4 = -2.5;\ M_4 = -2.5x + 4.5\ (12' \le x \le 18')$

3.6 $V_1 = 2.7;\ M_1 = 2.7x - 28.35\ (0 \le x \le 6')$
$V_2 = 0.3x - 4.5;\ M_2 = -1.5x^2 + 4.5x - 33.75\ (6' \le x \le 15')$

3.11 V: plateaus of 4.73, 1.73, −2.27, −7.27;
M: straight lines with peaks of 18.9, 25.8, 14.5.

3.15 V: flat 4.1, straight line to 2.9, jump to −0.1, straight line to −2.5 and flat; M: straight line to 12.3, parabola with peak of 22.8, straight line from 15 to 0.

3.16 V: flat 2.7, straight line to 0; M: straight line from −28.35 to −12.15, parabola to 0.

3.21 (a) $\sigma_t = \sigma_c = 25.8;\ \tau_{max} = 0.909$
(b) $\sigma_t = 14.12;\ \sigma_c = 6.64;\ \tau_{max} = 0.902$
(c) $\sigma_t = \sigma_c = 2.07;\ \tau_{max} = 0.389$
(d) $\sigma_t = 2.45;\ \sigma_c = 1.63;\ \tau_{max} = 0.406$

3.25 (a) $\sigma_t = \sigma_c = 22.8$; $\tau_{max} = 0.513$
(b) $\sigma_t = 12.48$; $\sigma_c = 5.87$; $\tau_{max} = 0.509$
(c) $\sigma_t = \sigma_c = 1.83$; $\tau_{max} = 0.220$
(d) $\sigma_t = 2.16$; $\sigma_c = 1.44$; $\tau_{max} = 0.229$

3.26 (a) $\sigma_t = \sigma_c = 28.35$; $\tau_{max} = 0.338$
(b) $\sigma_t = 2.85$; $\sigma_c = 6.06$; $\tau_{max} = 0.335$
(c) $\sigma_t = \sigma_c = 2.28$; $\tau_{max} = 0.145$
(d) $\sigma_t = 1.79$; $\sigma_c = 2.69$; $\tau_{max} = 0.151$

3.31 (a) 1 = 2.35″; (b) 1 = 1.55″

3.35 (a) 1 = 2.07″; (b) 1 = 1.29″

CHAPTER 4

4.1 max. shears of 25(*ab*), −18.33(*bc*), 15(*cd*);
max. moments of 156.25(*ab*), 108.33(*bc*), −75(*cd*);
axial forces of −1.67(*ab*), −15(*bc*), −18.33(*cd*).

4.5 max. shears of −12.5(*ab*), −16.25(*bc*), 12.5(*cd*);
max. moments of 75(*ab*), 41(*bc*), 37.5(*cd*);
axial forces of −3.75(*ab*), −12.5(*bc*), −16.25(*cd*).

4.8 max. shears of 23.75(*ab*), 10(*bc*), 6.25(*cd*);
max. moments of −325(*ab*), 100(*bc*), 37.5(*cd*);
axial forces of −10(*ab*), −6.25(*bc*), −10(*cd*).

4.11 max. shears of −24(*ab*), 25.1(*bc*);
max. moments of −160(*ab*), −160(*bc*);
axial forces of −40(*ab*), −39.4 and −21.5(*bc*).

CHAPTER 5

5.1 $\delta_b = 8092/(EI)$; $\theta_b = -503.4/(EI)$ clockwise.

5.2 $\delta_b = 1782/(EI)$; $\theta_b = -66/(EI)$ clockwise.

5.3 $\delta_b = 4000/(EI)$; $\theta_b = 66.7/(EI)$ counterclockwise.

5.4 $\delta_b = 1440/(EI)$; $\theta_b = 41.3/(EI)$ counterclockwise.

5.11 $\delta_b = 6145.9/(EI)$; $\theta_b = 1333.3/(EI)$ counterclockwise.

5.13 $\delta_{max} = 9375.8/(EI)$ at 14.92′.

5.18 reaction at *b* is 18.25(↑); max. shear 10.25; max. moment −30.

5.19 reaction at *b* is 20(↑); max. shear 10; max. moment −18.

5.20 reaction at b is 21.5($\uparrow$); max. shear -20.75; max. moment 34.24.

5.21 reaction at b is 16.88($\uparrow$); max. shear -14.94; max. moment 28.24.

5.28 reaction at b is 29.5($\uparrow$); max. shear -10.5; max. moment 11.28.

5.30 $\theta_b = -2281/(EI)$ clockwise; $\delta_{bx} = \delta_{cx} = 98{,}953/(EI)$ right; $\theta_c = \theta_d = 1281.5/(EI)$ counterclockwise; $\delta_{dx} = 124{,}583/(EI)$ right.

5.36 reaction $d_x = 10.99$ left; max. shears of 9.01(ab), -41.67(bc), 10.99(cd); max. moments of 180.2(ab), -219.8(bc), -219.8(cd).

5.42 $\sigma_{max} = 20.25$ (determinate) vs. 5.40 (indeterminate).

5.44 $d = 7.15''$ (det.) vs. $4.80''$ (indet.).

CHAPTER 6

6.1 $\delta_{ay} = 34{,}558/(AE)$; $\delta_{ax} = 2025/(AE)$ left.

6.3 $\delta_{ay} = 2014/(AE)$; $\delta_{ax} = 606/(AE)$ right.

6.5 $\delta_{ay} = 8728/(AE)$; $\delta_{ax} = 10{,}567/(AE)$ right.

6.6 $\delta_{dy} = 1.054''$.

6.11 $P_{ab} = -108(c)$; $P_{ac} = 177(t)$; $P_{bc} = 157(t)$; $P_{ad} = 424(t)$; $P_{cd} = -300(c)$.

6.14 $P_{ab} = -67(c)$; $P_{af} = -92.7(c)$; $P_{bc} = -32.5(c)$; $P_{bf} = 37(t)$; $P_{cd} = -74.3(c)$; $P_{ce} = 56(t)$; $P_{cf} = -67.1(c)$; $P_{ef} = -40.6(c)$; $P_{de} = -50(c)$.

6.17 $P_{ab} = 15(t)$; $P_{ac} = -21.2(c)$; $P_{ae} = -45(c)$; $P_{bc} = 45(t)$; $P_{be} = 21.2(t)$; $P_{cd} = 42.4(t)$; $P_{ce} = -15(c)$; $P_{de} = -30(c)$.

6.22 $\delta_{by} = 51{,}328/(EI)$.

6.24 $\delta_{by} = 5758/(EI)$.

6.28 $\theta_b = 5812.5/(EI)$ clockwise.

6.31 $\theta_a = 2315/(EI)$ clockwise.

6.33 (a) $\sigma_t = \sigma_c = 83.1$; $\tau_{max} = 1.91$;
(b) $\sigma_t = 44.29$; $\sigma_c = 36.0$; $\tau_{max} = 0.95$;
(c) $\sigma_t = \sigma_c = 33.0$; $\tau_{max} = 1.09$.

6.34 (a) $\sigma_t = \sigma_c = 42.5$; $\tau_{max} = 2.51$;
(b) $\sigma_t = 18.4$; $\sigma_c = 22.6$; $\tau_{max} = 1.25$;
(c) $\sigma_t = \sigma_c = 16.9$; $\tau_{max} = 1.43$.

6.38 (a) $\delta_{bx} = 337{,}500/(EI)$ right;
(b) $\theta_b = 22{,}500/(EI)$ clockwise.

CHAPTERS 7 AND 8

joint moments stated

7.1/8.1 $M_{ab} = 10.59$; $M_{ba} = -M_{bc} = 21.18$; $M_{cb} = -M_{cd} = 93.10$.

7.5/8.5 $M_{ab} = -284.34$; $M_{ba} = -M_{bc} = 106.32$; $M_{cb} = -M_{cd} = -34.63$.

7.6/8.6 $M_{ab} = -461.7$; $M_{ba} = -M_{bc} = -402.5$; $M_{cb} = -M_{cd} = 283.1$.

7.11/8.11 $M_{ab} = 47.6$; $M_{ba} = 95.2$; $M_{bd} = 57.2$; $M_{be} = 47.6$.

7.15/8.15 $M_{da} = -85.66$; $M_{eb} = -116.28$; $M_{fc} = -118.99$;
$M_{ab} = -M_{ad} = 50.39$; $M_{ba} = 189.15$; $M_{be} = -111.63$;
$M_{bc} = -77.52$; $M_{cb} = -M_{cf} = -117.05$.

7.17/8.17 $M_{ab} = -M_{ad} = -79.07$; $M_{ba} = -44.63$; $M_{da} = 56.13$.

7.18/8.18 $M_{ab} = -M_{ac} = 133.3$; $M_{ca} = -266.7$.

7.25/8.25 $M_{ea} = -89.0$; $M_{ab} = -M_{ae} = 74.1$; $M_{ba} = 81.1$;
$M_{bf} = -64.7$; $M_{bc} = -16.4$; $M_{fb} = -84.3$;
$M_{cb} = -M_{cd} = -31.8$; $M_{dc} = -M_{dg} = 35.8$; $M_{gd} = -43.7$.

CHAPTER 9

9.1 (a) $\bar{D} = \begin{bmatrix} \theta_b \\ \theta_c \end{bmatrix}$; (b) $\bar{\bar{S}} = \begin{bmatrix} .8 & .2 \\ .2 & .4 \end{bmatrix} EI.$

9.4 (a) $\bar{D} = \begin{bmatrix} \theta_b \\ \theta_c \\ \delta_d \\ \theta_d \end{bmatrix}$; (b) $\bar{\bar{S}} = \begin{bmatrix} 85.33 & 21.33 & 0 & 0 \\ 21.33 & 106.66 & 12 & 32 \\ 0 & 12 & 3 & 12 \\ 0 & 32 & 12 & 64 \end{bmatrix} EI/(128).$

9.6 (a) $\bar{D}_G = \begin{bmatrix} \delta_{b\bar{X}} \\ \delta_{b\bar{Y}} \\ \theta_b \end{bmatrix}$; (b) $\bar{\bar{S}}_G = \begin{bmatrix} 13.35 & 0 & -90 \\ 0 & 81.45 & 0 \\ -90 & 0 & 3600 \end{bmatrix} EI/(15)^3.$

9.9 (a) $\bar{D}_G = \begin{bmatrix} \delta_{b\bar{X}} \\ \delta_{b\bar{Y}} \\ \theta_{ba} \\ \theta_{bc} \end{bmatrix}$; (b) $\bar{\bar{S}}_G = \begin{bmatrix} 23.52 & 0 & 0 & -144 \\ 0 & 23.52 & 144 & 0 \\ 0 & 144 & 2304 & 0 \\ -144 & 0 & 0 & 2304 \end{bmatrix} EI/(24)^3.$

9.13 (a) $\bar{D}_G = \begin{bmatrix} \delta_{b\bar{X}} \\ \delta_{b\bar{Y}} \\ \delta_{c\bar{X}} \\ \delta_{d\bar{X}} \end{bmatrix}$; (b) $\bar{\bar{S}}_G = \begin{bmatrix} 1 & 0 & -0.5 & 0 \\ 0 & 2 & 0.5 & 0 \\ -0.5 & 0.5 & 1.5 & -1 \\ 0 & 0 & -1 & 2 \end{bmatrix} EA/L.$

9.16 (a) $\theta_b = 38.69/(EI)$ clockwise; $\theta_c = -92.27/(EI)$ counterclockwise; (b) $V_{ab} = 7.68$; $M_{ab} = -8.93$; $V_{ba} = 12.32$; $M_{ba} = -M_{bc} = 32.15$; $V_{bc} = 18.21$; $V_{cb} = 11.79$; $M_{cb} = 0$.

9.19 (a) $\theta_a = -192/(EI)$ counterclockwise; $\theta_b = 312/(EI)$ counterclockwise; $\delta_d = -4202.6/(EI)$ left; $\theta_d = 610.6/(EI)$ clockwise; (b) $V_{ab} = 8$; $M_{ab} = -32$; $V_{ba} = -8$; $M_{ba} = -64$; $V_{bc} = 7$; $M_{bc} = -36$; $V_{cb} = 17$; $M_{cb} = -M_{cd} = 96$; $V_{cd} = 20$; $V_{dc} = -4$; $M_{dc} = 0$.

9.21 (a) $\delta_{b\underline{X}} = 2739.7/(EI)$ right; $\delta_{bY} = -725.1/(EI)$ down; $\theta_b = 31.38/(EI)$ clockwise; (b) $P_{ab} = -0.55$; $V_{ab} = 14.32$; $M_{ab} = -55.98$; $P_{ba} = 0.55$; $V_{ba} = 0.68$; $M_{ba} = -12.2$; $P_{bc} = 0.55$; $V_{bc} = -0.58$; $M_{bc} = 39.39$; $P_{cb} = -0.55$; $V_{cb} = 20.58$; $M_{cb} = 66.45$; $P_{db} = 0.1$; $V_{db} = 18.9$; $M_{db} = -106.37$; $P_{bd} = -0.1$; $V_{bd} = 1.1$; $M_{bd} = -27.19$.

9.24 (a) $\delta_{b\underline{X}} = -4462.8/(EI)$ left; $\delta_{bY} = -5950.4/(EI)$ down; $\theta_{ba} = 11.9/(EI)$ clockwise; $\theta_{bc} = -8.93/(EI)$ counterclockwise; (b) $P_{ab} = 3.72$; $V_{ab} = 15.04$; $M_{ab} = -120.99$; $P_{ba} = -3.72$; $V_{ba} = 4.96$; $M_{ba} = 0$; $P_{cb} = 4.96$; $V_{cb} = -11.28$; $M_{cb} = 90.75$; $P_{bc} = -4.96$; $V_{bc} = -3.72$; $M_{bc} = 0$.

9.28 (a) $\delta_{b\underline{X}} = 290L/(EA)$ right; $\delta_{bY} = -70L/(EA)$ down; $\delta_{c\underline{X}} = 180L/(EA)$ right; $\delta_{d\underline{X}} = 90L/(EA)$ right; (b) $P_{ab} = 155.6(t)$; $P_{bc} = -127.2(c)$; $P_{ac} = P_{cd} = 90(t)$; $P_{bd} = -70(c)$.

CHAPTER 10

10.1 The base moments are: $M_a = 26.1$; $M_b = 73.8$; $M_c = 63.9$; $M_d = 16.2$.

10.7 The base moments are: $M_a = 23.1$; $M_b = 54.42$; $M_c = 60.9$; $M_d = 29.58$.

10.8 The base moments are: $M_a = 28.02$; $M_b = 55.98 = M_c$; $M_d = 28.02$.

10.9 Top floor joint moments are: $M_{mn} = -43.74$; $M_{nm} = M_{no} = -43.74$; $M_{on} = -43.74$; $M_{op} = -46.62$; $M_{po} = -44.13$.

CHAPTER 11

11.2 (a) Influence line for R_a is a constant = 1.0; influence line for M_a linear from 0 to −30; (b) Influence line for $V_{10'}$ two plateaus: 0 to 10′, 1.0 to 30′; $V_{15'}$ two plateaus: 0 to 15′, 1.0 to 30′; (c) $M_{10'}$ = 0 to 10′, linear to 30′ (0 to −20); $M_{15'}$ = 0 to 15′, linear to 30′ (0 to −15).

11.5 (a) $R_{a|max} = 150$; $M_{a|max} = -2250$;
(b) $V_{10'|max} = 100$; $V_{15'|max} = 75$;
(c) $M_{10'|max} = -1000$; $M_{15'|max} = -562.5$.

11.8 (a) $R_{a|max} = 80$; $M_{a|max} = -1500$;
(b) $V_{10'|max} = 70$; $V_{15'|max} = 60$;
(c) $M_{10'|max} = -800$; $M_{15'|max} = -550$.

11.11 (a) Influence line for R_a parabolic: 1.0 at 0′, 0.8519 at 10′, 0.4815 at 20′, 0 at 30′; influence line for R_b parabolic: 0 at 0′, 0.1481 at 10′, 0.5185 at 20′, 1.0 at 30′; influence line for M_a parabolic: 0 at 0′ and 30′, −5.555 at 10′, −4.444 at 20′;
(b) influence line for $V_{10'}$ made of two parabolic curves: 0 at 0′, −0.1481 at 10′, jump to 0.8519 at 10′, 0.4815 at 20′ and 0 at 30′;
(c) influence line for $M_{10'}$ made of two parabolic curves: 0 at 0′, cusp of 2.962 at 10′, 0.3704 at 20′, 0 at 30′.

11.14 $R_{a|max} = 93.75$; $R_{b|max} = 56.25$; $M_{a|max} = -562.5$.

INDEX